Electrochemistry: Science and Technology

Electrochemistry: Science and Technology

Edited by **Heather Wright**

Willford Press
New York

Published by Willford Press,
118-35 Queens Blvd., Suite 400,
Forest Hills, NY 11375, USA
www.willfordpress.com

Electrochemistry: Science and Technology
Edited by Heather Wright

International Standard Book Number: 978-1-68285-322-1 (Hardback)

Printed in China.

Contents

	Preface	**VII**
Chapter 1	**Inhibition of corrosion of carbon steel by heptane sulphonic acid - Zn^{2+} system** C. Mary Anbarasi and Susai Rajendran	**1**
Chapter 2	**In-situ activated hydrogen evolution by molybdate addition to neutral and alkaline electrolytes** John Gustavsson, Christine Hummelgård, Joakim Bäckström, Inger Odnevall Wallinder, Seikh Ohammad Habibur Rahman, Göran Lindbergh, Sten Eriksson and Ann Cornell	**19**
Chapter 3	**Voltammetric determination of hydroxylamine in water and waste water samples using a NiO nanoparticle/new catechol derivative modified carbon paste electrode** Mahbobeh Moazampour, Fahimeh Tahernejad-Javazmi, Maryam Salimi-Amiri, Hassan Karimi-Maleh and Mehdi Hatami	**35**
Chapter 4	**Electrochemical studies of nevirapine, an anti-HIV drug, and its assay in tablets and biological samples** Nagappa laxman teradal, shankar narayan prashanth, jaldappagari Seetharamappa	**45**
Chapter 5	***p*-toluene sulfonic acid doped polyaniline carbon nanotube composites: synthesis via different routes and modified properties** Ashok K. Sharma and Yashpal Sharma	**54**
Chapter 6	**Ni-MoO_2 cathodes for hydrogen evolution in alkaline solutions. Effect of the conditions of their electrodeposition** Gvozden Tasić, Borka Jović, Uroš Lačnjevac, Nedeljko Krstajić and Vladimir Jović	**64**
Chapter 7	**Electrochemical investigations of unexplored anthraquinones and their DNA binding** Afzal Shah, Abdur Rauf, Asad Ullah, Zeema Munir, Rumana Qureshi, Iftikhar Ahmad, Muhammad Tahir Soomro and Zia-ur-Rehman	**72**
Chapter 8	**Ir-Ni oxide as a promising material for nerve and brain stimulating electrodes** Joan Stilling, Nehar Ullah and Sasha Omanovic	**81**
Chapter 9	**Novel Mannich bases bearing pyrazolone moiety Synthesis, characterization and electrochemical studies** Krishna Naik, Aluru Raghavendra Guru Prasad, Yadati Narasimha Spoorthy and Lakshmana Rao Krishna Rao Ravindranath	**93**
Chapter 10	**Binder-less activated carbon electrode from gelam wood for use in supercapacitors** Nirwan Syarif, Ivandini A. Tribidasari and Widayanti Wibowo	**123**
Chapter 11	**On the suppression of cathodic hypochlorite reduction by electrolyte additions of molybdate and chromate ions** John Gustavsson, Gongzhuo Li, Christine Hummelgård, Joakim Bäckström and Ann Cornell	**132**

Chapter 12 **Morphology and cyclic voltammetry analysis of in situ polymerized polyaniline/graphene composites** 146
Devesh Kumar Mahla, Subhendu Bhandari, Mostafizur Rahaman and Dipak Khastgir

Chapter 13 **Electrochemical behaviour of PES ionomer and Pt-free catalyst for PEMFCs** 156
Stefania Giordano, Mariangela Longhi, Leonardo Formaro, Hermes Farina and Giuseppe Di Silvestro

Chapter 14 **Electrochemical determination of an antitumour platinum(IV) complex: trans-[$PtCl_2(OH)_2$(dimethylamine)(isopropylamine)]. Application to biological samples** 165
Carmen S. Hernandez Domínguez and Pedro Hernández

Chapter 15 **Electrochemical treatment of leachates from sanitary landfills** 174
Annabel Fernandes, Edite Catalão, Lurdes Ciríaco, Maria J Pacheco and Ana Lopes

Chapter 16 **Formation of the honeycomb-like electrodes by the regime of pulsating overpotential in the second range** 185
Nebojša D. Nikolić, Goran Branković and Miomir G. Pavlović

Chapter 17 **Electrochemical treatment of trypan blue synthetic wastewater and its degradation pathway** 193
Anantha N. Subba Rao and Venkatesha T. Venkatarangaiah

Chapter 18 **Determination of nevirapine in the presence of cucurbit(7)uril with a gold electrode** 211
Ana M. Esteva, Elías Blanco, Juan J. Piña, Abel I. Balbin, Carmen Quintana and Pedro Hernández

Chapter 19 **Anodic formation of nanoporous crystalline niobium oxide** 219
Leonid Skatkov, Larisa Lyashok, Valeriy Gomozov, Irina Tokareva and Boris Bayrachniy

Chapter 20 **Synthesis of $CuInSe_2$ thin films from electrodeposited $Cu_{11}In_9$ precursors by two-step annealing** 228
Tsung-Wei Chang, Shao-Yu Hu and Wen-Hsi Lee

Chapter 21 **The effect of cetyltrimethylammonium bromide on size and morphology of ZnO and CuO** 237
Anantha N. Subba Rao and Venkatesha T. Venkatarangaiah

Chapter 22 **Novel electroanalysis of hydroxyurea at glassy carbon and gold electrode surfaces** 251
Keerti M. Naik and Sharanappa T. Nandibewoor

Chapter 23 **Comparative voltammetric study and determination of carbamate pesticide residues in soil at carbon nanotubes paste electrodes** 262
Thommandru Raveendranath Babu, Sarvareddy Rajasekhar Reddy and Puchakayala Sujana

Permissions

List of Contributors

Preface

Electrochemistry has a wide range of applications, especially in industries for processes like electrodeposition, redox reactions, galvanization, etc. This book includes some of the vital pieces of work being conducted across the world, on various topics related to electrochemistry. It presents a thorough understanding of concepts such as electrochemical materials science, energy conversion catalysis, molecular electrochemistry, electrochemical synthesis, etc. This book includes contributions of experts and scientists which will provide innovative insights into this field. This book also contains many topics apt for research, which will be beneficial for research scholars as well as professionals in this field.

This book is the end result of constructive efforts and intensive research done by experts in this field. The aim of this book is to enlighten the readers with recent information in this area of research. The information provided in this profound book would serve as a valuable reference to students and researchers in this field.

At the end, I would like to thank all the authors for devoting their precious time and providing their valuable contributions to this book. I would also like to express my gratitude to my fellow colleagues who encouraged me throughout the process.

Editor

1

Inhibition of corrosion of carbon steel by heptane sulphonic acid – Zn^{2+} system

C. MARY ANBARASI and SUSAI RAJENDRAN*✉

PG Department of Chemistry, Jayaraj Annapackiam College for Women, Periyakulam-625601, India, E-mail: anbuc_m@yahoo.co.in
**Corrosion Research Centre, PG and Research Department of Chemistry, GTN Arts College, Dindigul-624005 and Department of Chemistry, RVS School of Engineering and Technology, Dindigul-624005, India*

✉Corresponding author: E-mail: susairajendran@gmail.com

Abstract

Corrosion inhibition of carbon steel in dam water by sodium heptane sulphonate (SHS) and zinc ion system was investigated using weight loss and potentiodynamic polarization methods. Results of weight loss method indicated that inhibition efficiency (IE) increased as the inhibitor concentration increased. A synergistic effect existed between SHS and Zn^{2+}. The influence of sodium potassium tartrate (SPT) on the IE of the SHS-Zn^{2+} system was evaluated. As the immersion period increased, the IE decreased. Polarization study revealed that SHS-Zn^{2+} system functioned as a cathodic inhibitor. AC impedance spectra revealed that a protective film was formed on the metal surface. The nature of the metal surface was analyzed by FTIR spectra, SEM and AFM analyses.

Keywords

Carbon steel; Corrosion; Electrochemical techniques; FTIR; SEM; AFM.

Introduction

Natural water is frequently used in cooling systems. Cooling water circuits may present several problems. Corrosion, scale and fouling by microorganisms can appear when natural water is used as thermal fluid. These problems can occur jointly, reducing the thermal efficiency of the circuits with significant economic repercussions. To eliminate or to reduce these problems, water used in cooling circuits is treated with inhibitive formulations. Inhibitors are generally used in these processes to control metal dissolution.

Organic compounds are recognized as effective inhibitors of the corrosion of many metals and alloys. The efficiency of an organic compound as a corrosion inhibitor is closely associated with the chemical adsorption [1-5]. Most of these organic compounds contain nitrogen, sulphur, oxygen and multiple bonds in the molecules, which are adsorbed on the metal surface and the organic compound [6, 7]. Much interest has been devoted to organic inhibitors such as polyethylene glycols [8], 1,2,3 benzotriazole [9], 2 mercaptobenzimidazole [10], ethoxylated fatty acids [11] and meta substituted aniline-N-sallicylidenes [12]. Protecting efficiency was also found to be improved with increasing length of alkyl chain and with the increase of the organic compound concentration [13]. A survey of the available literature reveals that the corrosion inhibition of 2-naphthalenesulfonic acid, 2,7-naphthalenedisulfonic acid, and 2-naphthol-3,6-disulfonic acid on Armco-iron electrode in sulfuric acid has been investigated by Vračar and Dražić [14]. The inhibition efficiency changes with the number of functional groups substituted on benzene ring and increases with the concentration of inhibitors [14]. The inhibitory action of 2-mercaptobenzoxazol, 2-mercaptobenzimidazole, N-cetylpyridiniumbromide, and propargylbenzene-sulphonate on the corrosion of carbon steel in acid media has also been studied by Prakash Rajesh Kumar Singh and Ranju Kumar [15]. The corrosion inhibitors are used to reduce corrosion damage in sub-surface equipments in oil well field. The corrosion inhibition activity was studied by gravimetric and potentiostatic polarization methods in the presence of 20% HCl [15]. Manickavasagam *et al.* [16] reported the corrosion inhibition of poly-(styrene sulphonic acid)-doped polyaniline on carbon steel in acid media. The polymer acts as an anodic inhibitor. The adsorption of the compound on the metal surface obeys Temkin's adsorption isotherm [16]. Aliev [17] has described the influence of salts of alkyl phenol sulphonic acid on the corrosion of ST3 steel. The protective effect increases with temperature. The investigated compounds inhibit corrosion of ST3 steel as a result of chemical adsorption [17]. Shakkthivel and Vasudevan [18] have studied acrylic acid-diphenylamine sulphonic acid copolymer threshold inhibitor for sulphate and carbonate scales in cooling water systems. The results showed that the polymer acts as a very good antiscaling inhibitor both in the carbonate and sulphate brines. Copolymer of acrylic acid-diphenylamine sulphonic acid can be used safely in cooling water industries [18].

The present work is undertaken to evaluate the inhibition efficiency (*IE*) of sodium heptane sulphonate (SHS) – Zn^{2+} system in controlling corrosion of carbon steel immersed in dam water in the absence or presence of Zn^{2+} by weight loss method; to investigate the influence of immersion period on the IE of the system; to analyze the protective film by FTIR spectra, fluorescence spectra, Scanning Electron Microscope (SEM), and Atomic Force Microscope (AFM); and to propose the mechanism of corrosion inhibition based on the above results.

Experimental

Preparation of the Specimen

Carbon steel specimens of size 1.0 cm × 4.0 cm × 0.2 cm and chemical composition 0.026 % sulphur, 0.06 % phosphorous, 0.4 % manganese, 0.1 % carbon, and the rest iron were polished to a mirror finish, degreased with trichloroethylene, and used for the weight loss method and surface examination studies.

Weight-loss Method

Carbon steel specimens were immersed in 100 ml of the Sothuparai dam water in Tamil Nadu, India, containing various concentrations of the inhibitor sodium heptane sulphonate (Scheme 1) in the absence and presence of Zn^{2+} for 3 days. The weights of the specimens before and after the immersion were determined using a Digital Balance (Model AUY 220 SHIMADZU). The corrosion products were cleaned with Clarke's solution prepared by dissolving 20 g of Sb_2O_3 and 50 g of $SnCl_2$ in one litre of concentrated HCl of specific gravity 1.9 [19]. The corrosion IE was then calculated using the equation

$$IE = 100\ [1-(W_2/W_1)]\ \% \quad (1)$$

where W_1 is the weight loss value in the absence of inhibitor and W_2 is the weight loss value in the presence of inhibitor. Corrosion rate (in mils per year) was calculated using the formula

$$CR = 534\ W / DAT \quad (2)$$

W = weight loss in milligrams
D = density of specimen = 7.87 g cm^{-3}
A = area of specimen = 10.4 cm^2
T = exposure time in hours = 72 hrs.

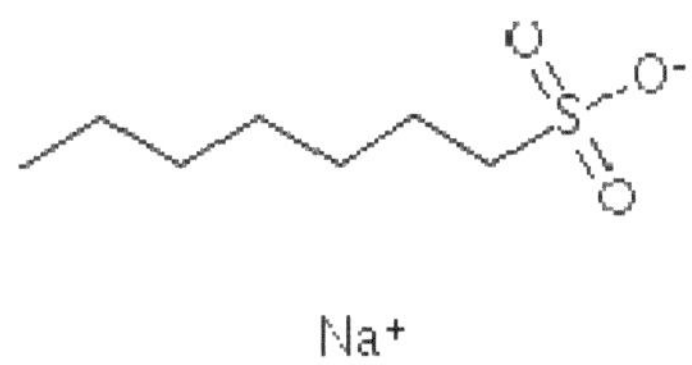

Scheme 1. *Molecular structure of sodium heptane sulphonate*

Surface Examination Study

The carbon steel specimens were immersed in various test solutions for a period of one day. After one day, the specimens were taken out and dried. The nature of the film formed on the surface of the metal specimen was analyzed by various surface analysis techniques.

Fourier Transforms Infrared Spectra

These spectra were recorded in a Perkin-Elmer-1600 spectrophotometer using KBr pellets. The FTIR spectrum of the protective film was recorded by carefully removing the film, mixing it with KBr and making the pellet.

UV visible absorbance spectra and fluorescence spectra

The instrument UV Spetord S-100 Analytic Jena was used for recording UV visible absorbance spectra. Fluorescence spectra were recorded in a Glaxo F-6300 Spectro fluorimeter.

Scanning Electron Microscopic Studies (SEM)

The carbon steel samples immersed in blank solution and in the inhibitor solution for a period of one day were removed, rinsed with double distilled water, dried and observed under a scanning electron microscope to examine the surface morphology. The surface morphology measurements of carbon steel were examined using JEOL MODEL6390 computer-controlled scanning electron microscope.

Atomic Force Microscopy Characterization (AFM)

The carbon steel specimen immersed in blank and in the inhibitor solution for a period of one day were removed, rinsed with double distilled water, dried and subjected to the surface examination. Atomic force microscopy (Veeco dinnova model) was used to observe the samples' surface in tapping mode, using cantilever with linear tips. The scanning area in the images was 5 µm × 5 µm and the scan rate was 0.6 Hz s^{-1}.

Potentiodynamic Polarization

Polarization studies were carried out with a CHI-electrochemical work station with impedance model 660A. It was provided with IR compensation facility. A three-electrode cell assembly was used. The working electrode was carbon steel. An SCE was the reference electrode. Platinum was the counter electrode. From polarization studies, corrosion parameters such as corrosion potential (E_{corr}), corrosion current (I_{corr}), Tafel slopes anodic = b_a, and cathodic = b_c were calculated and linear polarization study (*LPR*) was done. The scan rate (v) was 0.01 V s^{-1}. Hold time at E_f was 0 s and quiet time t_q was 2 s.

AC Impedance Spectra

The instrument used for polarization study was also used to record AC impedance spectra. The cell set up was also the same. The real part (Z') and imaginary part (Z'') of the cell impedance were measured in ohms at various frequencies. Values of charge transfer resistance (R_t) and the double layer capacitance (C_{dl}) were calculated. AC impedance spectra were recorded with initial $E_{(v)}$ = 0 V, high frequency limit was 1×10^5 Hz, low frequency limit was 1 Hz, amplitude = 0.005 V and quiet time t_q = 2 s.

Determination of the Biocidal Efficiency

The biocidal efficiency of the system was determined using a Zobell medium and calculating the number of colony forming unit per ml using a bacterial colony counter. The biocidal efficiency of sodium dodecyl sulphate (SDS) was determined. Various concentrations of SDS, namely 50 ppm, 100 ppm, 150 ppm, 200 ppm, and 250 ppm, were added to the formulation consisting of the inhibitor system. Polished and degreased carbon steel specimens in duplicate were immersed in these environments for a period of three days. After three days, one ml of each test solution from the above environment was pipetted out into sterile Petri dishes each containing about 20 ml of the sterilized Zobell medium. The Petri dishes were then kept in a sterilized environment inside the laminar flow system fabricated for 48 hours. The total viable hydrotropic bacterial colonies were counted using a bacterial colony counter. The biocidal efficiencies of the formulation consisting of the inhibitor in the presence of various concentrations of SDS were also calculated.

Results and Discussion

Weight-loss Study

The physicochemical parameters of dam water are given in Table 1. The corrosion inhibition efficiencies of sodium 1-heptane sulphonate (SHS)-Zn^{2+} systems and the corresponding corrosion rates (mils per year) are given in Table 2.

The inhibition efficiencies of SHS-Zn^{2+} systems are given in Table 2. It was found that the *IE* increases as the concentration of SHS increases. As the concentration of Zn^{2+} increases, *IE* also increases. A synergistic effect exists between SHS and Zn^{2+}. For example, 250 ppm of SHS has 28 % *IE*. 30 ppm of Zn^{2+} has 21% *IE*. However, interestingly, the formulation consisting of 250 ppm of SHS and 30 ppm of Zn^{2+} has 75% *IE*. This is the mixture of inhibitors that shows better inhibition efficiency than the individual inhibitors [20].

***Table 1.** Water Analysis (Sothuparai dam water, Tamil Nadu, India)*

Parameter	**Result**
Appearance	Brownish
Total dissolved solids	100 ppm
Electrical conductivity	140 $\mu S\ cm^{-1}$
pH	8.25
Total hardness as $CaCO_3$	50 ppm
Calcium	10 ppm
Magnesium	6 ppm
Iron	1.2 ppm
Nitrate	10 ppm
Chloride	10 ppm
Sulphate	2 ppm

***Table 2.** The corrosion inhibition efficiencies and the corresponding corrosion rates of SHS-Zn^{2+} systems*

c_{SHS} / ppm	$c_{Zn^{2+}}$ / ppm					
	0		15		30	
	IE / %	*CR* / mpy	*IE* / %	*CR* / mpy	*IE* / %	*CR* / mpy
0	-	4.4384	12	3.9058	21	2.5063
50	12	3.9058	33	2.9737	47	2.3523
100	17	3.6839	31	3.0625	4748	2.3079
150	21	3.5063	36	2.8406	53	2.0860
200	22	3.4620	39	2.7074	65	1.5534
250	28	3.1956	41	2.6187	75	1.1096

Synergism Parameter (S_I)

Synergism parameters are indications of synergistic effect existing between inhibitors [21, 22]. S_I value is found to be greater than the one indicating synergistic effect existing between Zn^{2+} of concentrations 15ppm and 30 ppm and various concentrations of SHS. The results are given in Table 3.

$$S_I = (1-I_{1+2})/(1-I'_{1+2}) \qquad [3]$$

where $I_{1+2} = (I_1+I_2)-(I_1 \times I_2)$

I_1 = surface coverage of inhibitor (SHS)

I_2 = surface coverage of inhibitor (Zn^{2+})

I'_{1+2} = combined surface coverage of inhibitors (SHS) and (Zn^{2+})

Surface coverage = (*IE* /100) / %.

I_2 for Zn^{2+} (15 ppm) = 0.12 and I_2 for Zn^{2+} (30 ppm) = 0.21

Table 3. *Synergism parameter (S_I).*

c_{SHS} / ppm	I_1	SHS-Zn^{2+}(15 ppm) I'_{1+2}	S_I	SHS-Zn^{2+}(30 ppm) I'_{1+2}	S_I
50	0.12	0.33	1.1558	0.47	1.3117
100	0.17	0.31	1.058	0.48	1.2610
150	0.21	0.36	1.086	0.53	1.3279
200	0.22	0.39	1.125	0.65	1.7606
250	0.28	0.41	1.073	0.75	2.2752

Analysis of Variance (F- Test)

F-test was used to investigate whether the synergistic effect existing between two inhibitors is statistically significant [23, 24]. If F-value was above 5.32 for 1,8 degrees of freedom, the synergistic effect proved to be statistically significant. If it was below 5.32 for 1, 8 degrees of freedom, it was statistically insignificant at a 0.05 level of significance. The results are given in Table 4.

Table 4. *Analysis of Variance (F- Test)*

$c_{Zn^{2+}}$ /ppm	Source of variances	Sum of squares	Degrees of freedom	Mean square	F-value	*p* value
15	between	128	1	128	4.8762	< 0.05
	within	210	8	26.25		
30	between	706.88	1	706.88	7.7979	>0.05
	within	725.20	8	90.65		

The obtained F- value of 4.8762 for 15 ppm of Zn^{2+} was not statistically significant. Therefore, it was concluded that the influence of 15 ppm of Zn^{2+} on the inhibition efficiencies of various concentrations of SHS was not statistically significant.

The obtained F- value of 7.7979 for 30 ppm of Zn^{2+} was statistically significant. Therefore, it was concluded that the influence of 30 ppm of Zn^{2+} on the inhibition efficiencies of various concentrations of SHS was statistically significant.

Influence of Immersion Period on the IE of SHS (250 ppm) - Zn^{2+} (30 ppm) System

The influence of immersion period on *IE* of SHS (250 ppm) - Zn^{2+} (30 ppm) is shown in Table 5.

Table 5. *Influence of immersion period on the IE of SHS (250 ppm) - Zn^{2+} (30 ppm) system.*

System	Immersion period, days				
	1	3	5	7	10
Dam water (DW) *CR* / mpy	7.8431	4.4384	6.8400	4.0914	4.4327
DW+SHS (250 ppm)+Zn^{2+}(30 ppm) *CR* /mpy	1.4902	1.1096	2.1888	1.6365	2.3050
IE / %	81	75	68	60	48

It was found that, as the immersion period increases, the inhibition efficiency decreases. This may be due to the fact that, as the period of immersion increases, the protective film Fe^{2+}- SHS complex formed on the metal surface is broken by the continuous attack of other ions present in the environment; hence, the IE decreases as the immersion period increases. A similar behavior was observed in the corrosion prevention of carbon steel by methyl orange [25].

Influence of pH on the IE of SHS (250 ppm) - Zn^{2+} (30 ppm) System

The influence of pH (addition of H_2SO_4 or addition of NaOH) on the *IE* of SHS -Zn^{2+} system is given in Table 6.

Table6. *Influence of pH on the IE of SHS (250 ppm) - Zn^{2+} (30 ppm) system; Immersion period three days*

System	pH				
	3	5	7	9	11
Dam water (DW) *CR* / mpy	6.1535	5.9589	4.4384	4.8036	3.3443
DW+SHS (250 ppm)+Zn^{2+}(30 ppm) *CR* /mpy	2.8306	2.2644	1.1096	0.9127	0.4682
IE / %	54	62	75	81	86

All the reactions involved in corrosion process can be classified into a few general reactions [26]. The anodic reaction in every corrosion reaction is the oxidation of a metal to its ion.

$M \rightarrow M^{n+} + ne^-$ (4)

The number of electrons produced is the valence of the ion.

There are several different cathodic reactions that are frequently encountered in metallic corrosion. The most common are:

Hydrogen evolution $2H^+ + 2e^- \rightarrow H_2$ (5)

Oxygen reduction (acid solution) $O_2 + 4H^+ + 4e^- \rightarrow 2H_2O$ (6)

Oxygen reduction (neutral or basic solution) $O_2 + 2H_2O + 4e^- \rightarrow 4OH^-$ (7)

In the present study, it was observed that the *IE* decreased when the pH of the solution was lowered (pH=5 and pH=3). This is acidic medium and hence the anodic reaction is dissolution of metal (equation 4) and the cathodic reactions are evolution of hydrogen and oxygen reduction (equations 5 and 6). Because of the release of hydrogen gas from the metal surface, blistering of the protective film takes place and the protective film formed is broken [21].

When the pH is increased (alkaline medium), oxygen is reduced to hydroxide ion (equation 7). The Zn^{2+} ion combines with OH^- ion to form $Zn(OH)_2$ as an insoluble film on the metal surface. Thus, cathodic reaction is controlled and hence the corrosion potential shifts to more negative direction.

Influence of Sodium Potassium Tartrate (SPT) on the IE of SHS (50 ppm) - Zn^{2+} (15 ppm) System

The *IE* and *CR* of carbon steel immersed in solutions containing SHS (50 ppm) - Zn^{2+} (15 ppm) and various concentrations of SPT are presented in Table 7.

***Table 7.** Influence of sodium potassium tartrate (SPT) on the IE of SHS (50 ppm) - Zn^{2+} (15 ppm) system*

c_{SHS} / ppm	$c_{Zn^{2+}}$/ ppm	c_{SPT} / ppm	*CR* / mpy	*IE* / %
0	0	0	4.4384	-
50	15	0	2.9737	33
50	15	50	0.3551	92
50	15	100	0.2663	94
50	15	150	0.0888	98
50	15	200	0.5326	88
50	15	250	0.6214	86

It is noted that for SPT, *IE* increases up to the concentration of 150 ppm and then it decreases. As the concentration of inhibitors (SPT) increases, more of the inhibitors are transported towards the metal surface through the formation of Zn^{2+}-inhibitor complex. On the metal surface, Zn^{2+}-inhibitor complex is converted into Fe^{2+}- inhibitor complex. This complex is insoluble, stable, and more compact and, therefore, *IE* increases up to 150 ppm concentration. On further increasing the concentration of the inhibitors, the complex formed on the metal surface goes into solution. Thus, the *IE* decreases. Another reason may be that when more inhibitors are added, the Zn^{2+}- inhibitor complex formed is precipitated in the bulk of the solution. Also, the inhibitors are not transported towards the metal surface; hence, the *IE* decreases. Similar observations have been made by Rajendran et al. while studying the inhibition efficiency of Henna extract [27] and also by Arockia Selvi *et al.* [28] who investigated the inhibition efficiency of diethylenetriamine penta-methylene-phosphonic acid-Zn^{2+} system.

Biocidal Efficiency of SDS - SHS (250 ppm)- Zn^{2+} (30 ppm) System

***Table 8.** Influence of biocide (SDS) on SHS (250 ppm)- Zn^{2+} (30 ppm) system*

c_{SHS} / ppm	$c_{Zn^{2+}}$/ ppm	c_{SDS} / ppm	Colony forming unit	Biocidal efficiency, %
0	0	0	26×10^4	-
0	0	150	18×10^4	30
250	30	50	59×10^3	77
250	30	100	22×10^3	91
250	30	150	14×10^3	94
250	30	200	43×10^2	98
250	30	250	nill	100

The biocidal nature of SDS (an anionic surfactant) has also been studied [29]. The inhibitor system that offered the best corrosion inhibition efficiency was selected. Biocidal efficiency of SDS at various concentrations on the best system in dam water is determined. It is interesting to note that the SHS (250 ppm)- Zn^{2+} (30 ppm) system has some biocidal efficiency (BE).

In Table 8, 150 ppm of SDS alone has 30 % BE. When 50 ppm of SDS is added to the SHS (250 ppm)- Zn^{2+} (30 ppm) system, BE increases from 30% to 77%. On increasing the concentration of SDS from 50 ppm to 100 ppm, 150 ppm, 200 ppm, BE continues to increase and 100% BE is achieved when the concentration of SDS is 250 ppm.

Analysis of polarization curves

The potentiodynamic polarization curves of carbon steel immersed in various test solutions are shown in Fig.1. The corrosion parameters are given in Table 9.

***Table 9**. Corrosion parameters of carbon steel immersed in dam water in the presence and absence of inhibitor obtained by polarization method*

c_{SHS} / ppm	$c_{Zn^{2+}}$/ ppm	E_{corr} / mV *vs* SCE	I_{corr} / A cm^{-2}	b_a / mV dec^{-1}	b_c / mV dec^{-1}	*LPR* / Ω cm^2
0	0	-494	2.66×10^{-6}	166	203	2.053×10^4
250	30	-573	4.96×10^{-7}	181	152	7.701×10^4

When carbon steel is immersed in dam water, the corrosion potential, E_{corr}, is -494 mV *vs* SCE. The formulation consisting of 250 ppm SHS + 30 ppm Zn^{2+} shifts the corrosion potential to -573 mV *vs* SCE. This suggests that the cathodic reaction is controlled predominantly. The corrosion current value and linear polarization resistance (*LPR*) value for dam water are $I_{corr} = 2.66\times10^{-6}$ A cm^{-2} and $LPR = 2.053\times10^4$ Ω cm^2.

For the formulation of SHS (250 ppm) + Zn^{2+} (30 ppm), the corrosion current value (I_{corr}) has decreased to 4.96×10^{-7} A cm^{-2}, and the LPR value has increased to 7.701×10^4 Ω cm^2. This indicates that a protective film is formed on the metal surface. When a protective film is formed on the metal surface, *LPR* value increases and corrosion current value decreases. The fact that the *LPR* value increases with decrease in corrosion current indicates that inhibition of corrosion takes place by the formation of a protective film. There is an increase in the value of anodic Tafel slope. Similar observations have been reported by Joseph Rathish *et al.* [30] while studying the corrosion behavior of metals in artificial sweat; Yesu Thangam *et al.* [31] while studying inhibition of corrosion of carbon steel in a dam water by sodium molybdate-Zn^{2+} system, and also by Agnesia Kanimozhi and Rajendran [29] while studying the corrosion inhibition sodium tungstate-Zn^{2+}-ATMP system.

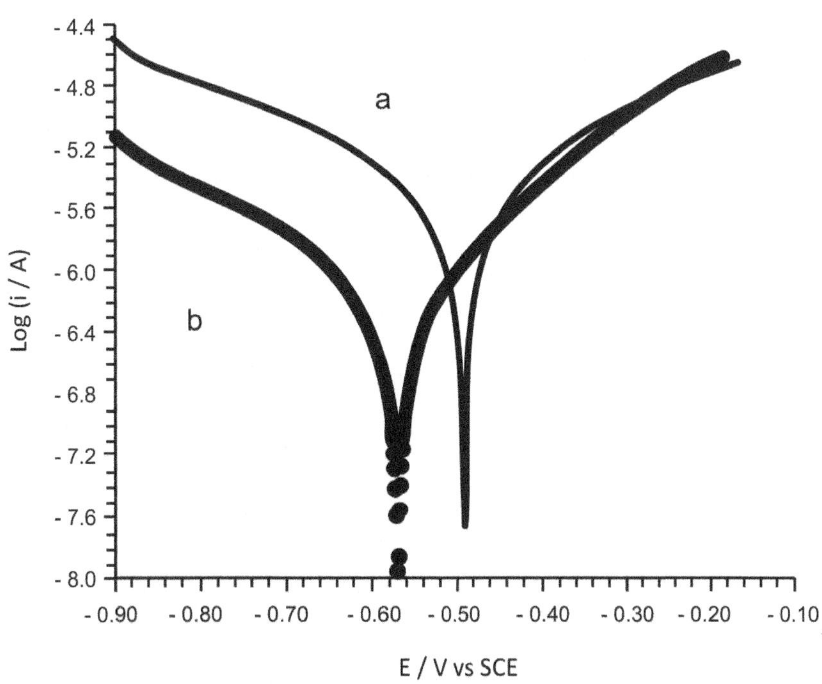

***Fig 1.** Polarization curves of carbon steel immersed in various test solutions **a)** dam water **b)** dam water containing 250 ppm of SHS and 30 ppm of Zn^{2+}*

Analysis of AC impedance spectra

AC impedance spectra have been used to detect the formation of a film on the metal surface. If a protective film is formed, the charge transfer resistance (R_t) increases and double layer capacitance (C_{dl}) value decreases [32, 33]. The AC impedance spectra of carbon steel immersed in various solutions are shown in Fig. 2.

Table 10. *Impedance parameters of carbon steel in dam water in the presence and absence of inhibitor obtained by AC impedance method*

c_{SHS} / ppm	$c_{Zn^{2+}}$/ ppm	R_t/ Ω cm^2	C_{dl} / F cm^{-2}	log (Z / Ω)
0	0	1.084×10^4	8.24×10^{-10}	4.062
250	30	1.8837×10^4	4.72×10^{-10}	4.276

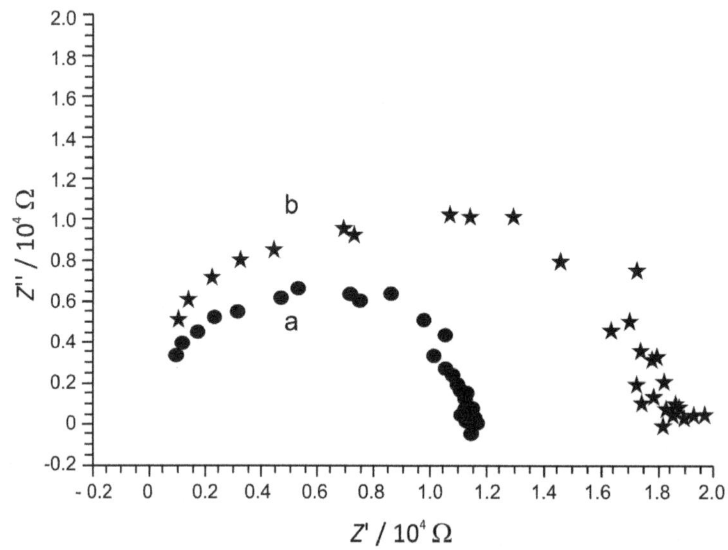

Fig. 2. *AC impedance spectra of carbon steel immersed in various test solutions* ***a)*** *dam water* ***b)*** *dam water containing 250 ppm of SHS and 30 ppm of* Zn^{2+}

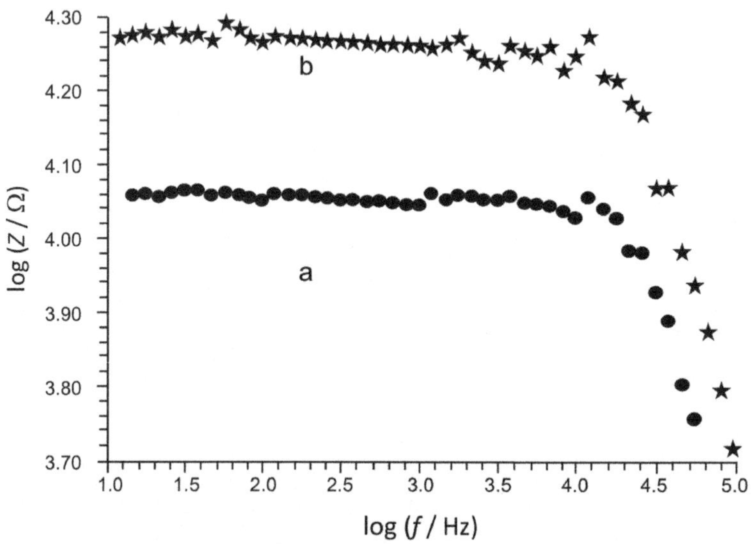

Fig. 3. *AC impedance spectra of carbon steel immersed in various test solutions. (frequency Bode plots)* ***a)*** *dam water* ***b)*** *dam water containing 250 ppm of SHS and 30 ppm of* Zn^{2+}

The AC impedance parameters, *i.e.*, charge transfer resistance (R_t) and double layer capacitance (C_{dl}), are given in Table 10. When carbon steel is immersed in dam water, R_t value is 1.084×10^4 Ω cm^2 and C_{dl} value is 8.24×10^{-10} F cm^{-2}. When SHS and Zn^{2+} are added to dam water, R_t value increases from 1.084×10^4 Ω cm^2 to 1.8837×10^4 Ω cm^2 and the C_{dl} decreases from 8.24×10^{-10} F cm^{-2} to 2.13×10^{-10}. This suggests that a protective film is formed on the surface of the metal. This is further supported by the increase in impedance value log*Z* from 4.062 Ω to 4.276 Ω (Fig. 3). This accounts for the very high IE of SHS-Zn^{2+} system.

Analysis of FTIR spectra

Earlier researchers have confirmed that FTIR spectrometer is a powerful instrument that can be used to determine the type of bonding for organic inhibitors adsorbed on the metal surface [34]. FTIR spectra were used to analyze the protective film formed on metal surface. FTIR spectrum of pure SHS is given in Fig. 4a. The FTIR spectrum of the film formed on the metal surface after immersion in the dam water for 3 days containing 250 ppm of SHS and 30 ppm of Zn^{2+} is shown in Fig. 4b. The S=O stretching frequency has decreased from 1193 cm^{-1} to 1118 cm^{-1}. This indicates that the oxygen atom of S=O group has coordinated with Fe^{2+} formed on the metal surface, resulting in the formation of Fe^{2+}-SHS complex on the anodic sites of the metal surface. The peak at 1399 cm^{-1} is due to Zn-O stretching. The stretching frequency due to –OH appears at 3398 cm^{-1}. Therefore, it is concluded that Zn $(OH)_2$ is formed on cathodic sites of the metal surface [35].

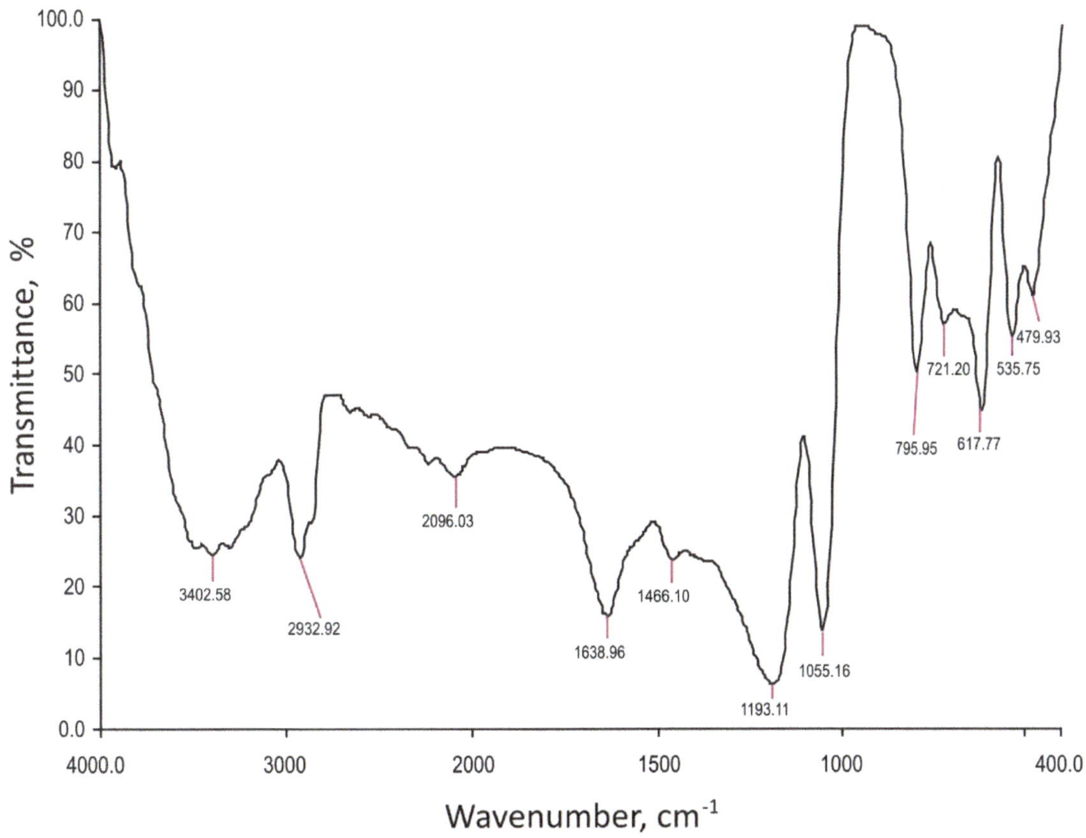

Fig. 4a. *FTIR spectrum of pure sodium heptane sulphonate*

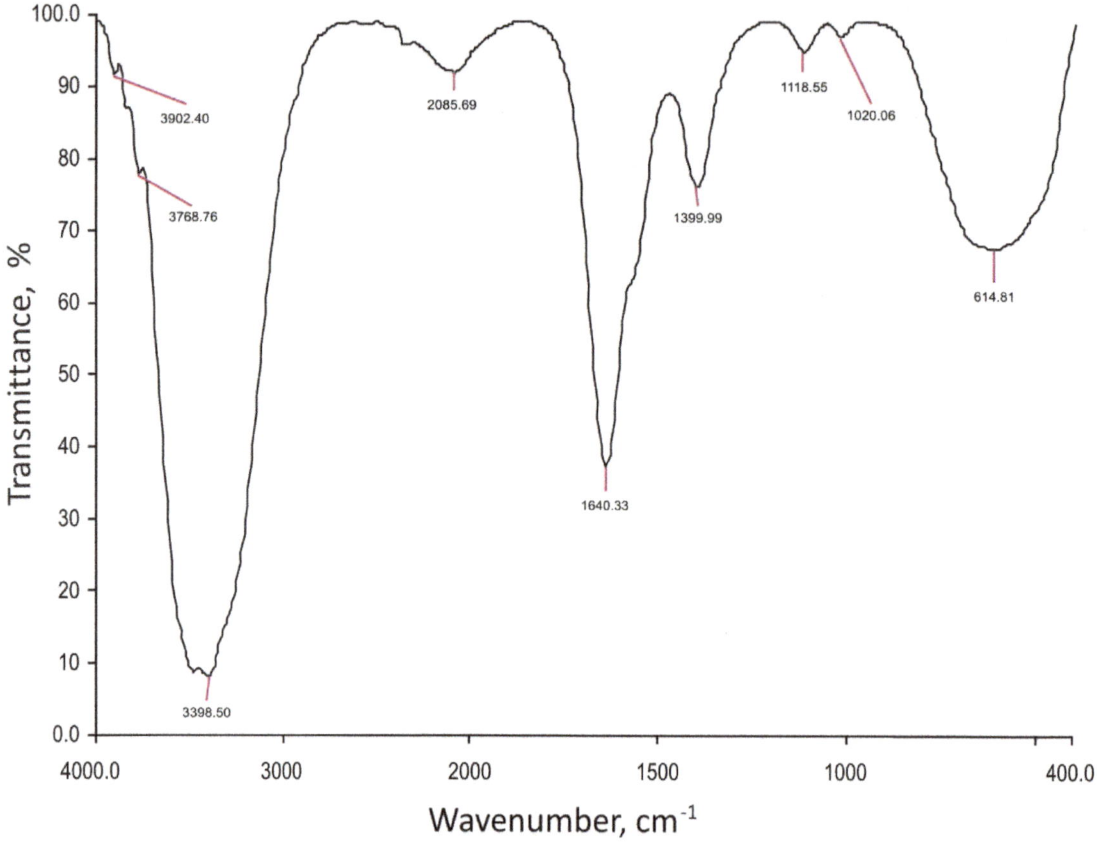

***Fig. 4b**. FTIR spectrum of film formed on metal surface after immersion in dam water containing 250 ppm of SHS -30 ppm of Zn^{2+}*

UV-visible absorbance spectra and fluorescence spectra.

UV-visible absorption spectra and fluorescence spectra can be used to confirm the protective film formed on the metal surface [36, 37].

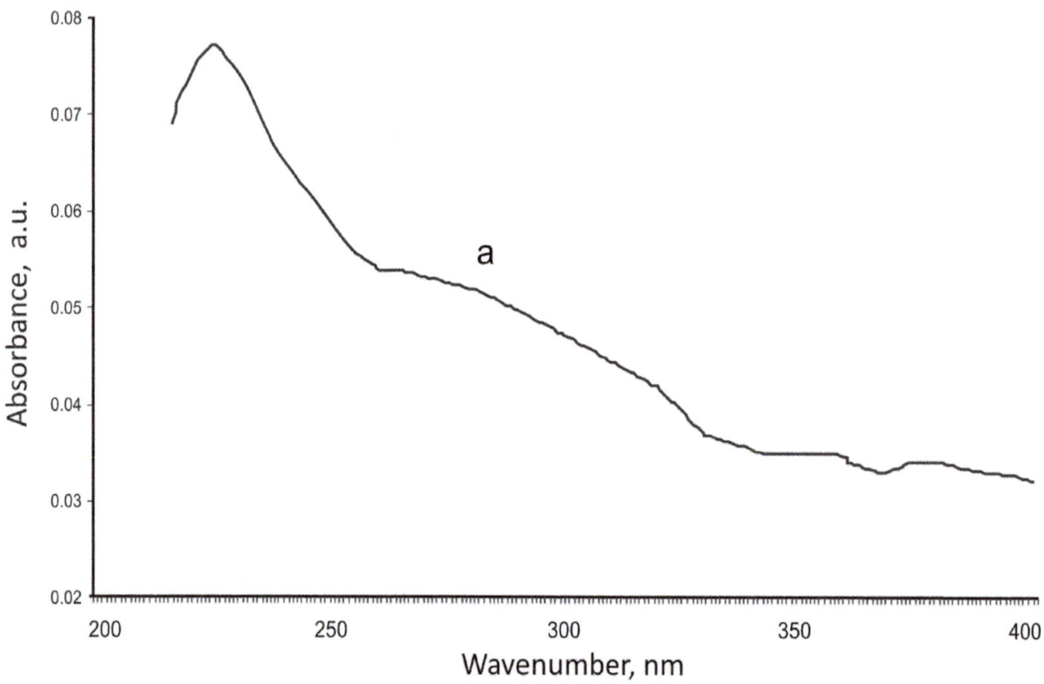

***Fig. 5a.** UV visible absorption spectra of SHS solution*

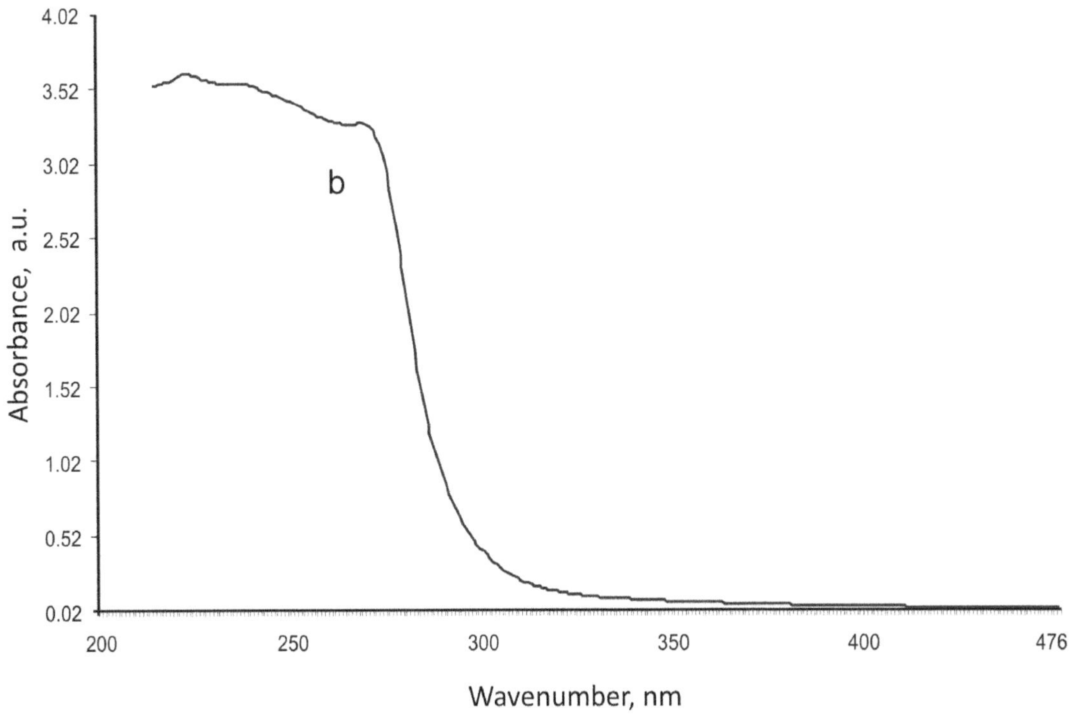

Fig. 5b. *UV visible absorption spectra of SHS+Fe^{2+} solution*

UV-visible absorption spectrum of an aqueous solution of sodium heptane sulphonate (SHS) is shown in Fig. 5 curve (a). The peak appears at 226 (0.077) nm (the peak intensity is given in parentheses). The corresponding emission spectrum (λ_{ex} = 226 nm) is shown in Fig. 6 curve (a). Peaks appear at 396 (3.71), 450 (1.41), and 505 (1.91) nm (the peak intensities are given in parentheses). Fe^{2+}-SHS complex in solution was prepared by mixing SHS solution and Fe^{2+} ion (as $FeSO_4{\cdot}7H_2O + H_2O$ freshly prepared). Its UV-visible absorption spectrum is shown in Fig. 5 curve (b). Peaks appear at 226 (3.62), 242 (3.56) and 278 (3.29) nm. The corresponding emission spectrum (λ_{ex} = 226 nm) is shown in Fig. 6 curve (b). Peaks appear at 397 (4.35) and 510 (2.30) nm. These peaks correspond to Fe^{2+} - SHS complex.

The emission spectrum (λ_{ex} = 226 nm) of the film formed on the carbon steel metal surface after immersion in the solution containing 30 ppm of Zn^{2+} and 250 ppm of SHS is shown in Fig. 6 curve (c). Peaks appear at 395 (3.88) and 511 (1.79) nm. These peaks correspond to Fe^{2+}-SHS complex. Thus, fluorescence spectral study leads to the conclusion that protective film consists of Fe^{2+}-SHS complex formed on the metal surface.

SEM Analysis of Metal Surface

The SEM image of ×500 magnification of carbon steel specimen immersed in dam water for one day in the absence and presence of inhibitor system are shown in Figs.7b and 7c, respectively.

The SEM micrographs of polished carbon steel surface (control) in Fig. 7a shows the smooth surface of the metal. This implies the absence of any corrosion product formed on the metal surface.

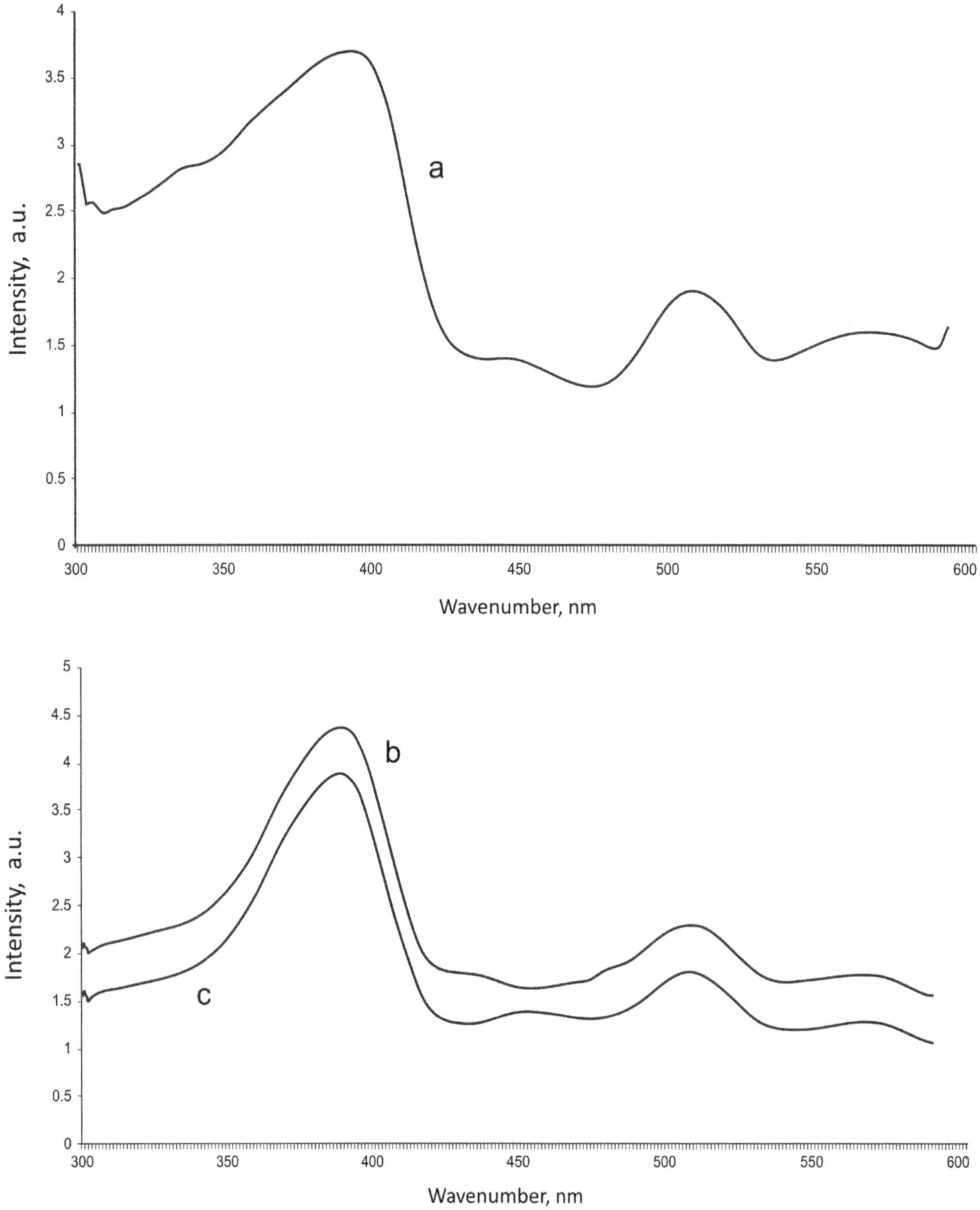

Fig. 6: *Fluorescence Spectra* ***a)*** *SHS solution;* ***b)*** *SHS+Fe^{2+} solution ,**c)** film formed on metal surface after immersion in solution containing dam water + 250 ppm of SHS and 30 ppm of Zn^{2+}*

The SEM micrographs of carbon steel surface immersed in dam water in Fig. 7b show the roughness of the metal surface, which indicates the corrosion of carbon steel in dam water. Fig. 7c indicates that, in the presence of 250 ppm of SHS and 30 ppm of Zn^{2+} mixture in dam water, the surface coverage increases, which in turn results in the formation of insoluble complex on the surface of the metal (SHS-Zn^{2+} inhibitor complex) and the surface is covered by a thin layer of inhibitors, which control the dissolution of carbon steel. Such results were reported earlier [38].

a b c

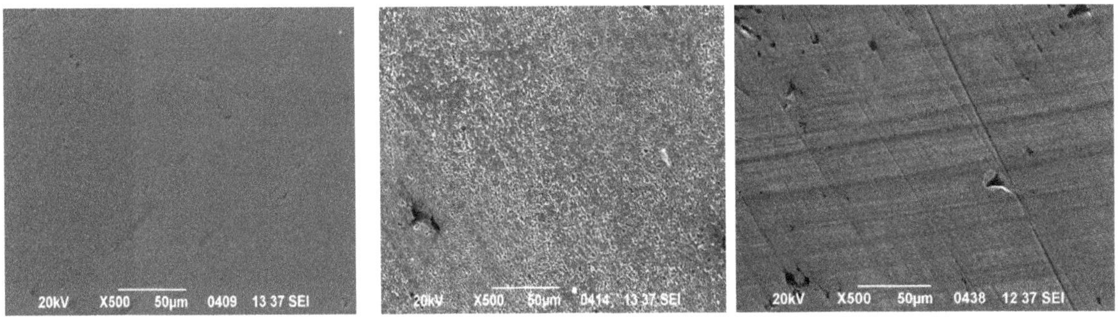

Fig. 7. *SEM micrographs (magnification-×500) of* ***a)*** *polished carbon steel (control) -*
b) *carbon steel immersed in dam water*
c) *carbon steel immersed in dam water containing 250 ppm of SHS and 30 ppm of* Zn^{2+}

Atomic Force Microscopy Characterization

AFM is a powerful technique to investigative the surface morphology at nano- to micro-scale and has become a new choice to study the influence of inhibitor on the generation and progress of the corrosion at the metal/solution interface [38-40]. The three dimensional (3D) AFM morphologies and the AFM cross-sectional profile for polished carbon steel surface (reference sample), carbon steel surface immersed in dam water (blank sample), and carbon steel surface immersed in dam water containing the formulation of 250 ppm of SHS and 30 ppm of Zn^{2+} are shown as Figs. 8a and 8d; 8b and 8e; and 8c and 8f, respectively.

Root– Mean-Square Roughness, Average Roughness and Peak-to-Valley Value

AFM image analysis was performed to obtain the average roughness, R_a (the average deviation of all points roughness profile from a mean line over the evaluation length), root-mean-square roughness, R_q (the average of the measured height deviations taken within the evaluation length and measured from the mean line), and the maximum peak-to-valley, *P-V*, height values (largest single peak-to-valley height in five adjoining sampling heights) [38]. Table 11 is a summary of R_q, R_a, *P-V* values for carbon steel surface immersed in different environment.

Figs. 8a and 8d displays the surface topography of un-corroded metal surface. The value of R_q, R_a and *P-V* height for the polished carbon steel surface (reference sample) are 4.3 nm, 3.41 nm, and 35.28 nm respectively. The slight roughness observed on the polished carbon steel surface is due to atmospheric corrosion.

Figs. 8b and 8e displays the corroded metal surface with few pits in the absence of the inhibitor immersed in dam water. The R_q, R_a and *P-V* height values for the carbon steel surface are 31.9 nm, 24.9 nm, and 420.3 nm, respectively. These data suggest that carbon steel surface immersed in dam water has a greater surface roughness than the polished metal surface, which shows that the unprotected carbon steel surface is rougher due to the corrosion of the carbon steel in dam water environment.

Figs. 8c and 8f displays the steel surface after immersion in dam water containing 250 ppm of SHS and 30 ppm of Zn^{2+}. The R_q, R_a and *P-V* height values for the carbon steel surface are 1.17 nm, 0.89 nm, and 10.37 nm, respectively. The R_q, R_a and *P-V* height values are considerably lower in the inhibited environment compared to the uninhibited environment. These parameters confirm that the surface is smoother. The smoothness of the surface is due to the formation of a compact protective film of Fe^{2+}- SHS complex and $Zn(OH)_2$ on the metal surface, thereby inhibiting the corrosion of carbon steel [38].

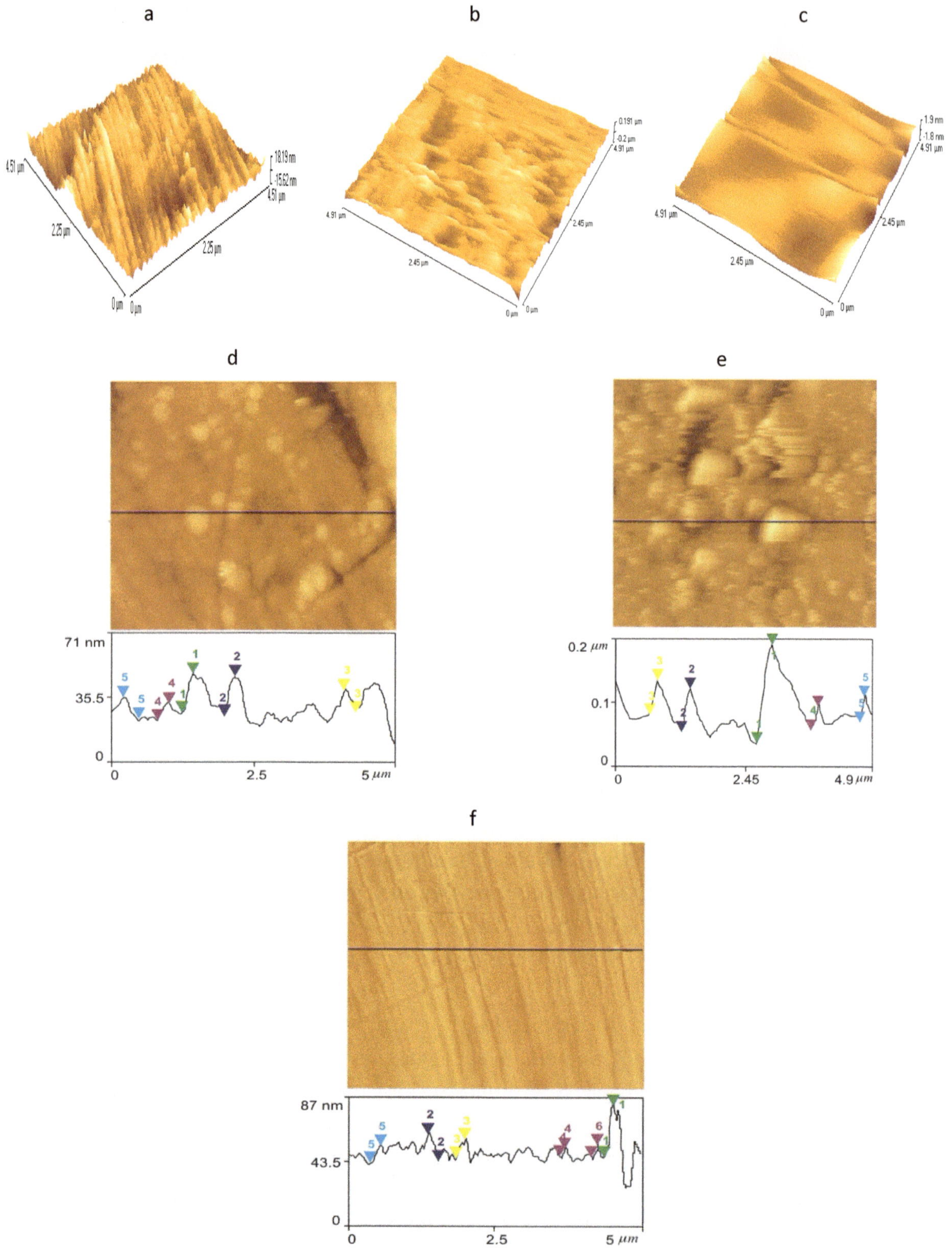

Fig. 8. *Three dimensional AFM images of:* ***a)*** *polished carbon steel (control);* ***b)*** *carbon steel immersed in dam water (blank);* ***c)*** *carbon steel immersed in dam water containing SHS (250 ppm) + Zn^{2+} (30ppm); AFM cross-sectional images of the surface of:* ***d)*** *polished carbon steel (control);* ***e)*** *carbon steel immersed in dam water (blank);* ***f)*** *carbon steel immersed in dam water containing SHS (250 ppm) + Zn^{2+} (30ppm).*

Table 11. *AFM data for carbon steel surface immersed in inhibited and uninhibited environment*

Samples	R_q/ nm	R_a /nm	Maximum peak-to-valley height, nm
Polished carbon steel	4.33	3.41	35.28
Carbon steel immersed in dam water (blank)	31.9	24.9	420.3
Carbon steel immersed in dam water + SHS (250ppm) + Zn^{2+}(30ppm)	1.17	0.89	10.37

Mechanism of corrosion inhibition

From the above discussion, a mechanism may be proposed for the corrosion inhibition of carbon steel immersed in dam water containing 250 ppm SHS and 30 ppm Zn^{2+} system.

When the formulation consists of 250 ppm of SHS and 30 ppm of Zn^{2+} in dam water, there is formation of SHS – Zn^{2+} complex in solution.

When carbon steel is immersed in this solution SHS– Zn^{2+} complex diffuses from the bulk of the solution towards the metal surface.

SHS Zn^{2+} complex is converted into SHS Fe^{2+} complex on the anodic sites of the metal surface with the release of Zn^{2+} ion.

$$Zn^{2+}SHS + Fe^{2+} \rightarrow Fe^{2+}SHS + Zn^{2+} \quad (8)$$

The released Zn^{2+} combines with OH^- to form $Zn(OH)_2$ on the cathodic sites of the metal surface

$$Zn^{2+} + 2\ OH^- \rightarrow Zn(OH)_2\downarrow \quad (9)$$

Thus the protective film consists of SHS Fe^{2+} complex and $Zn(OH)_2$.

This accounts for the synergistic effect of SHS Zn^{2+} system.

Conclusions

The present study leads to the following conclusions:

The inhibition efficiency (IE) of SHS in controlling corrosion of carbon steel immersed in dam water in the absence and in presence of Zn^{2+} has been evaluated by weight loss method. The formulation consisting of 250 ppm SHS and 30 ppm Zn^{2+} has 75% IE. As the immersion period increases, the IE decreases. Polarization study reveals that SHS – Zn^{2+} system functions as a cathodic inhibitor. AC impedance spectra reveal that a protective film is formed on the metal surface. FTIR spectra reveal that the protective film consists of Fe^{2+} SHS complex and $Zn(OH)_2$.

Acknowledgements: *The authors are thankful to their respective management and University Grants Commission, New Delhi.*

References

[1] A. Spinelli, R.S. Gonclaves, *Corros.Sci.* **30** (1990) 1235-1246
[2] W.X. Oliver, R.S.Gonclaves, *J. Braz .Chem. Soc.* **3** (1992) 92-94
[3] Z.B. Vendrame, R.S. Gonclaves, *J. Braz .Chem. Soc.***9** (1998) 441-448
[4] L.D. Melloo, R.S. Gonclaves, *Corros.Sci.* **43** (2001) 457-470
[5] A.M. Lucho, R.S. Gonclaves, D.S. Azambuja, *Corros.Sci.* **44** (2002) 467-479.
[6] S. Martinez, I. Stern, *Appl. Surf. Science* **199** (2002) 83-89
[7] S. Manov, A.M. Lamazouere, L. Aries, *Corros.Sci.* **42** (2000) 1235-1248
[8] J. Dobryszycki, S. Biallozor, *Corros.Sci.* **43** (2001) 1309-1319
[9] K. Aramaki, *Corros.Sci.* **43** (2001) 1985-2000

[10] L. Wang, J.X. Pu, H.C. Luo, *Corros.Sci.* **45** (2003) 677-683
[11] E.E.F. El-Sherbin, S.M.A. Wahaab, M. Deyab, *Mat. Chem. Phys.* **89** (2005) 83-191
[12] J.D. Talti, M.N. Desai, N.K. Shah, *Mat. Chem. Phys.* **93** (2005) 54-64
[13] A. Srhiri, M.Etman, F.Dabosi, *Electrochim. Acta* **41** (1996) 429-437
[14] Lj.M. Vracar , D.M .Drazic, *Corros.Sci.* **44** (2002) 1669-1680
[15] D. Prakash Rajesh Kumar Singh, R.Kumar, *Indian J. Chem.Techn.* **13** (2006) 555-560
[16] R. Manickavasagam, K. Jeya Karthik, M. Paramasivam, S. Venkatakrishna Iyer, *Anti- Corros. Methods M.,* **49** (2002) 19-26
[17] T.A. Aliev, *Mat. Sci.* **44** (2008) 69-74
[18] P. Shakkthivel , T. Vasudevan, *Desalination* **197** (2006) 179-189
[19] G. Wranglen, Chapman & Hall, London (1985) 236
[20] T. Umamathi, J. A. Selvi, S.Agnesia Kanimozhi, S. Rajendran, A. John Amalraj, *Indian J. Chem.Techn.* **15** (2008) 560-565
[21] S. Rajendran, S. Shanmugapriya, T. Rajalakshmi , A.J. Amalraj, *Corrosion* **61** (2005) 685-692.
[22] K.A.Radha, R.Vimala, B.Narayanasamy, J.A.Selvi , S.Rajendran, *Chem. Eng. Commun.* **195** (2008) 352-366
[23] S. Rajendran, A. Raji, J. A. Selvi, A. Rosaly , S. Thangasamy, *J.Mater. Ed.* **29** (2007) 245-258
[24] S. Rajendran, A. Raji, J. Arockia Selvi, A. Rosaly, Thangasamy, *Edutracks.* **6** (2007)30-33
[25] J. Sathiyabama, S. Rajendran, J. A. Selvi, A. J. Amalraj, *Indian J. Chem.Techn.* **15** (2008)462-466
[26] M.G. Fontana, Corrosion Engineering, TATA McGraw-Hill Publishing Company Limited, New Delhi, Third edition, 2006, p.16
[27] S. Rajendrran, M. Agasta, R. B. Devi, B. Sh.Devi, K. Rajam, J. Jeyasundari, *Zastita Materijala.* **50** (2009) 77-84
[28] J. A. Selvi, S.Rajendran, J. Jeyasundari, *Zastita Materijala.* **50** (2009) 91-98
[29] S. A.Kanimozhi, S. Rajendran, *Arab. J. Sci. Eng.* **34** (2009) 37-47
[30] R.J. Rathish, S. Rajendran, J.L. Christy J, *et al. Open Corrosion Journal* **3** (2010) 38-44
[31] Y.Y. Thangam, M.Kalanithi, C.M.Anbarasi, S.Rajendran, *Arab. J. Sci. Eng.* **34** (2009) 49-60
[32] J. Sathiyabama, S. Rajendran, J. A.Selvi, *B. Electrochem.* **22** (2006) 363-370
[33] S. Rajendran M. Manivannan, J.W. Sahayaraj, J.A. Selvi, J. Sathiyabama, A. Amalraj, N. Palaniswamy, *Trans. SAEST.* **41** (2006) 63-67
[34] A. Lalitha, S. Ramesh, S. Rajeswari, *Electrochim. Acta* **51** (2005) 47–55
[35] R. M. Silverstein, F. X. Webster: *Spectrometric Identification of Organic Compounds*, VI Edition, Wiley Student Editon, 2007 p-108
[36] S. Rajendran, B.V. Apparao, N. Palaniswamy, *B. Electrochem* **12** (1996) 15-19
[37] S. Rajendran, S. M.Reenkala, N. Anthony, R. Ramaraj, *Corros. Sci.* **44** (2002) 2243-2252
[38] B. Sherine, A. J.A. Nasser, S. Rajendran, *Int. J. Eng. Sci. Technol.* **24** (2010) 341-357
[39] A.K. Singh, M.A. Quraishi, *Corros.Sci.* **53** (2011) 1288-1297
[40] B. Wang, M.Du, J. Zhang, C.J. Gao, *Corros.Sci.* **53** (2011) 353-361

2

In-situ activated hydrogen evolution by molybdate addition to neutral and alkaline electrolytes

JOHN GUSTAVSSON✉, CHRISTINE HUMMELGÅRD*, JOAKIM BÄCKSTRÖM*, INGER ODNEVALL WALLINDER**, SEIKH MOHAMMAD HABIBUR RAHMAN***, GÖRAN LINDBERGH, STEN ERIKSSON*** and ANN CORNELL

Applied Electrochemistry, School of Chemical Science and Engineering, KTH Royal Institute of Technology, SE-100 44 Stockholm, Sweden
**Department of Natural Sciences, Engineering and Mathematics, Mid Sweden University, SE-851 70 Sundsvall, Sweden*
***Department of Surface and Corrosion Science, KTH Royal Institute of Technology, SE-100 44 Stockholm, Sweden*
****Department of Chemical and Biological Engineering, Chalmers University of Technology, SE-412 96 Gothenburg, Sweden*

✉Corresponding Author

Abstract
Activation of the hydrogen evolution reaction (HER) by in-situ addition of Mo(VI) to the electrolyte has been studied in alkaline and pH neutral electrolytes, the latter with the chlorate process in focus. Catalytic molybdenum containing films formed on the cathodes during polarization were investigated using scanning electron microscopy (SEM), energy-dispersive X-ray analysis (EDS), X-ray photoelectron spectroscopy (XPS), and X-ray fluorescence (XRF). In-situ addition of Mo(VI) activates the HER on titanium in both alkaline and neutral electrolytes and makes the reaction kinetics independent of the substrate material. Films formed in neutral electrolyte consisted of molybdenum oxides and contained more molybdenum than those formed in alkaline solution. Films formed in neutral electrolyte in the presence of phosphate buffer activated the HER, but were too thin to be detected by EDS. Since molybdenum oxides are generally not stable in strongly alkaline electrolyte, films formed in alkaline electrolyte were thinner and probably co-deposited with iron. A cast iron-molybdenum alloy was also investigated with respect to activity for HER. When polished in the same way as iron, the alloy displayed a similar activity for HER as pure iron.

Keywords
Molybdate, molybdenum dioxide, electrodeposition, electrolysis

Introduction

Sodium chlorate is produced in an energy-intensive process where sodium chloride is oxidized into sodium chlorate with hydrogen gas as a product (reaction 1). Industrial, dimensionally stable, anodes (DSA®) are used as anodes while the cathodes are normally made from low-alloyed carbon steel or titanium. Steel cathodes corrode, which shortens their lifetime and may result in short circuiting of the electrolysis cells and furthermore corrosion products that are formed can contaminate the product. Steel cathodes have overpotentials of around 800 mV for hydrogen evolution (reaction 2) [1].

$$NaCl + 3\,H_2O \rightarrow NaClO_3 + 3\,H_2 \quad (1)$$

$$2\,H_2O + 2e^- \rightarrow H_2 + 2\,OH^- \quad (2)$$

Titanium cathodes are more corrosion resistant than steel cathodes. Drawbacks are that titanium hydride forms over time and that Ti shows even higher overpotentials compared to steel. The electric energy used for electrolysis accounts to up to 80% of the variable production cost of chlorate, and with rising costs for electricity a reduction of the cathode overpotential is of utmost importance. Another major concern is the search for an alternative to the carcinogenic Cr(VI) electrolyte additive. At present, Cr(VI) is necessary to obtain high current efficiency in the process. During operation Cr(VI) is electrodeposited on the cathode, forming a film of chromium hydroxide that hinders the electroreduction of chlorate and hypochlorite (HClO and ClO^-), an intermediate in the chlorate process [2]. Chromate also functions as a buffer to keep the electrolyte pH between 5.9 and 6.7, which is optimal for the process [3].

Many attempts have been made to replace the cathode material in the chlorate process by electrodes with coatings containing ruthenium compounds as electrocatalyst, see for example Refs [4] and [5]. Although they show a high catalytic activity for hydrogen evolution, their long-term stability has not yet been resolved. The chlorate cathode is subject to stringent requirements– it should be stable both during vigorous hydrogen evolution and during production stops in the corrosive electrolyte. Also, its corrosion products should not catalyze the decomposition of hypochlorite in the bulk electrolyte and cathodes containing for example nickel or cobalt are therefore excluded [1].

In-situ activation, the addition of a compound to the electrolyte that enhances the electrocatalytic properties of the cathodes, has several advantages over the above mentioned cathode replacement – fast, simple implementation in existing plants, a constant renewal of the cathode surface and the possibility to use substrate materials that can survive the harsh conditions in the chlorate process.

Molybdenum (VI) has shown some promising features as an in-situ activator for hydrogen evolution, both in chlorate electrolyte [6][7] and in strongly alkaline solutions [8]-[13]. What makes it even more interesting for the chlorate application is the ability of Mo-containing films formed in-situ to suppress cathodic oxygen reduction [14] (and thereby also possibly the reduction of hypochlorite). Molybdenum (VI) is regarded as an environmentally friendly alternative to Cr(VI) in corrosion applications [14].

Li *et al.* [6] have investigated the possibility to replace Cr(VI) with Mo(VI) in the chlorate electrolyte. They concluded that Mo(VI) addition decreased the overpotential for hydrogen evolution by 100-130 mV on steel cathodes, that the two compounds had comparable buffer capacity, and that the Mo(VI) addition increased the levels of unwanted oxygen in the cell gas. Later studies [7] have indicated that the effect on oxygen levels depends on the Mo(VI) concentration, and that low concentrations of MoO_3 (1-10 mg dm^{-3} (7-70 μM Mo(VI))) can activate the hydrogen evolution reaction (HER) without any increased oxygen production. In the absence of Cr(VI), such low Mo(VI) levels require an additional buffer such as phosphate to stabilize the electrolyte pH. These studies in chlorate electrolyte did not include any surface analyses of possible electrode films.

At strongly alkaline conditions, 30 wt.% KOH at 70°C [8]-[13], it was found that films containing molybdenum and iron, the latter present as an impurity in the electrolyte, co-deposited on the metal cathodes and resulted in hydrogen overpotentials at -1 kA m^{-2} that were virtually independent of the substrate material. Typically, after 17 h of polarization in the presence of 4 mM Mo(VI), the observed overpotentials for hydrogen evolution on metal substrates of Co, Cu, Fe, Mo, Nb, Ni, Pd, Pt, V, W, and Zr varied within 40 mV compared to a 650 mV variation in the absence of Mo(VI) [11]. In the absence of electrolyte impurities such as iron, no molybdenum-containing deposits were found [13]. Elemental molybdenum cannot be electrodeposited from aqueous electrolytes, but as alloys with iron group metals [14] or as molybdenum oxides [14].

The ability of electrolyte additions of Mo(VI) to suppress the cathodic oxygen reduction has been investigated on copper [14]. The inhibiting effect was most efficient at pH 8.2 and ended at pH 11, probably due to differences in the cathodic surface films formed in the electrolytes of varying pH. In particular the oxide MoO_2, a probable constituent of a cathode film formed by electroreduction of Mo(VI), has limited stability in alkaline solutions [14]4,[16]. The study did not include surface analyses of cathode films that formed.

The aim of the present study is to demonstrate how Mo(VI) additions to the electrolyte can activate the HER. As the properties of the cathode films formed depend on the electrolyte pH, films formed at pH 6.5 (relevant for the chlorate process) have been compared to films formed at alkaline conditions, 1 M NaOH. Polarization curves for hydrogen evolution have been combined with surface analyses by scanning electron microscopy (SEM), energy-dispersive X-ray analysis (EDS), X-ray photoelectron spectroscopy (XPS), X-ray fluorescence (XRF), optical microscopy and X-ray diffraction (XRD). Interesting results on films of co-precipitated Mo and Fe initiated the casting of iron-molybdenum alloy. This cast material could potentially be polished to a smooth surface and allow electrocatalytic effects to be separated from effects of increased surface area, which can be difficult when studying electrodeposited films.

The long-term goal is a chromate-free chlorate process with stable, energy-efficient cathodes. Selectivity aspects of replacing Cr(VI) with Mo(VI) will be presented in a later communication.

Experimental

Electrochemistry

The working electrodes were rotating-disc electrodes (RDEs) of commercially pure titanium (Ø 4 mm), platinum (Ø 4 mm), gold (Ø 4 mm), molybdenum (Ø 6 mm), iron (Ø 5 mm), and iron-molybdenum alloy (Ø 6mm), all embedded in Teflon, and a titanium RDE (Ø 11.3 mm) in a titanium holder shielded by epoxy and silicon tubing. The RDEs were run with an electrode rotator Model 616 from Pine Instrument Company. A platinum grid was used as the counter electrode and the reference electrodes were a red rod electrode (REF201 from Radiometer) with saturated KCl, a saturated calomel electrode (REF401 from Radiometer), or an Hg/HgO/1 M KOH electrode (XR400 and XR440 from Radiometer), all connected to a Luggin capillary.

The platinum RDE was polished with alumina paste (Alpha micropolish No. 1 C, particle size 1.0 μm, from Buehler), washed with MilliQ water, cleaned in an ultrasonic bath with acetone and thoroughly washed with MilliQ water. The molybdenum, iron, iron-molybdenum alloy and titanium RDEs were polished with grit 4000 SiC grinding paper. When the experiments began, the iron electrode was immersed in the electrolyte under cathodic polarization to avoid uncontrolled corrosion.

The electrolytes were made from MilliQ water; sodium chloride, sodium hydroxide, hydrochloric acid, iron(III) chloride hexahydrate, sodium molybdate dihydrate, sodium dihydrogen phosphate, all from Merck, and of pro analysi grade. When we use the terms "phosphate concentration" or "total phosphate concentration", we are referring to the sum of phosphate species, for example $[PO_4^{3-}]_{tot}=[H_3PO_4]+ [H_2PO_4^{-}]+ [HPO_4^{2-}]+ [PO_4^{3-}]$. In the present study, we excluded $NaClO_3$ from the electrolytes as trace amounts of Cr(VI) present in the chlorate salt may form cathodic surface films that influence the kinetics for the HER. As model systems, either 2 M NaCl (around neutral pH) or 1 M NaOH (pH≈13, 70°C) were used. Nitrogen purging was used 15 min prior and during the electrochemical experiments to deaerate the electrolyte

Galvanostatic polarization experiments were recorded with a PAR 273A potentiostat controlled by LabView. Each point was recorded for 15 seconds and a current interrupt technique [17] was used for the IR-correction. Each experiment was repeated at least once with good repeatability.

Current efficiency measurements were made to evaluate how large a proportion of the applied current that was related to this side reaction. A divided cell was used [23] with stationary titanium electrodes and electrolytes of 2 M NaCl with and without 100 mM Mo(VI) polarized at -3 kA m^{-2} at 70°C. Both an electrolyte of 2 M NaCl, pH 6.5 and an alkaline electrolyte containing 2 M NaCl, 1 M NaOH were investigated. This relatively high concentration was chosen to see any effect on current efficiency more clearly.

Surface analysis

RDEs polarized for SEM measurements were removed from the electrolyte without breaking the polarization current and were carefully dipped into MilliQ water for a few seconds to remove excess electrolyte. For the SEM analysis, LEO 1450 EP SEM and Zeiss EvO 50 SEM with EDS were used with accelerator voltages of 20 kV. XRF measurements were performed using a Niton XLT898 instrument with an accelerator voltage of 35 kV and a silver anode.

The molybdenum iron alloy sample was molded in epoxy and polished in four steps: (i) SiC grit 220 grinding paper, (ii) 9 µm diamond suspension, (iii) 3 µm diamond suspension, and (iv) 0.04 µm colloid silica suspension. All polishing material was purchased from Struers. The polished sample was metallographically etched in a freshly prepared 1:1 mixture of 10 wt.% KOH (aq) and 10 wt.% $K_3Fe(CN)_6$ (aq) [18][20] and inspected using a Zeiss Axiotech Vario optical microscope with a Kappa digital camera.

XPS compositional analyses were performed using an UltraDLD spectrometer (Kratos Analytical, Manchester, UK) with a monochromatic Al x-ray source (1486.5 eV) on two separate areas of approximately 700 × 300 µm for each sample. Binding energies are given as the average of the peak positions of the two areas. The chamber pressure was approximately 2×10^{-9} mbar during analysis. Wide spectra (pass energy of 160 eV) and core level high resolution spectra (20 eV pass energy) were acquired for Mo 3d, Ti 2p, O 1s, and C 1s. All binding energies were corrected by setting the hydrocarbon C1s peak to 285.0 eV.

XRD measurements were carried out at ambient temperature using a Bruker AXS D8 ADVANCE VARIO powder diffractometer. The sample disc surface was polished gradually using grit 500, 1000, 2400 to 4000 SiC paper and rinsed with acetone in an ultrasonic cleaner. The diffractometer was operated with CuK_{α} radiation generated at 40 kV acceleration voltage and 40 mA current. High resolution X-ray diffraction patterns were collected in order to identify the phases using monochromatic Cu$K_{\alpha 1}$ (1.5406 Å) radiation obtained with a germanium primary monochromator, using a solid-state rapid LynxEye detector in a Bragg-Brentano geometry. Scans were performed for the 2θ range 20-100°, with a step size of 0.0092° and 3.9 s collection time per step. The resulting pattern was imported into the evaluation software package DIFFRAC^{plus}EVA with SEARCH, where phases were identified, after background subtraction with the aid of the ICDD database [21].

Iron-molybdenum alloy casting

Iron pieces 99.99% and molybdenum slug 6.35 mm diameter × 6.35mm length, 99.95% were ordered from Alfa Aesar for the alloy production. The alloy was melted in a casting form with 40 mm × 40 mm sides and the mass was 500 g total alloy with nominal composition 67 at.% (54.2 wt.%) iron and 33 at.% (45.8 wt.%) molybdenum by Swerea KIMAB AB, Stockholm, Sweden. A small part of the molybdenum formed a separate layer and was later removed. Since the melting point increases with increasing molybdenum content [22] alloys with higher molybdenum content could not be cast. According to Laser Ablation Inductively Coupled Plasma Mass Spectrometry (LA-ICP-MS) analysis, the alloy produced had a composition of 72.4 at.% (60.4 wt.%) iron and 27.6 at.% (39.5 wt.%) molybdenum.

Results and discussion

The section is organized as follows: first, we present the electrochemical measurements showing an in-situ activation effect for HER of adding Mo(IV) salt to NaCl and NaOH electrolytes. We also investigated how addition of phosphate buffer modified the results. In the next subsection, we present our results from surface analyses of the films formed during electrolysis in the preceding subsection. In the final subsection, we address the issue of synergetic effects when combining iron and molybdenum by investigating the electrochemistry of a well-characterized cast alloy sample during HER.

In-situ activation

In Figure 1, polarization curves for hydrogen evolution are shown for molybdenum and titanium rotating disc electrodes in 2 M NaCl at pH 6.5 in both absence and presence of 4 mM Mo(VI). Without Mo(VI) in the electrolyte, both titanium and molybdenum were less active than iron with Tafel slopes of -340 mV dec^{-1} and -180 mV dec^{-1}, respectively. When reversing the polarization curves (not shown), titanium showed a more active electrode, indicating non-stationary conditions. In the presence of Mo(VI), the two substrates gave similar potentials for the HER and Tafel slopes in the range -170 to -180 mV dec^{-1}. At technical current densities (between -2 and -3 kA m^{-2}), the in-situ activated electrodes were approximately as active as iron. Thus, as found earlier for strongly alkaline conditions [11], the addition of Mo(VI) changed the electrode kinetics and seemed to make the cathode potential relatively independent of the substrate material.

The Mo(VI) compounds in the electrolyte may be cathodically reduced to, for example, Mo(IV) species. Current efficiency measurements, made under vigorous magnetic stirring and hydrogen gas bubble generation, showed current efficiencies for hydrogen production of 95-100 % regardless of whether 100 mM Mo(VI) was present or not and irrespective of electrolyte pH. Any reduction of Mo-species was thus only a minor part of the total applied current and the altered kinetics were not due to a major change in electrode reaction from hydrogen evolution to Mo(VI) reduction. Again, addition of Mo(VI) did not lead to a significant loss in current efficiency, which is important for the energy efficiency of the process.

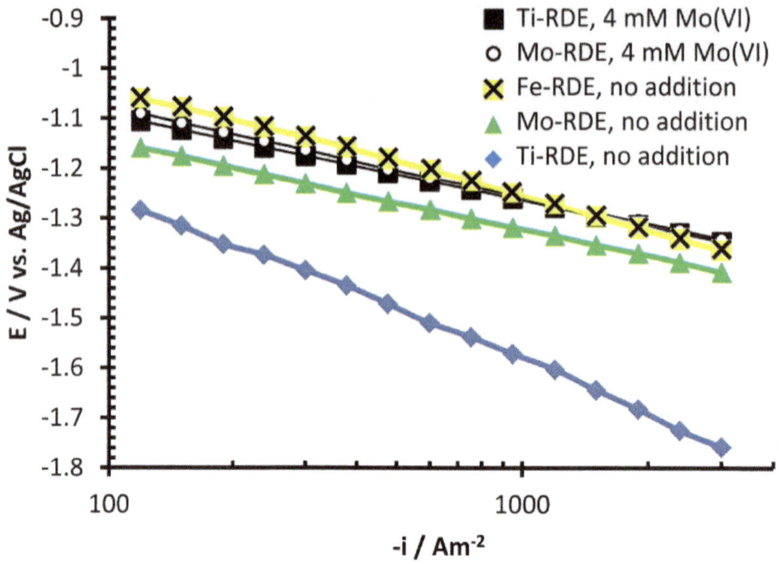

Figure 1. *IR-corrected polarization curves recorded in 2 M NaCl, pH≈6.5, 70°C at a rotation rate of 5000 rpm. All polarization curves were recorded in the anodic direction. The electrodes were pre-polarized at -3 kA m^{-2} for 15 min prior to the polarization curves.*

Electrolysis in electrolytes of near neutral pH may benefit from the addition of buffers, as is the case of chlorate electrolysis. Mo(VI) can act as a pH buffer with a pK_a value of 6.0 [5], but an additional buffer is required if low Mo(VI) concentrations are desired. Phosphoric acid, dihydrogen phosphate and hydrogen phosphate with pK_a values of 2.15, 7.20 and 12.38 at 25°C [23] could for example be used. We have investigated the impact of addition of 40 mM of total phosphate to the Mo(VI)-containing electrolyte using both Mo and Ti electrodes. As seen in Figure 2, there were no negative effects of phosphate addition on the Mo(VI) activation of the HER at high current densities. There was even a positive effect on the cathode potential at lower current densities. The in-situ activation by phosphate is a buffer effect, seen as an increased limiting current for $H_2PO_4^-$ and/or HPO_4^{2-}, that buffers the surface pH and possibly electrochemically deprotonates at the cathode surface [24]. Note that the polarization curves for the Mo-activated hydrogen evolution in Figures 1 and 2 show no limiting currents, indicating a different activation mechanism than for phosphate. In the case of Mo(VI)-addition, a catalytic film was probably formed while the phosphate species resulted in an activation by replacing water as a reactant in the HER. To verify the role of Mo(VI), the following experiment was performed: two titanium electrodes were pre-polarized for 30 min at -3 kA m^{-2} and 3000 rpm in 2 M NaCl at 70°C and pH 6.5 with and without the addition of 100 mM Mo(VI). The electrodes were then rinsed and transferred to an Mo(VI) free electrolyte of identical composition, and the electrode potentials were measured at -3 kA m^{-2}. Electrodes pre-polarized in an Mo(VI)-containing electrolyte had about 200 mV lower overpotential for hydrogen evolution compared with electrodes pre-polarized in the absence of Mo(VI). These results indicate that the activation of hydrogen evolution relates to a catalytic film formed on the cathode surface and not primarily to a mass-transport controlled deprotonation of electrolyte species, as found earlier in pH-neutral electrolytes for catalysis of the HER by for example phosphate [24] or yttrium ions [25]. Since the 100 mM Mo(VI) was present in the electrolyte, the overpotential for HER increased with time after the initial activation. At the end of the 30 min period, the activation was only 100 mV compared to titanium without addition of Mo(VI) to the electrolyte (without IR-correction the electrode with molybdenum -containing film with time became even worse than the pure titanium). When electrodes with deposited molybdenum-containing films were transferred to a Mo(VI)-free electrolyte, the overpotential decreased again. Obviously, there is an optimal Mo(VI) concentration to achieve the best activation of the HER. In this study, we have chosen to work mostly with additions of 4 mM Mo(VI), the same concentration as Huot and Brossard [9] recommend for alkaline water electrolysis. In Figure 1, the observed activation of HER at 400 mV on titanium with 4 mM Mo(VI) is significantly higher than the activation at 100 mV with 100 mM Mo(VI) (not shown).

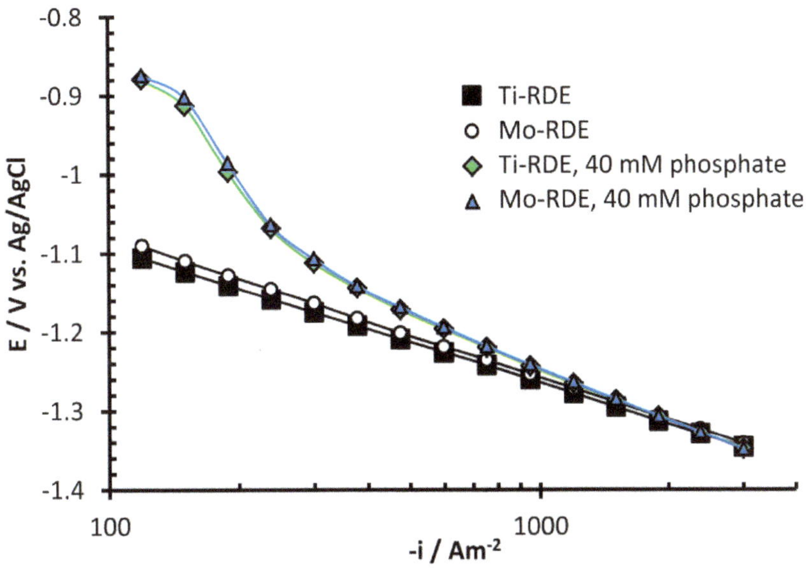

Figure 2. *IR-corrected polarization curves recorded in 2 M NaCl, 4 mM Mo(VI), pH≈6.5, 70°C at a rotation rate of 5000 rpm. All polarization curves were recorded in the anodic direction. The electrodes were pre-polarized at -3 kA m^{-2} for 15 min prior to the polarization curves.*

As already mentioned, electrodeposited MoO_2 is not stable at alkaline conditions [14][16]. To investigate if this would affect activation for hydrogen evolution, cathodic polarization curves were recorded in an electrolyte of 1 M NaOH. As can be seen in Figure 3, without additions of Mo(VI) the Tafel slopes were -160 mV dec^{-1} for titanium as cathode material and -100 mV dec^{-1} for molybdenum. After polarization in the presence of 4 mM Mo(VI), the Tafel slopes were the same for the two electrode substrates, -130 mV dec^{-1}. Furthermore, the electrodes were more active in the presence of Mo(VI) in the electrolyte. Both titanium and molybdenum substrates showed a similar potential for HER. When comparing the results in Figure 1 at near neutral pH with Figure 3 at 1 M NaOH and the data at 30 wt.% KOH, 80°C, in ref [11], the activation seems to follow the same pattern irrespective of the electrolyte pH although different films were formed. The addition of Mo(VI) to the electrolyte results in film formation on the cathode surfaces and the films, which, rather than the substrate, will to a large extent determine HER activity.

Surface characterization

In Figure 4, secondary electron mode SEM images are shown of titanium electrodes that had been polarized with and without Mo(VI) in the electrolyte. A film, that contained molybdenum according to EDS data, can be seen after polarization in the presence of Mo(VI) as well as crystals of high Mo content and of Na and Cl. In Figure 5, back-scatter mode SEM images show titanium electrodes polarized with Mo(VI) in both absence and presence of 20 mM total phosphate.

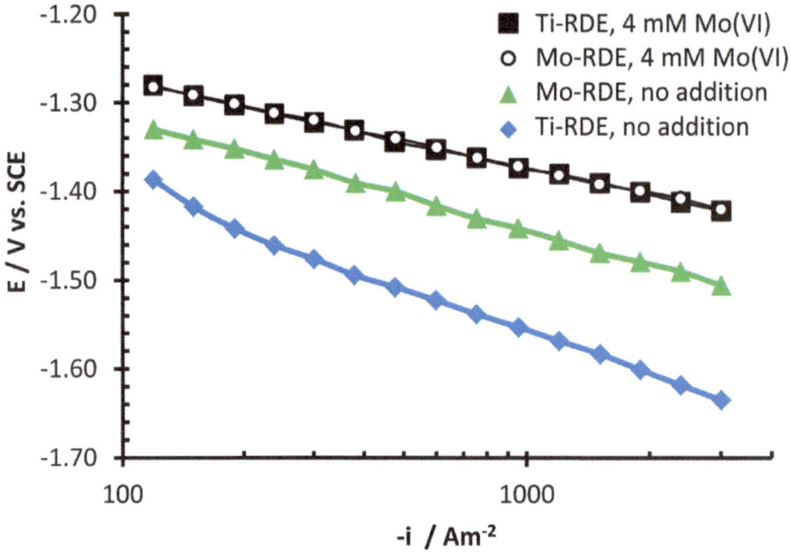

Figure 3. *IR-corrected polarization curves recorded in 1 M NaOH, 70°C at a rotation rate of 5000 rpm. All polarization curves were recorded in the anodic direction. The electrodes were pre-polarized at -3 kA m^{-2} for 15 min prior to the polarization curves.*

The molybdenum-containing film can be seen only on the electrode polarized in the absence of phosphate. No molybdenum-containing film was observed when the total phosphate concentration was varied between 10 mM and 40 mM. This may be due to competitive adsorption of molybdate and phosphate, which has earlier been found on gibbsite in soil [26]. This may explain why molybdenum could not be detected with EDS on the titanium surface in the presence of phosphate. Another possible explanation is the formation of some metal phosphate interfering with the film formation. Activation of the HER by Mo(VI) was also found in the presence of phosphate, as shown in Figure 2. Obviously, a catalytic film was formed which dissolved, detached or was too thin to be detected by EDS.

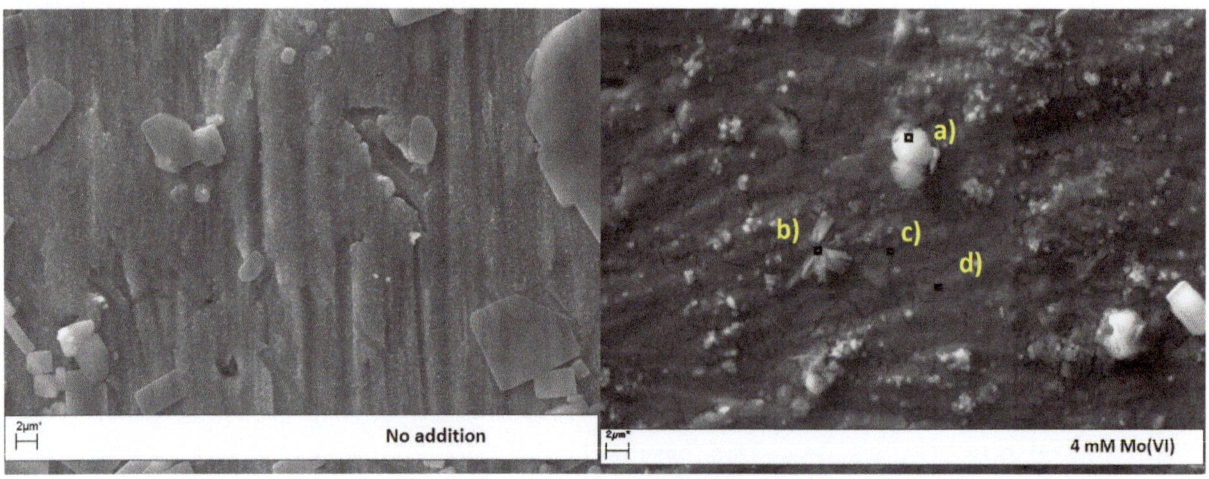

Figure 4. *SEM images on Ti-RDEs polarized for 30 min at -3 kA m^{-2} in 2 M NaCl, pH≈6.5 and 70°C with and without 4 mM Mo(VI). The rotation rate during polarization was 3000 rpm. The composition was analyzed with EDS in atomic % and the relative Mo/(Mo+Ti) atomic ratio was in the marked points: a.) 23 %, b) 26 %, c.) 6 %, d.) 9 %*

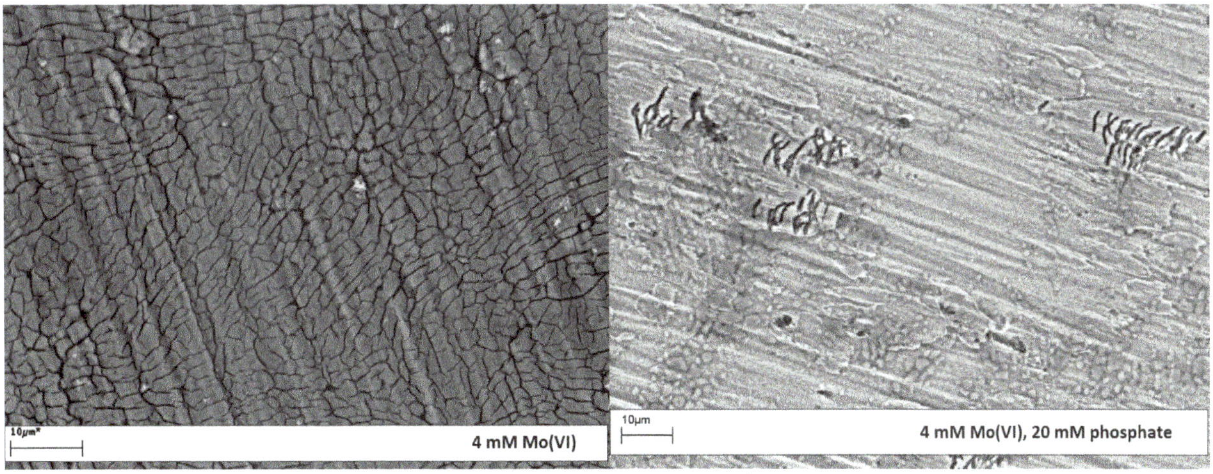

Figure 5. *SEM images on Ti-RDEs polarized for 30 min at -3 kA m^{-2} in 2 M NaCl, 4 mM Mo(VI), pH≈6.5 and 70°C with and without 20 mM phosphate. The rotation rate during polarization was 3000 rpm.*

Further surface analyses were performed on titanium cathode samples that had been polarized at -3 kA m^{-2} for 30 min in either 2 M NaCl at near neutral pH or in 1 M NaOH, both with addition of 4 mM Mo(VI). Analyses by means of XRF showed that there was significantly more molybdenum on electrodes polarized in neutral media compared with alkaline electrolyte (Figure 6). The Mo K-L_3 and K-M_3 peaks were in good agreement with data available in open literature [27].

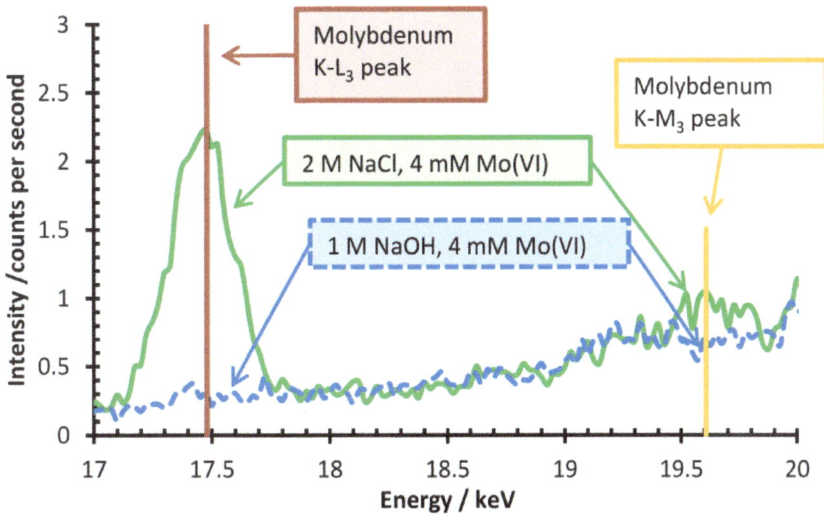

Figure 6. *XRF-spectra for Ti-RDEs polarized for 30 min at -3 kA m^{-2}, 70°C and a rotation rate of 3000 rpm. The pH was adjusted to 6.5 for the electrolyte containing 2 M NaCl, 4 mM Mo(VI).*

XPS analysis shows, in agreement with XRF, a relatively higher intensity of the molybdenum peaks for samples polarized in neutral electrolyte (Figure 7) compared with alkaline electrolyte (Figure 8). Evidently, there was more molybdenum on the sample polarized in neutral electrolyte. The sample polarized in neutral electrolyte displayed XPS spectra which were more complex with several overlapping peaks Mo($3d_{5/2}$: 229.6 eV; 230.7 eV; 232.5 eV; 233.8 eV; 235.0 eV), indicating multiple oxidation states of molybdenum. Unambiguous assignments of the peaks were difficult due to overlap between molybdenum oxides and chlorides, and too large variations in reported binding energies also for a given molybdenum oxide, e.g. between 228.3 and 232.7 eV for MoO_2 and between 231.6 and 235.6 eV for MoO_3 [28]. The sample polarized in alkaline electrolyte revealed only two different oxidation states, of which the predominant Mo($3d_{5/2}$) peak, located at

232.6 eV, was most probably assigned to MoO_3 and the peak at 230.4 to MoO_2[28]. The XPS measurements indicate that the film formed in neutral electrolyte was more complex and probably consisted of polyvalent molybdenum oxides. However, the presence of molybdenum chlorides cannot be excluded. Furthermore, signals from the titanium substrate were only observed when the electrode had been polarized in alkaline electrolyte, indicating that the surface oxide layer was either very thin or formed locally. Relative Mo/Ti atomic ratios determined from XPS were 0.08 and 0.21 as measured in two different areas on the electrode surface. The molybdenum-containing film thus appeared to be formed locally and was unevenly distributed.

XPS measurements also revealed the presence of oxidized iron and copper in the outermost surface of the electrode polarized in 1 M NaOH with 4 mM Mo(VI). We suspect that the iron and copper were present as low level impurities in the p.a. sodium hydroxide and that the 30 min of cathodic polarization was long enough for them to codeposit together with molybdenum on the surface. Iron and molybdenum are known to codeposit [8]-13,15,29,[30] and for example KOH can contain some iron [8]-[13].

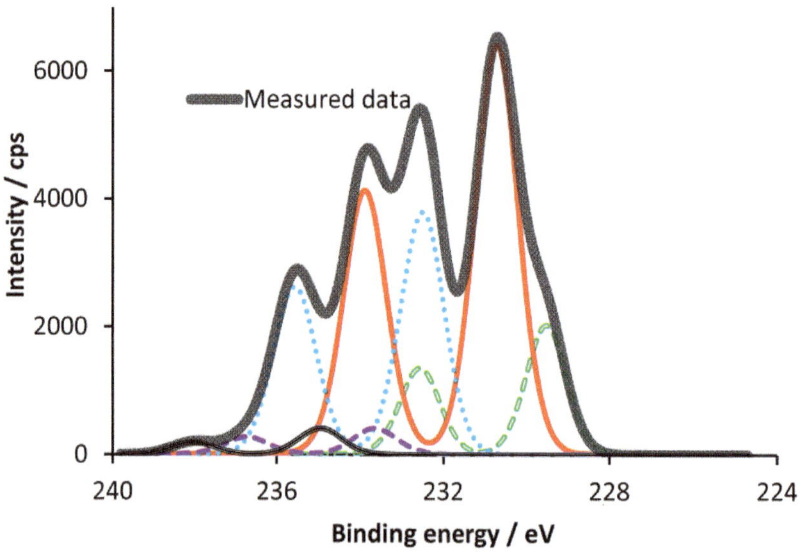

Figure 7. *XPS spectrum of the Mo 3d signal with resolved peaks observed in the outermost surface layer of a Ti-RDE polarized for 30 min in 2 M NaCl, 4 mM Mo(VI), pH≈7, 70°C at -3 kA m^{-2} and a rotation rate of 3000 rpm.*

To investigate the molybdenum-iron codeposition, potential-sweep experiments were performed on a platinum rotating disc electrode after 30 min of pre-polarization at -3 kA m^{-2}. The results are shown in Figure 9. With added molybdenum, there was an oxidation peak at -0.65 V, peak 2. Gold and titanium were also used as electrode materials to exclude substrate effects. A similar peak was confirmed on gold and titanium electrodes that had been polarized in 1 M NaOH, 4 mM Mo(VI), 70°C at -3 kA m^{-2} and Au-RDE (Ti-RDE not tested) polarized in 2 M NaCl, 1 M NaOH, 4 mM Mo(VI), 70°C at -3 kA/m^2. The peak did not appear when the electrolyte was 2 M NaCl, 4 mM Mo(VI), pH 7. According to these findings, pH rather than chlorides affected the appearance of the peak. Considering that on a Pt-RDE when both Fe(III) and Mo(VI) were added to the electrolyte, an even larger peak appeared, it is likely that the peak was connected to the oxidation of an iron-molybdenum deposit. The charge of the oxidation peak corresponds to a film significantly thicker than 10 times the approximately 5-10 nm information depth of XPS.

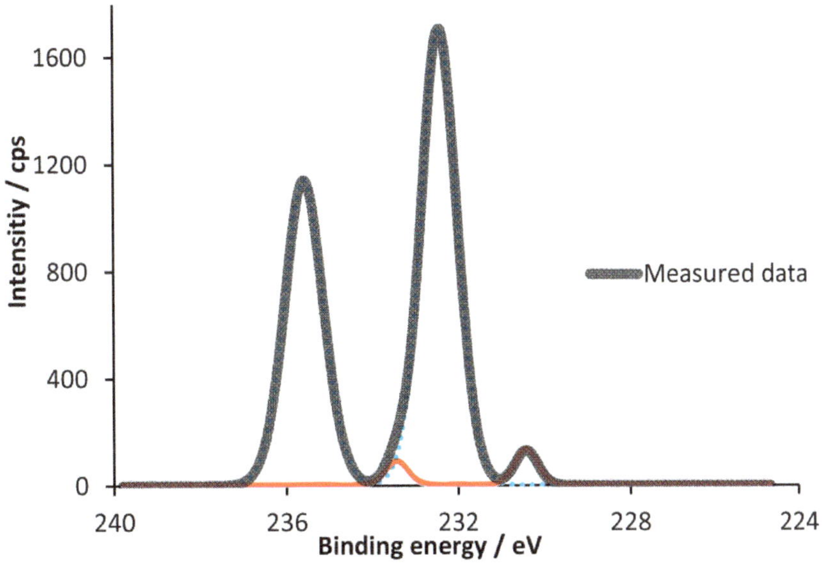

Figure 8. *XPS spectrum of the Mo 3d signal with resolved peaks in the outermost surface layer of a Ti-RDE polarized for 30 min in 1 M NaOH, 4 mM Mo(VI), 70°C at -3 kA m^{-2} and a rotation rate of 3000 rpm.*

The surface was probably not completely covered by film at the time of XPS analysis since information from the titanium substrate was also observed. When only Fe(III) was added the peak did not appear, again suggesting that the oxidation peak was caused by oxidation of the deposit formed by induced codeposition. A comparison with cyclic voltammograms on iron in 1 M NaOH at room temperature [31], indicates that Peak 1 at -0.98 V may be the oxidation of Fe(0) to Fe(II). In pH neutral electrolytes, the solubility of Fe(III) is quite low without chelating agents, so it is unlikely that there will be any significant amount of iron and molybdenum codeposition.

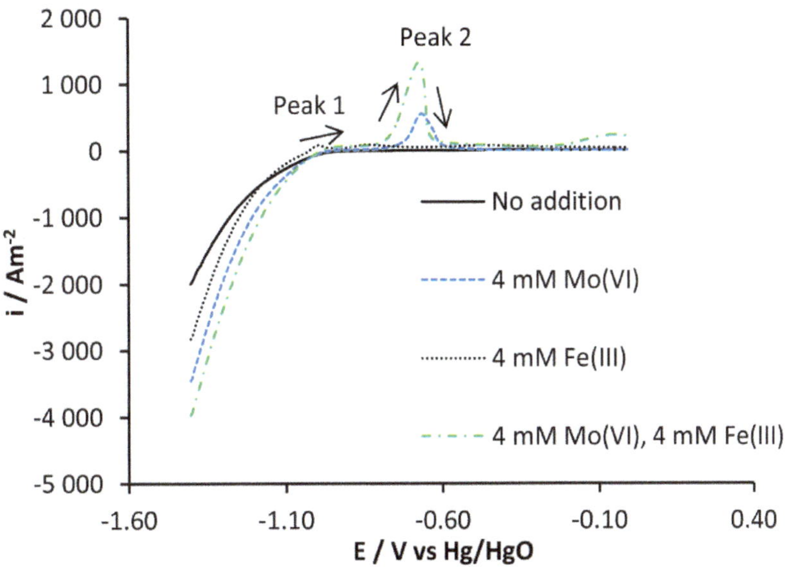

Figure 9. *Potential sweeps recorded on a platinum RDE in 1 M NaOH, 70°C at a sweep rate of 50 mV s^{-1} and a rotation rate of 5000 rpm. The electrode was pre-polarized at -3 kA m^{-2} for 30 min prior to the polarization curves. Concentrations given for Fe(III) correspond to the added amount since some precipitation was observed.*

Thus, the film formed at near neutral pH contained polyvalent Mo-oxides, whereas the film formed at alkaline conditions was much thinner and contained less molybdenum and a considerable amount of iron impurities. This is consistent with the findings by [8]-[13] and with the limited pH range of stability of MoO_2 [14]4,[16]. Most likely, MoO_2 is precipitated in the near neutral electrolyte, whereas in alkaline conditions an Mo-Fe alloy is the obtained deposit.

Combining iron and molybdenum

As seen in Figure 9 the combination of added Mo(VI) and Fe(III) resulted in the largest cathodic currents at negative potentials, which is consistent with an activated HER. Intermetallic phases based on metals from different sides of a volcano plot over the hydrogen binding energy would, according to Brewer-Engel valance bond theory, have a higher activity for HER than the individual metals [31],[33]. Iron and molybdenum are on different sides of the volcano plot and have been codeposited and investigated in the literature for hydrogen evolution activity [9]-[13],[29][30][34]. In all these studies, the alloys were electrodeposited, and thus resulted in films with rough surfaces.

Elezović *et al.* [34] investigated the activity for HER for Mo-Fe films of varying composition: 34 at.% (46.9 wt.%) Mo, 41.8 at.% (55.2 wt.%) Mo and 59.2 at.% (71.4 wt.%) Mo. The higher the Mo content, the lower the overpotential, but when normalizing with surface roughness factors as determined from double layer capacitances, there was no significant difference between the films. The trend in overpotential between the alloys was therefore mainly attributed to variations in real surface area [34]. In a recent paper [35], a commercial $Fe_{80}Mo_{20}$ alloy with a polished surface was used as reference sample when investigating HER on $Fe_{75}Mo_{20}R_5$ (R=Rare earth metal) alloys. Although the Brewer-Engel theory is discussed in both references [34][35], no comparisons were made with the catalytic activity of the pure metals, iron and molybdenum. Both studies were made in 1 M NaOH at 25°C.

An alloy was cast to investigate true electrocatalytic effects by alloying Fe and Mo. Rotating disc electrodes, which could be polished to a smooth surface, were prepared to enable comparisons with electrodes of the pure metals without major differences in surface area. As described earlier, the nominal composition of the cast alloy was 72.4 at.% (60.4 wt.%) iron and 27.6 at.% (39.5 wt.%) molybdenum. Optical microscopy on the alloy showed several phases mixed on the micrometer scale (Figures 10 and 11), and some lighter and darker features. The different features were probably some inhomogeneity of the alloy. The main phases as identified by XRD (Figure 12) were $MoFe_2$ "PDF 04-003-4286" with hexagonal structure, space group *P63/mmc* (194) and $Mo_{5.08}Fe_{7.92}$ "PDF 04-003-7152" with rhombohedral structure, space group *R*-3*m* (166) [21]. There were also some minor phases such as $MoFe_2O_4$, MoO_3 and Mo. Weight fractions of constituents semi-quantitatively determined by comparing the integrated intensities of the diffraction peaks from each of the known phases are shown in Figure 12. These data indicates that $Mo_{5.08}Fe_{7.92}$ was the most common phase (around 50 %) in the alloy. To our knowledge, cast iron-molybdenum alloys have not been investigated for HER in neutral electrolytes before.

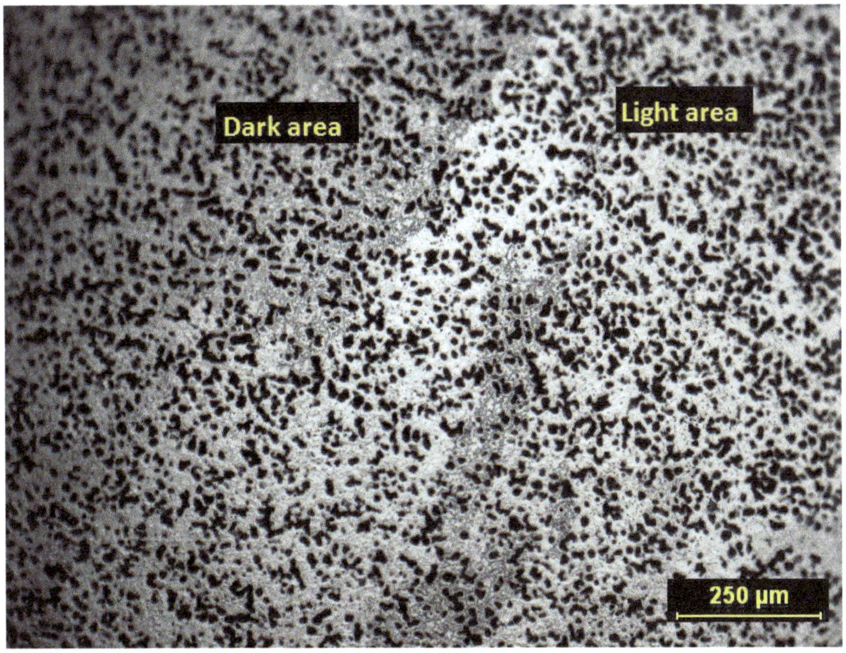

Figure 10. *Optical microscopy image of the cast iron-molybdenum alloy.*

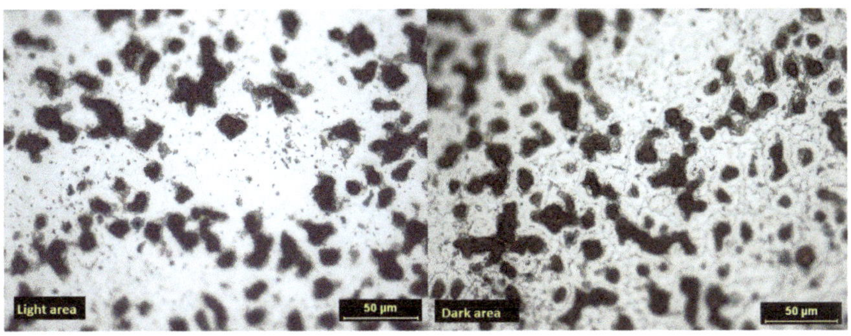

Figure 11. *Optical microscopy images of the cast iron-molybdenum alloy. To the left is an image of a lighter area and to the right an image of a darker area on the surface, see Figure 10.*

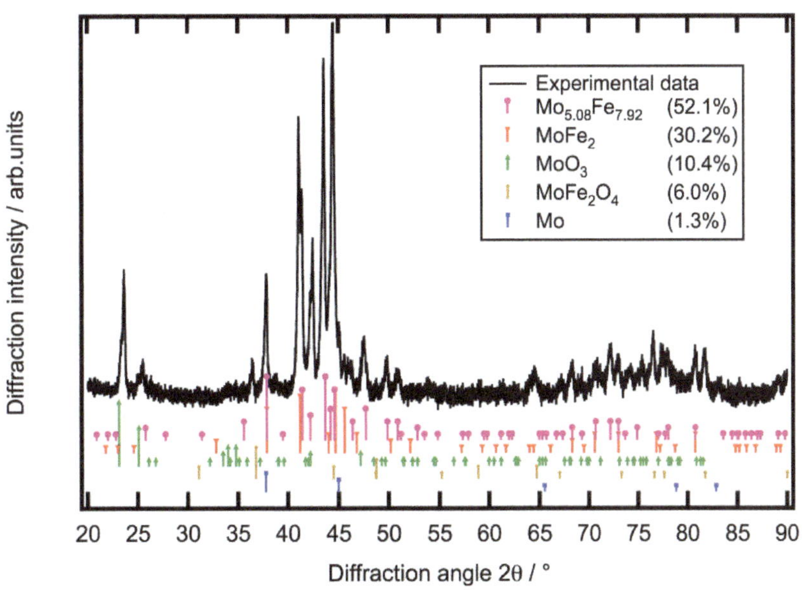

Figure 12. *X-ray diffraction pattern of the cast Fe-Mo alloy. The positions of the known diffraction peaks of the identified phases are marked, excluding the peaks with negligible strengths. The percentage numbers in brackets displayed in the legend show the fractions (by mass) determined via the semi-quantitative analysis of the sample composition.*

Polarization curves recorded on Mo-RDE, Fe-RDE and RDE made from the alloy showed that iron and the alloy were both more active for the HER compared with Mo-RDE (Figure 13). The polarization curve of the alloy had a Tafel slope for HER of -170 mV dec^{-1}. The activity of the alloy for the HER was approximately the same as for pure iron; no synergistic catalytic effect of the alloying could therefore be seen.

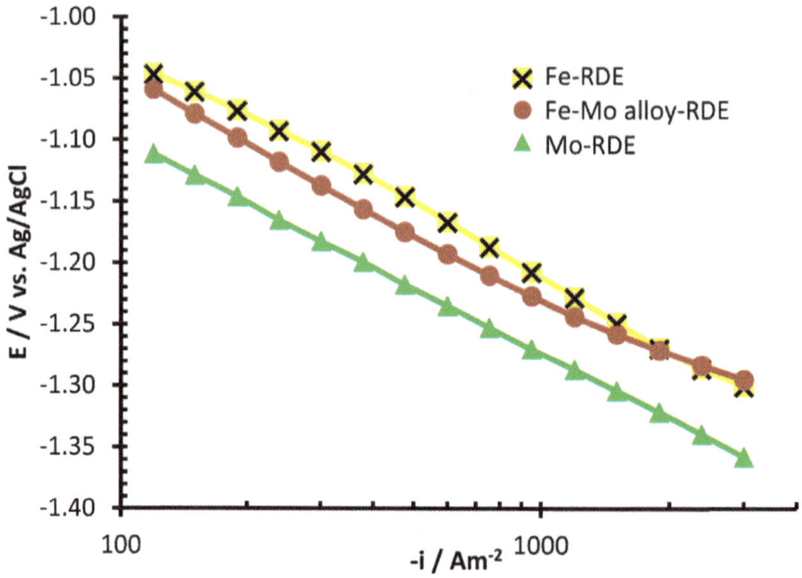

Figure 13. *IR-corrected polarization curves recorded in 2 M NaCl, pH 6.5, 70°C at a rotation rate of 5000 rpm. All polarization curves were recorded in the anodic direction. The electrodes were pre-polarized at -3 kA m^{-2} for 1 min prior to the polarization curves.*

Conclusions

The addition of Mo(VI) can activate the HER on titanium and molybdenum electrodes in both neutral NaCl and in alkaline NaOH electrolytes. In the presence of Mo(VI), the overpotentials for hydrogen evolution are similar for the two materials in a wide current density range (between -100 and -3000 kA m^{-2}). Cathodic Mo-containing films are formed during polarization, and any electroreduction of Mo(VI) species does not affect the current efficiency for hydrogen evolution by more than a few percent at most. Films formed at pH 6.5 are active for hydrogen evolution also when transferred to an Mo(VI)-free electrolyte and therefore the mechanism of activation is concluded to originate from a catalytically active film.

At technical current densities in neutral electrolytes, the formed films gave the cathode a similar activity to an iron electrode. It thus seems possible to use a corrosion resistant, robust cathode material in the chlorate process and still get a continuously renewing surface with similar activity as steel cathodes in use today. Phosphate, a possible alternative to chromate for buffering the chlorate electrolyte, does not appear to have any negative effect on this activation at technical current densities. Instead, at a low current density and a good mass-transport, phosphate can further reduce the overpotential for hydrogen evolution.

The cathode films formed during polarization at neutral pH are much thicker than those formed at alkaline pH. XPS analyses showed that the former have more complex chemical composition, most probably consisting of several polyvalent molybdenum oxides.

The investigated cast alloy of molybdenum and iron was approximately as active as iron for hydrogen evolution. This indicates that electrodeposited iron-molybdenum alloys containing

$MoFe_2$ and $Mo_{5.08}Fe_{7.96}$ will only give a stronger activation of the HER than iron if the real surface area is increased during deposition.

Acknowledgements *Grants from Knut and Alice Wallenberg foundation for the XPS instrument are acknowledged. We thank Staffan Palovaara at SCA R&D Centre AB for helping with the SEM and EDS measurements. Bernth Nordin at Permascand is acknowledged for helping with metallographic specimen preparation. The Swedish Energy Agency, the Swedish Research Council, Permascand, Eka Chemicals and the European Regional Development Fund through the project "EnergyWise" are acknowledged for financing this study.*

References

[1] J. E. Colman, *American Inst. Chem. Eng. Symp. Ser.* **77** (1981) 244-263
[2] G. Lindbergh and D. Simonsson, *J. Electrochem. Soc.* **137** (1990) 3094-3099
[3] K. Viswanathan and B. V. Tilak, *J. Electrochem. Soc.* **131** (1984) 1551-1559
[4] A. Cornell and D. Simonsson, *J. Electrochem. Soc.* **140** (1993) 3123-3129
[5] L. Roué, M. E. Bonneau, D. Guay, M. Blouin and R. Schulz, *J. Appl. Electrochem.* **30** (2000) 491-498
[6] M. Li, Z. Twardowski, F. Mok, N. Tam, *J. Appl. Electrochem.* **37** (2007) 499–504
[7] M. Rosvall, K. Hedenstedt, A. Sellin, J. Gustavsson and A. Cornell, (Akzo Nobel Chemicals International B.V.), WO 2010/130546A1 (2010)
[8] J. Y. Huot and L. Brossard, *Int. J. Hydrogen Energ.* **12** (1987) 821-830
[9] J. Y. Huot and L. Brossard, *Surf. Coat. Tech.* **34** (1988) 373-382
[10] J. Y. Huot and L. Brossard, *J. Appl. Electrochem.* **18** (1988) 815-822
[11] J. Y. Huot and L. Brossard, *Int. J. Hydrogen Energ.* **14** (1989) 229-232
[12] J. Y. Huot and L. Brossard, *J. Appl. Electrochem.* **20** (1990) 281-288
[13] L. Brossard and J. Y. Huot, *J. Appl. Electrochem.* **21** (1991) 508-515
[14] F. J. Presuel-Moreno, M. A. Jakab and J. R. Scully, *J. Electrochem. Soc.* **152** (2005) B376-B387
[15] A. Brenner, *Electrodeposition of Alloys*, 1-2, Academic Press Inc., New York, USA, 1963
[16] P. Wang, L. L. Wilson, D. J. Wesolowski, J. Rosenqvist and A. Anderko, *Corros. Sci.* **52** (2010) 1625-1634
[17] A. Cornell, B. Håkansson, and G. Lindbergh, *Electrochim. Acta* **48** (2003) 473-481
[18] A. Cornell, G. Lindbergh and D. Simonsson, *Electrochim. Acta* **37** (1992) 1873-1881
[19] S. Modin and H. Modin, *Handbok i metallmikroskopering*, Meritförlaget, Johanneshov, Sweden, 1977, p. 366
[20] G. Petzow, *Metallographic etching*, 2nd ed., ASM International, Materials Park, USA, 1999, p. 86
[21] ICDD (**2010**). PDF-4+ 2010 (Database), edited by Dr. Soorya Kabekkodu, International Centre for Diffraction Data, Newtown Square, PA, USA
[22] R. Morales, D. U. Sichen and S. Seetharaman I. Arvanitides,, *Metall. Mater. Trans.* **33B** (2002) 589-594
[23] G. Aylward and T. Findlay, *SI Chemical Data* 4th edition, Jacaranda Wiley Ltd, Milton, Australia, 1998, p. 137
[24] V. Marinović and A. Despić, *J. Serbian Chem. Soc.* **63** (1998) 545-553
[25] J. Gustavsson, A. Cornell and G. Lindbergh, *Int. J. Hydrogen Energy* 37 (2012) 9496-9503
[26] S. Goldberg, *Soil Sci.* **175** (2010) 105-110
[27] J. A. Bearden, *Rev. Mod. Phys.* **39** (1967) 78-124
[28] R. A. Walton, *J. Less Common Met.* **54** (1977) 71-80
[29] N.R. Elezović , V.D. Jović, N.V. Krstajić, *Electrochim. Acta* **50** (2005) 5594-5601

[30] C. Norman, S. Julia, O. Klaus and R. Glen, (Canexus Chemicals Canada Ltd), WO 2006/039804 A1 (2006)
[31] B. Müller-Zülow, S. Kipp, R. Lacmann and M. A. Schneeweiss, *Surf. Sci.* **311** (1994) 153-158
[32] M. M. Jakšić, *Electrochim. Acta* **29** (1984) 1539-1550
[33] M. M. Jakšić, *Int. J. Hydrogen Energ.* **12** (1987) 727-752
[34] N. R. Elezović, V. D. Jović and N. V. Krstajić, *Electrochim. Acta* **50** (2005) 5594-5601
[35] F. Rosalbino, D. Macciò, A. Saccone. E. Angelini and S. Delfino, *Int. J. Hydrogen Energ.* **36** (2011) 1965-1973

3

Voltammetric determination of hydroxylamine in water and waste water samples using a NiO nanoparticle/new catechol derivative modified carbon paste electrode

Mahbobeh Moazampour, Fahimeh Tahernejad-Javazmi, Maryam Salimi-Amiri*, Hassan Karimi-Maleh✉ and Mehdi Hatami**

Department of Chemistry, Graduate University of Advanced Technology, Kerman, Iran
**Department of Physics, Sari Branch, Islamic Azad University, Sari, Iran*
***Polymer Research Laboratory, University of Bonab, Bonab, Iran*

✉Corresponding author

Abstract

A (9,10-dihydro-9,10-ethanoanthracene-11,12-dicarboximido)-4-ethylbenzene-1,2-diol (DED) modified NiO/NPs carbon paste electrode "(DED/NiO nanoparticle (NiO/NPs)/CPE) was constructed for determination of hydroxylamine (HX). The cyclic voltammogram showed that the electrocatalytic oxidation of HX at the surface of DED/NiO/NPs/CPE occurs at a potential of about 800 mV less positive than with an unmodified electrode. Square-wave voltammetry results presented that the electrocatalytic oxidation peak currents of HX in pH 8.0 had two linear dynamic ranges in the range of 0.1 to 2.0 and 2.0 to 400.0 μM HX, with a detection limit of 0.07 μM. The kinetic parameters such as electron transfer coefficient α (0.47) and rate constant ($2.454 \times 10^3\ M^{-1}\ s^{-1}$) were determined for the chemical reaction between HX and DED. Finally, this method was evaluated for the determination of HX in water and waste water samples.

Keywords

Hydroxylamine; NiO nanoparticle; water and waste water analysis; sensor; voltammetry

Introduction

Hydroxylamine (HX) is known as a type of reducing agent and is widely used in industrial and pharmaceutical applications. It has been identified as a key intermediate in nitrogen cycles and nitrous oxide production [1]. The quantitative determination of HX is very important in both studies of biological processes and for industrial purposes. It has been confirmed that HX is produced during the reduction of nitrates by *Escherichia coli* and Torula yeast [2].

Electrochemical analysis is gaining significance within industrial process control, environmental monitoring and various pharmaceutical and biotechnology applications [3-7]. The use of unmodified electrodes for electrochemical detection has a number of limitations, such as low selectivity and sensitivity, poor reproducibility, slow electron transfer reaction, low stability over a wide range of solution compositions and the high overpotential at which the electron transfer process occurs [8-10]. Chemical modification of inert substrate electrodes with redox active thin films offers significant advantages in the design and development of electrochemical sensors. In operation, the redox active sites shuttle electrons between the analyte and the electrodes with a significant reduction in activation overpotential [11]. A further advantage of chemically modified electrodes is that they are less prone to surface fouling and oxide formation, compared to inert substrate electrodes [12-14]. A wide variety of compounds have been used as electron transfer mediators for the modification of electrode surfaces in various procedures [15-17].

Nanotechnology has become one of the most interesting disciplines in science and technology today. The intense interest in nanotechnology is being driven by various interesting fields and is creating a new industrial revolution [18]. Nano-materials such as nanoparticles, carbon nanotubes or nanocomposite connected with biomolecules are being used for several bioanalytical applications [19-21]. Electroanalysis is taking advantage of all the possibilities offered by nanomaterials that are easy to detect using conventional electrochemical methods. Nanocomposite of a variety of shapes, sizes and compositions continues to change the field of bioanalytical measurement.

In the present work, we describe the preparation and suitability of a DED modified NiO/NPs carbon paste electrode as a new electrode for electrocatalysis and determination of HX in an aqueous buffer solution. To demonstrate the catalytic ability of the modified electrode toward the electrooxidation of HX in real samples, we examined the utility of this method for the voltammetric determination of HX in water and waste water samples.

Experiment

Chemicals

All chemicals used were of analytical reagent grade purchased from Merck (Darmstadt, Germany), unless otherwise stated. Doubly distilled water was used throughout.

1.0×10^{-2} mol L^{-1} HX solution was prepared daily by dissolving 0.064 g HX in water and the solution was diluted to 100 mL with water in a 100 mL volumetric flask. The solution was kept in a refrigerator at $4^{\circ}C$ and in the dark. Further dilution was made with water.

Phosphate buffer solutions (sodium dihydrogen phosphate and disodium monohydrogen phosphate, plus sodium hydroxide, 0.1 mol L^{-1}) (PBS) with different pH values were used.

High viscosity paraffin ($d = 0.88$ kg L^{-1}) from Merck was used as the pasting liquid for the preparation of the carbon paste electrode. Spectrally pure graphite powder (particle size <50 μm) from Merck was used as the substrate for the preparation of the carbon paste electrode as a working electrode.

Apparatus

Cyclic voltammetry (CV), chronoamperometry and square wave voltammetry (SWV) were performed using an analytical system, Autolab, with PGSTAT 302N (Eco Chemie, The Netherlands). The system was run on a PC using GPES software. A conventional three-electrode cell assembly consisting of a platinum wire as an auxiliary electrode and an Ag/AgCl (KCl_{sat}) electrode as a reference electrode was used. The working electrode was either an unmodified carbon paste electrode (CPE) or a DED/NiO/NPs/CPE. X-ray powder diffraction studies were carried out using a STOE diffractometer with Cu–K$_\alpha$ radiation (l = 1.54 Å).

Preparation of the modified electrode

To prepare the modified electrode, 150.0 mg of NiO/NPs and 70.0 mg of DED was hand mixed with 780.0 mg of graphite powder using a mortar and pestle. Using a syringe, 15 drops of paraffin were added to the mixture and mixed well for 55 min until a uniformly wetted paste was obtained. The paste was then packed into a glass tube. By pushing a copper wire down the glass tube into the back of the mixture, electrical contact was created. When necessary, a new surface was obtained by pushing an excess of the paste out of the tube and polishing it on weighing paper. The unmodified carbon paste electrode (CPE) was prepared in the same way without NiO/NPs and DED to the mixture, to be used for comparison purposes.

Preparation of real samples

Water samples were stored in a refrigerator immediately after collection. Ten millilitres of the sample was centrifuged for 15 min at 1500 rpm. The supernatant was filtered using a 0.45 μm filter and then diluted three times with the PBS pH 8.0. The solution was transferred into the voltammetric cell to be analysed without any further pre-treatment. The standard addition method was used for the determination of HX in real samples.

Results and discussion

NiO/NPs characterisation

NiO/NPs were analysed by XRD analyses. The XRD pattern of NiO/NPs nanopowders in the 2θ range of 10-80° is shown in Fig. 1.

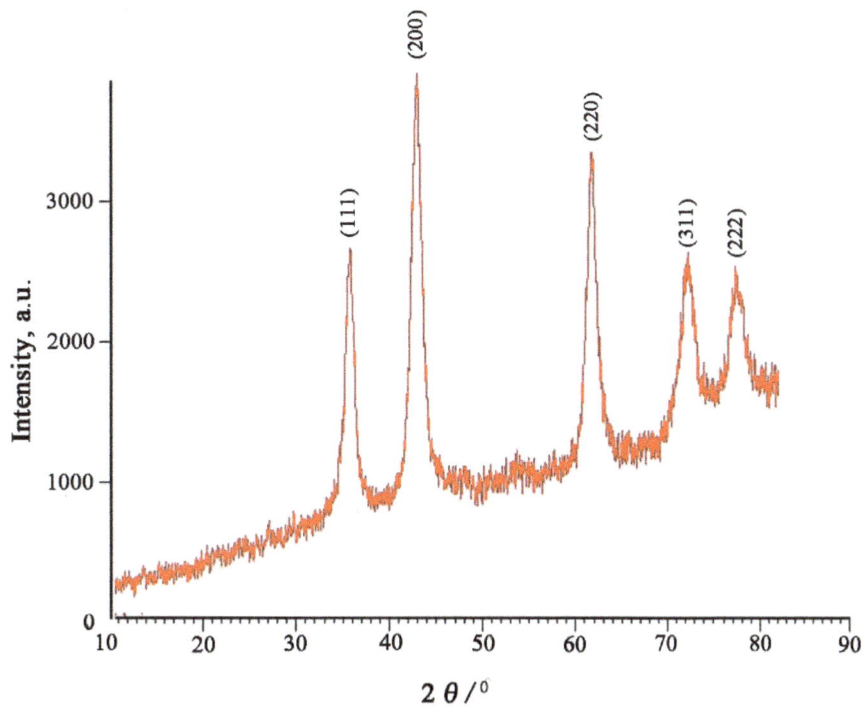

***Figure 1.** XRD patterns of as-synthesised NiO/NPs nanoparticles.*

Figure 1 clearly proves the presence of NiO/NPs. An average diameter of as-synthesised NiO/NPs was calculated from the broadness peak ($2\theta = 44°$) by using the Scherrer equation $D = K\lambda/\beta \cos\theta$, and measured about 25.0 nm.

Electrochemical investigation

Figure 2 depicts the cyclic voltammetry responses from the electrochemical oxidation of 400 µM HX at DED/NiO/NPs/CPE (curve c), DED/CPE (curve b), NiO/NPs/CPE (curve d) and unmodified CPE (curve e). As shown, the anodic peak potential for HX oxidation at DED/NiO/NPs/CPE (curve c) and at DED/CPE (curve b) was about 200 mV, while at NiO/NPs/CPE (curve d); the peak potential was about 1000 mV. At the unmodified CPE, the peak potential of HX was about 1050 mV (curve e). From these results, it was concluded that the best electrocatalytic effect for HX oxidation was observed at DED/NiO/NPs/CPE (curve c).

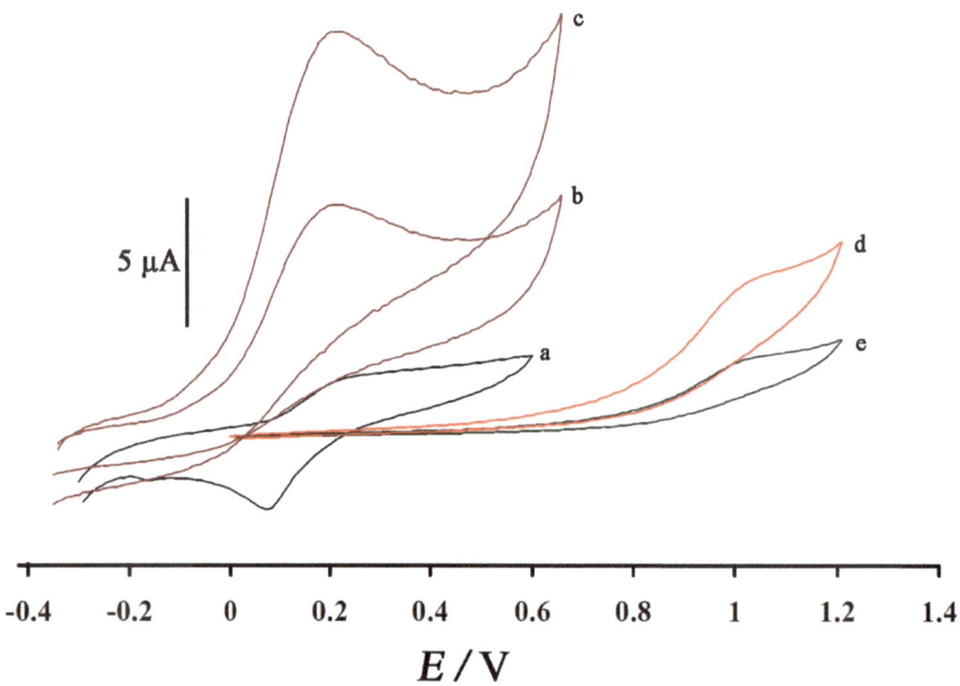

Figure 2. *Cyclic voltammograms of (a) the buffer solution at DED/NiO/NPs/CPE; (b) 400 µM HX at DED/CPE; (c) 400 µM HX at DED/NiO/NPs/CPE; (d) 400. µM HX at NiO/NPs/CPE; (e) 400 µM HX at CPE. Conditions: 0.1 mol L^{-1} PBS (pH 8.0), scan rate of 20 mV s^{-1}.*

For example, the results show that the peak potential of HX oxidation at DED/NiO/NPs/CPE (curve c) shifted by about 800 and 850 mV toward less positive values when compared with NiO/NPs/CPE (curve d) and unmodified CPE (curve e), respectively. Additionally, DED/NiO/NPs/CPE showed higher anodic peak current for the oxidation of HX compared to DED/CPE, indicating that the combination of NiO/NPs and the mediator significantly improved the performance of the electrode toward HX oxidation. In fact, DED/NiO/NPs/CPE in the absence of HX exhibited a well-behaved redox reaction (Figure 2a) in the buffer solution (pH 8.0). However, there was a drastic increase in the anodic peak current in the presence of 400 µM HX (curve c), which can be related to the electrocatalytic role of DED/NiO/NPs/CPE towards oxidation of HX.

We observed a linear variation of the peak current with the square root of scan rate ($v^{1/2}$) at scan rates ranging from 2-12 mV s^{-1} at pH 8.0 (Figure 3). This result clearly indicates a diffusion-controlled electrooxidation process [22].

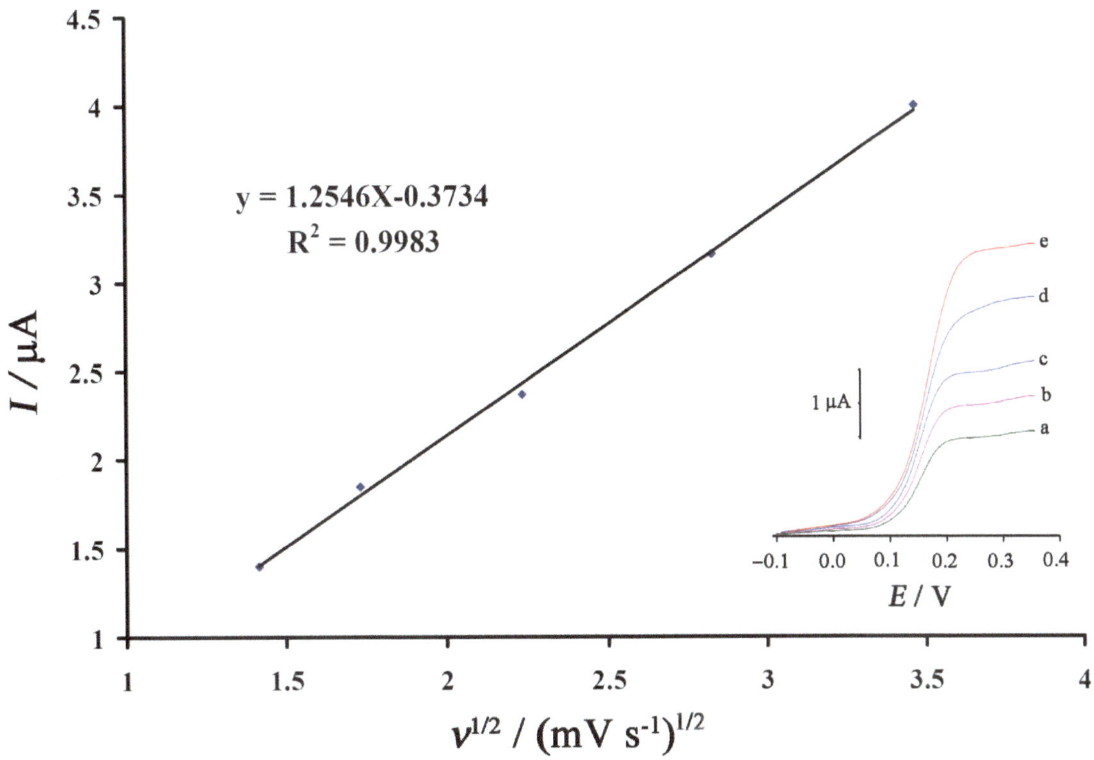

Figure 3 *Plot of I_{pa} versus $v^{1/2}$ for the oxidation of 100.0 µM HX at various scan rates of (a) 2.0; (b) 3.0; (c) 5.0; (d) 8.0; (e) 12.0 mV s^{-1} in 0.1 mol L^{-1} phosphate buffer solution (pH 8.0) at NiO/NPs/DED/CPE. Inset: Cyclic voltammograms of 100.0 µM HX at various scans.*

To obtain information about the rate-determining step, a Tafel plot was drawn, derived from points in the Tafel region of the linear sweep voltammogram (Figure 4). The slope of the Tafel plot was equal to $n(1-\alpha)F/2.3RT$, which resulted in 0.1115 V decade^{-1} [23]. Therefore, we obtained the value of α being equal to 0.47.

For further investigations, the value of α was calculated for the oxidation of HX at pH 8.0 for both the modified and unmodified paste electrodes using one other method (see equation 1):

$$\alpha n_{\alpha} = 0.048/(E_P - E_{P/2}) \quad (1)$$

where $E_{P/2}$ is the potential corresponding to $I_{P/2}$. The values for αn_{α} were found to be 0.47 and 0.12 at the surface of DED/NiO/NPs/CPE and CPE, respectively. This result was also confirmed by the larger I_{pa} values recorded during linear seep voltammetry at DED/NiO/NPs/CPE.

Chronoamperometric measurements of HX at DED/NiO/NPs/CPE were carried out for various concentrations of HX in buffered aqueous solutions (pH 8.0) by setting the working electrode potentials at 0.0 mV and 400 mV *vs.* Ag/AgCl/KCl_{sat} (Figure 5A). For an electroactive material (HX, in this case) with a diffusion coefficient of D, the current observed for the electrochemical reaction at the mass transport limited condition was described using the Cottrell equation. Experimental plots of I *vs.* $t^{-1/2}$ were employed, with the best fits for 300 µM of HX (Figure 5B). The slope of the resulting straight line was then plotted against HX concentration (Figure 5B). From the resulting slope and Cottrell equation, the mean value of the D was found to be 2.1×10^{-6} cm^2 s^{-1} [24].

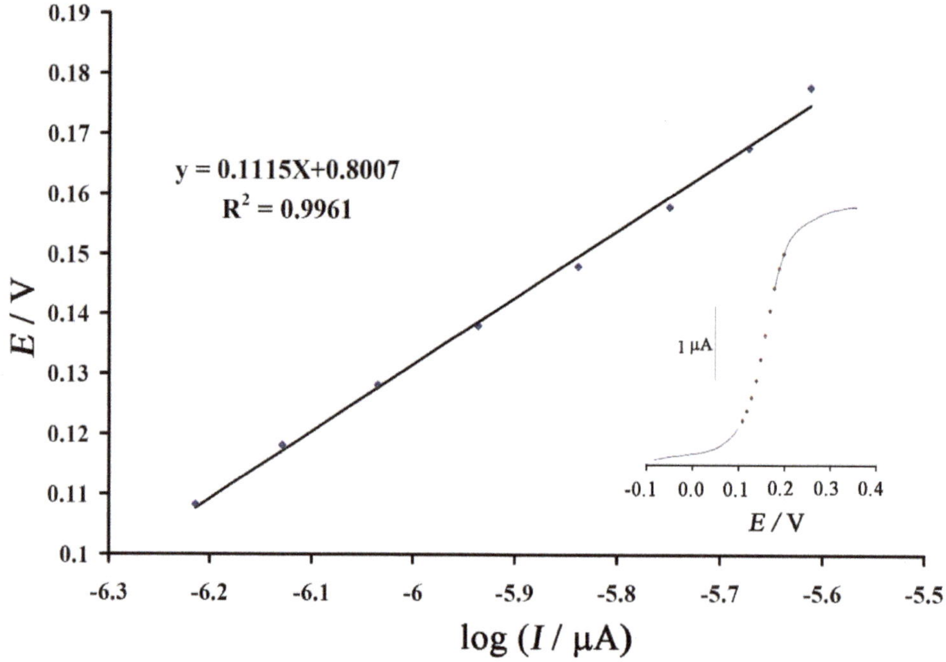

***Figure 4.** Tafel plot for DED/NiO/NPs/CPE in 0.1 mol L^{-1} PBS (pH 8.0) with a scan rate of 8 in the presence of 100.0 μM HX.*

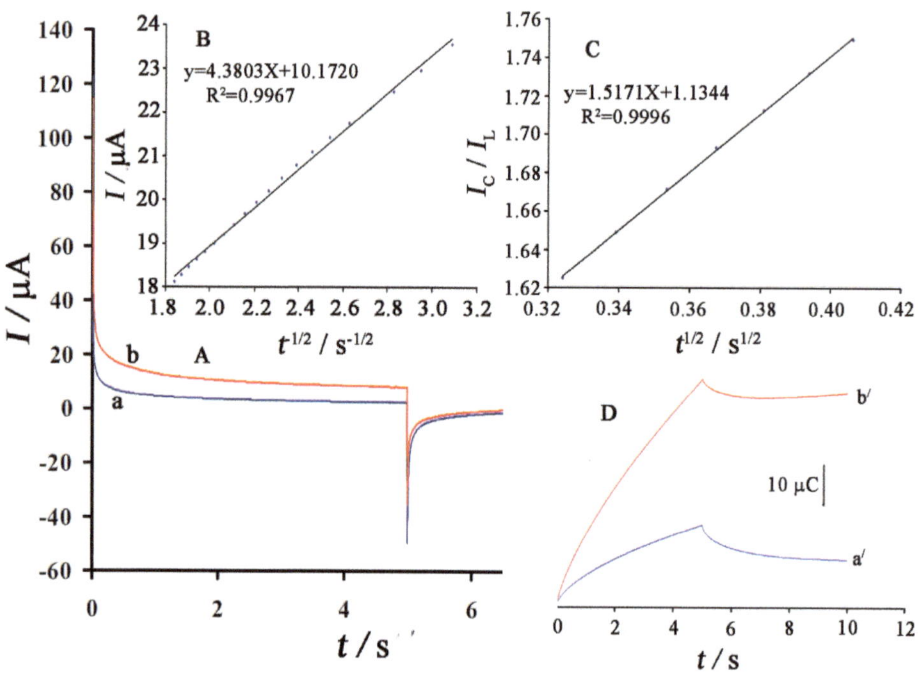

***Figure 5. A** – Chronoamperograms obtained at DED/NiO/NPs/CPE (a) in the absence and in the presence of (b) 300 μM HX at pH 8.0; **B** – Cottrell's plot for the data from the chronoamperograms; **C** – Dependence of IC/IL on $t^{1/2}$ derived from the chronoamperogram data; **D** – The charge-time curves (a') for curve (a) and (b') for curve (b).*

The rate constant for the chemical reaction between HX and redox sites in DED/NiO/NPs/CPE, k_h, can be evaluated by chronoamperometry according to the method set out by Galus [25]:

$$I_C/I_L = \pi^{1/2}\gamma^{1/2} = \pi^{1/2}(kC_bt)^{1/2} \tag{2}$$

Where I_C is the catalytic current of HX at DED/NiO/NPs/CPE, I_L the limited current in the absence of HX and t is the time elapsed (s). The above equation can be used to calculate the rate constant of the catalytic process k_h. Based on the slope of the I_C/I_L versus $t^{1/2}$ plots (Figure 5C), k_h can be obtained for a given HX concentration. From the values of the slopes, an average value of k_h was found to be k_h = 2.454×10^3 M^{-1} s^{-1}. The value of k_h also explains the sharp feature of the catalytic peak observed for catalytic oxidation of HX at the surface of DED/NiO/NPs/CPE.

Double potential step chronocoulometry, as well as other electrochemical methods, was in addition employed for the investigation of the electrode processes at DED/NiO/NPs/CPE. Forward and backward potential step chronocoulometry on the modified electrode in a blank buffer solution showed very symmetrical chronocoulograms. These had about an equal charge consumed for both oxidation and reduction of the DED_{Red}/DED_{Ox} redox system in DED/NiO/NPs/CPE. However, in the presence of HX, the charge value associated with forward chronocoulometry was significantly greater than that observed for the backward chronocoulometry (Figure 5D). This behaviour is typical of that expected for electrocatalysis at a chemically modified electrode [26].

Stability and reproducibility

The repeatability and stability of modified electrode was investigated using CV measurements of 400.0 μM HX in a buffer solution. The relative standard deviation (RSD) for five successive assays was 1.4 %. When seven different DED/NiO/NPs/CPEs were used, the RSD for ten measurements was 2.1 %. When the electrode was stored in the laboratory, the modified electrode retained 95 % of its initial response after a week and 92 % after 30 days (see Figure 6). These results indicate that DED/NiO/NPs/CPE has good stability and reproducibility, and could be used for HX measurements.

Determination of HX individually

Square wave voltammetry (SWV) was used to determine the concentration of HX. Since square wave voltammetry has a much higher current sensitivity and better resolution than cyclic voltammetry, the SWV was used for the determination of HX (Figure 7 inset). The plot of peak current vs. the HX concentration consisted of two linear segments with slopes of 5.9035 and 0.1498 μA/μM at the concentration ranges of 0.1-2.0 μM and 2.0-400.0 μM, respectively (Fig. 7).

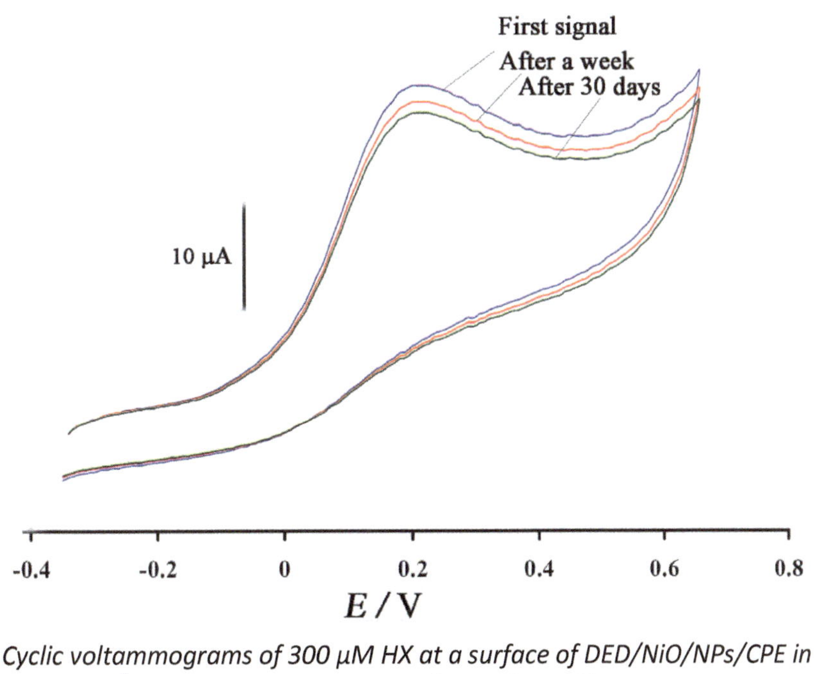

Figure 6. *Cyclic voltammograms of 300 μM HX at a surface of DED/NiO/NPs/CPE in a 0.1 mol L^{-1} phosphate buffer solution (pH 8.0) at different times.*

The decreasing of sensitivity (slope) of the second linear segment was likely due to kinetic limitation. The detection limit was determined as 0.07 μM for HX based on $Y_{LOD} = Y_B + 3\sigma$.

Interference study and real sample studies

Analytical selectivity was one of the important parameters that affected the accuracy of the analysis. In order to evaluate the selectivity of the proposed method for the determination of HX, the influence of various substances as potentially interfering compounds, which can be present in the water and waste water samples with the determination of HX, were studied under optimum conditions with 1.0 μM HX at pH 8.0. The tolerance limit was taken as the maximum concentration of the foreign substances, which caused an approximate 3% relative error (in potential or current) in the determination. The result of interfering studied for some of the various substances in oxidation current and oxidation peak potential of HX showed that 1000-fold of Ni^{2+}, CN^-, Ca^{2+}, Mg^{2+}, Mn^{+2}, K^+, Na^+, Cl^- and SCN^-, 800-fold of glucose, sucrose, lactose and fructose did not affect the selectivity.

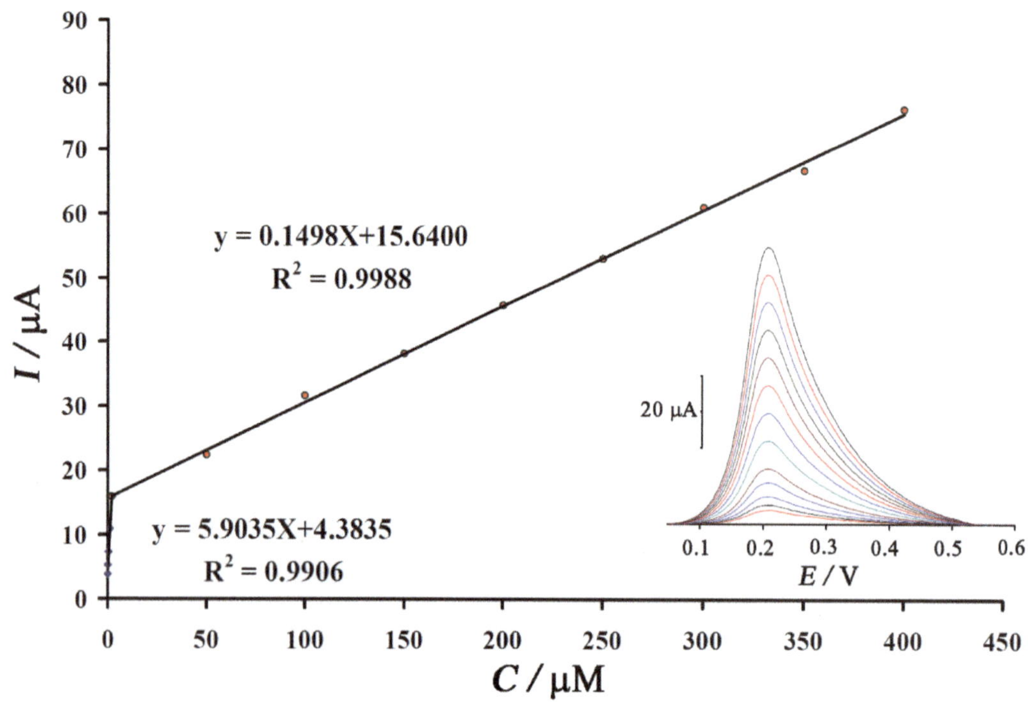

Figure 7. *The plots of the electrocatalytic peak current as a function of HX concentration. Inset shows the SWVs of DED/NiO/NPs/CPE in a 0.1 mol L^{-1} phosphate buffer solution (pH 8.0) containing different concentrations of HX. From bottom-up corresponds to 0.0, 0.1, 0.5, 1.0, 2.0, 50.0, 100.0, 150.0, 200.0, 250.0, 300.0, 350.0 and 400.0 μM of HX.*

In order to demonstrate the applicability of the new sensor in determining HX in real samples, we used the new sensor in determining HX in tap water, river water, wastewater and well water. The determinations of HX in samples were carried out using the standard addition method (Table 1). The accuracy of the method was examined by comparing the results obtained from this method with published methods for the determination of HX [2]. The results from the statistical calculation indicated good agreement between them for the mean values (t-test) and the precision (F-test) in the determination of HX in real samples for the three analyses. It was clear that a modified electrode was capable of voltammetric determination of HX, with high selectivity and good reproducibility.

Table 1: *Determination of HX in water samples (n=3).*

Sample	Added, µM	Founded, µM	Published method, µM	F_{ex}	F_{tab}	t_{ex}	t_{tab}
Tap water	1.00	1.05±0.10	0.98±0.11	6.3	19.0	1.2	3.8
Well water	5.0	4.85±0.16	5.22±0.32	7.8	19.0	1.5	3.8
River water[a]	10.0	10.45±0.65	10.55±0.76	9.6	19.0	2.1	3.8
Waste water	30.0	29.73±0.75	30.75±1.2	13.5	19.0	3.2	3.8

[a] *Tejan River, Sari, Iran*

F_{ex} - calculated *F* value; F_{tab} - reported *F* value from F-test table with 95 % confidence level and 2/2 degree of freedom; t_{ex} - calculated *t*; t_{tab} - reported *t* value from student t-test table with 98 % confidence level.

Conclusion

A carbon paste electrode modified with NiO/NPs and DED was used for electrocatalytic determination of HX. The results showed that the oxidation of HX was catalysed at pH 8.0, whereas the peak potential of hydrazine was shifted by 800 mV to a less positive potential at the surface of DED/NiO/NPs/CPE. In addition, it was shown that HX can be determined using the square wave voltammetry technique. The detection limit (3σ) was 0.07 according to the SWV method. The kinetic parameters, such as electron transfer coefficient, α (0.47) and a rate constant for the chemical reaction between HX and redox sites in DED/NiO/NPs/CPE, k_h (2.454×10^3 M^{-1} s^{-1}) were also determined using electrochemical approaches. Finally, the electrocatalytic oxidation of HX at the surface of this modified electrode can be employed as a new method for the voltammetric determination of HX in real samples such as tap water, river water, wastewater and well water.

Acknowledgements: *The authors wish to thank Gradate University of Advanced Technology, for their support.*

References

[1] M. Mazloum-Ardakani, H. Beitollahi, Z. Taleat, H. Naeimi, *Analytical Methods,* **2** (2010) 1764-1769.

[2] R. Sadeghi, H. Karimi-Maleh, M. A. Khalilzadeh, H. Beitollahi, Z. Ranjbarha, M. B. Pasha Zanousi, *Environmental Science and Pollution Research* **20** (2013) 6584-6593

[3] J. B. Raoof, R. Ojani, H. Karimi-Maleh, *Electroanalysis* 20 (2008) 1259-1262.

[4] E. Mirmomtaz, A. A. Ensafi, H. Karimi-Maleh, *Electroanalysis* **20** (2008) 1973-1979

[5] H. Beitollahi, H. Karimi-Maleh, H. Khabazzadeh, *Analytical Chemistry* 80 (2008) 9848–9851

[6] M. A. Khalilzadeh, F. Khaleghi, F. Gholami, H. Karimi-Maleh, *Analytical Letters,* 42 (2009) 584-599.

[7] H. Karimi-Maleh, A. A. Ensafi, H. R. Ensafi, *Journal of the Brazilian Chemical Society* **20** (2009) 880-887.

[8] H. Karimi-Maleh, A. A. Ensafi, A. R. Allafchian, *Journal of Solid State Electrochemistry* 14 (2010) 9–15.

[9] M. A. Khalilzadeh, H. Karimi-Maleh, A. Amiri, F. Gholami, R. Motaghed mazhabi, *Chinese Chemical Letters* **21** (2010) 1467-1470..

[10] A. A. Ensafi, E. Khoddami, B. Rezaei, H. Karimi-Maleh, *Colloids and Surfaces B: Biointerfaces* **81** (2010) 42-49.

[11] M. A. Khalilzadeh, H. Karimi-Maleh, *Analytical Letters* **43** (2010) 186–196.

[12] A. A. Ensafi, H. Karimi-Maleh, S. Mallakpour, M. Hatami, *Sensors and Actuators B* **155** (2011) 464–-72.

[13] A. A. Ensafi, H. Karimi-Maleh, S. Mallakpour, B. Rezaei, *Colloids and Surfaces B* **87** (2011) 480-488.

[14] A .A. Ensafi, H. Karimi-Maleh, M. Ghiaci, M. Arshadi, *Journal of Material Chemistry* **21** (2011) 15022-15030

[15] R. Moradi, S. A. Sebt, H. Karimi-Maleh, R. Sadeghi, F. Karimi, A. Bahari, H. Arabi, *Physical Chemistry Chemistry Physics* **15** (2013) 5888-5897.

[16] M. Keyvanfard, R. Shakeri, H. Karimi-Maleh, K. Alizad, *Materials Science and Engineering C* **33** (2013) 811-816

[17] M. Roodbari Shahmiri, A. Bahari, H. Karimi-Maleh, R. Hosseinzadeh, N. Mirnia, *Sensors and Actuators B* **177** (2013) 70-77.

[18] M. Elyasi, M. A. Khalilzadeh, H. Karimi-Maleh, *Food Chemistry* 141 (2013) 4311-4317.

[19] A. L. Sanati, H. Karimi-Maleh, A. Badiei, P. Biparva, A. A. Ensafi, *Materials Science and Engineering C* **35** (2014) 379–385.

[20] H. Karimi-Maleh, M. Moazampour, H. Ahmar, H. Beitollahi, A. A. Ensafi, *Measurement* **51** (2014) 91–99

[21] T. Tavana, M. A. Khalilzadeh, H. Karimi-Maleh, A. A. Ensafi, H. Beitollahi, D. Zareyee, *Journal of Molecular Liquids* **168** (2012) 69-74.

[22] H. Karimi-Maleh, M. Moazampour, H. Ahmar, H. Beitollahi, A.A. Ensafi, *Measurement* **51** (2014) 91-99.

[23] N. B. Salah, F. M. Mhalla, *Journal of Electroanalytical Chemistry* **485** (2000) 42-48.

[24] M. Mazloum Ardakani, M. A. Karimi, S. M. Mirdehghan, M. M. Zare, R. Mazidi, *Sensors and Actuators B* **132** (2008) 52-59.

[25] Z. Galus, Fundamentals of Electrochemical Analysis, Ellis Horwood, New York, 1976.

[26] M. Keyvanfard, S. Sami, H. Karimi-Maleh, K. Alizad, *Journal of the Brazilian Chemical Society* **24** (2013) 32-39.

4

Electrochemical studies of nevirapine, an anti-HIV drug, and its assay in tablets and biological samples

NAGAPPA LAXMAN TERADAL, SHANKAR NARAYAN PRASHANTH, JALDAPPAGARI SEETHARAMAPPA✉

Department of Chemistry, Karnatak University, Dharwad-580 003, India

✉Corresponding Author

Abstract

The electrochemical oxidation of nevirapine, an anti-HIV drug, at a glassy carbon electrode has been studied by voltammetric techniques. Nevirapine showed one well defined irreversible oxidation peak with a potential of 0.749 V in phosphate buffer at pH 10. The effects of different electrolytes, pH and scan rate on the electrochemical behaviour of nevirapine were examined to determine the optimum reaction conditions. The oxidation peak current was found to vary linearly with the concentration of nevirapine in the range of 5.0 – 350 μM. The limit of detection and limit of quantification values were calculated and found to be 1.026 μM and 3.420 μM, respectively. The low relative standard deviation values of inter-day and intra-day assays highlighted the good reproducibility of the proposed method for assay of nevirapine. Further, a sensitive and accurate differential pulse voltammetric method was developed for the determination of nevirapine concentrations in pharmaceutical formulations.

Keywords

Nevirapine, Non-nucleoside inhibitor, Voltammetric investigations, Pharmaceutical formulations.

Introduction

Antiretroviral drugs are used for the treatment of infections by retroviruses, primarily human immunodeficiency viruses (HIV) that can lead to acquired immunodeficiency syndrome (AIDS). Presently, four classes of antiretroviral drugs are available *viz.*, (i) nucleoside/tide reverse trans-criptase (NRTI) inhibitors, (ii) non-nucleoside (NNRTI) inhibitors, (iii) protease (PI) inhibitors and (iv) fusion inhibitors [1]. Nevirapine (NVP), (11-cyclopropyl-4-methyl-5, 11-dihydro-6H- dipyrido [3,2-b:2′,3′-e][1,4]diazepin-6-one) (Figure 1) is a NNRTI with activity against human immunodeficiency virus type 1 (HIV-1) [2]. NVP binds directly and reversibly to the catalytic site of the reverse transcriptase enzyme, and therefore interferes with viral RNA to DNA-directed polymerase activities [3]. It is recommended for treating HIV infections in combination with other reverse transcriptase inhibitors such as stavudine and lamivudine [4-7].

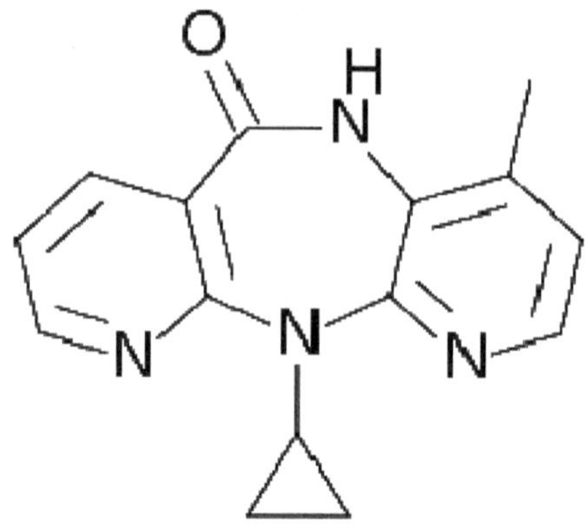

Figure 1. *Structure of nevirapine*

Electrochemical methods have proven to be sensitive and reliable for the determination of numerous electroactive compounds [8-10]. Under some circumstances, electrochemical methods can offer optimal solutions for drug analysis. Simplicity, low cost and relatively short analysis times make electrochemical techniques more useful for routine analytical applications.

A literature survey reveals that no attempt has been made up to the present date to investigate the electrochemical behavior of NVP, and determine it in pharmaceutical formulations by voltammetric methods. A few chromatographic methods *viz.*, HPLC, LC-MS or LC-MS/MS have been reported for the determination of NVP [11-16]. These methods require long analysis times, elaborate extraction and purification steps, or on-line sample extraction, and are relatively costly. Hence, the aim of the present work is to develop a simple, sensitive and accurate electrochemical method for the analysis of NVP, in bulk and dosage forms. The proposed method is more sensitive compared to reported methods [11,12].

Experimental

Reagents and Solution Preparations

A stock solution of NVP (2.5 mM) was prepared in methanol - water (20:80, v/v) and stored in a refrigerator at 4 °C. Working solutions of the drug were prepared daily by diluting the stock solution with the selected supporting electrolyte. In the present study, two different buffers *viz.*, Britton Robinson buffer (pH 2 - 10) and phosphate buffer (pH 3 - 10.6) were used as the supporting electrolytes. All other chemicals used in this investigation were of analytical grade.

Apparatus and procedure

Electrochemical studies were carried out on a CHI-1110a Electrochemical Analyser (CH Instruments Ltd. Co., USA, version 4.01) electrode system consisting of a glassy carbon electrode (GCE) (3 mm diameter) as the working electrode, a platinum wire as the counter electrode and an Ag/AgCl reference electrode. pH measurements were made on a EQ-610 pH meter (Equip-Tronics, India).

The GCE was polished using 0.3 micron Al_2O_3 before each measurement. After polishing, the electrode was rinsed thoroughly with water and then used for analysis.

The parameters for differential pulse voltammetry (DPV) were: pulse amplitude - 50 mV; pulse width - 30 ms; and scan rate - 20 mV s^{-1} maintained.

Assay of tablets

Tablets of NVP (each containing 200 mg of NVP) were obtained from local commercial sources. Ten tablets were finely powdered, and a portion of this powder equivalent to 2.5 mM of NVP was weighed and transferred into a 25 mL calibrated flask containing a methanol-water mixture (20:80, v/v). It was then sonicated for 15 min to effect complete dissolution, and diluted to volume with the same solvent. Suitable amounts of this solution were taken and analyzed. The amount of NVP in the tablet was calculated using a calibration graph or regression equation.

Determination of NVP in human urine and plasma samples

A spiked urine sample was obtained by treating a 1.8 ml aliquot of urine with 200 µl of NVP solution (2.5 mM). A suitable aliquot of the spiked urine was then diluted with phosphate buffer without any pre-treatment to prepare appropriate sample solutions, and their differential pulse voltammograms were recorded under optimized conditions.

For the determination of NVP in plasma, spiked serum samples were prepared following the procedure reported earlier [17]. Serum samples obtained from healthy individuals (after obtaining their written consent) were stored frozen until assay. For the assay of NVP in plasma, 1 ml of NVP solution (5 mM) was added to 1 ml of untreated plasma. The mixture was vortexed for 30 s. In order to precipitate the plasma proteins, the plasma samples were treated with 500 µl of 15% $HClO_4$. Afterward, the mixture was vortexed for 30 s and then centrifuged at 5000 rpm for 5 min. An appropriate volume of supernatant liquor was transferred to a voltammetric cell containing phosphate buffer at pH 10, and voltammograms were recorded. The voltammograms of blank samples (without NVP) did not show any signal that could interfere with this direct determination. The content of the drug in plasma was determined by referring to a calibration graph or regression equation.

Results and Discussion

Voltammetric behavior of NVP at a GCE

The electrochemical behavior of NVP at a GCE was investigated by CV. NVP showed one oxidation peak at 0.749 V in phosphate buffer at pH 10 with a scan rate of 100 mV s^{-1} (Figure 2). No peak was observed in the reverse scan, suggesting that the oxidation of NVP on the GCE was irreversible. Multi-sweep cyclic voltammograms of NVP (data not shown) revealed a significant decrease in peak current, indicating fouling of the electrode surface due to adsorption of NVP or its oxidation product.

Effect of different electrolytes and pH on electrooxidation of NVP

The influence of pH on the peak current (i_p) and peak potential (E_p) was investigated in phosphate (pH 3.0 – 10.6) and BR (pH 2 – 12) buffers. In both buffers, the oxidation peak current of NVP decreased in the pH range of 3 – 7 (data not shown). Above pH 7, the peak current increased up to pH 10.6 (Figure 3). Furthermore, a sharper, well defined peak was noted in phosphate buffer at pH 10; hence, we have used this buffer throughout the study.

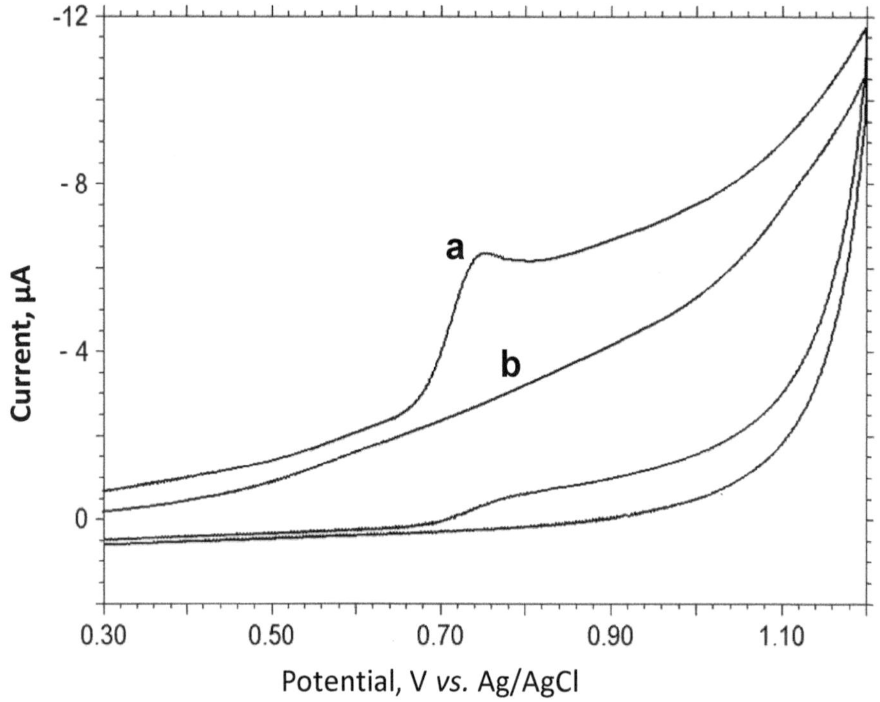

Figure 2. *Cyclic voltammogram of* ***a)*** *50 µM NVP in phosphate buffer at pH 10 and* ***b)*** *blank (buffer solution).*

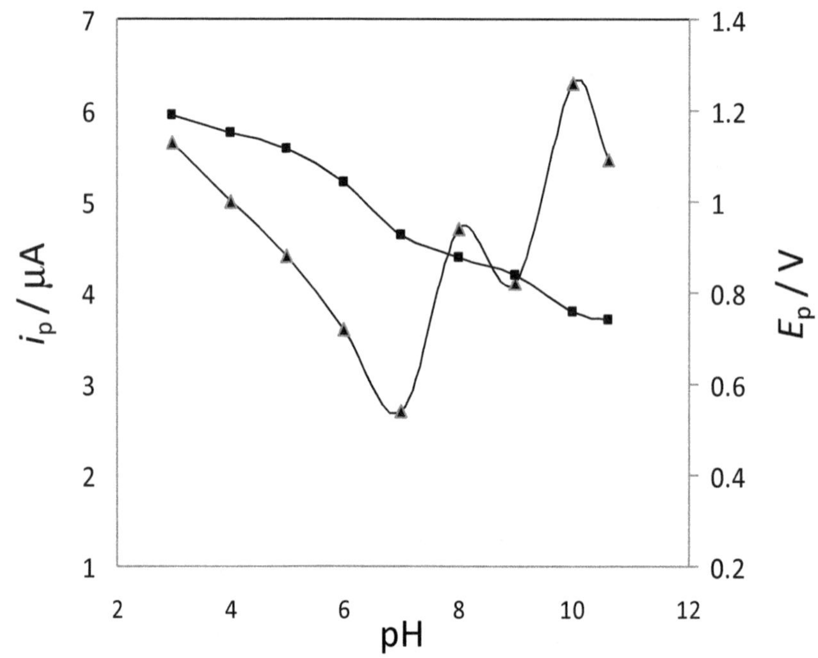

Figure 3. *Effect of pH on peak potential (■) and peak current (▲) for 50 µM NVP*

Figure 3 shows the effect of pH on peak potential (■) and peak current (▲) of 50 µM NVP in phosphate buffer. The peak current of NVP was observed to be maximal at pH values between 8 and 10. This might be attributed to the deprotonation of radical cations in the basic medium, leading to the formation of radical species. This in turn facilitated the oxidation of NVP. This was less pronounced at pH values 7 and 9; hence, the peak heights were found to be minimal. The anodic peak potential of NVP was shifted towards less positive potential with increasing pH of the buffer. This revealed that the pH of the supporting electrolyte exerted a significant influence on electrooxidation of NVP on the GCE, and involvement of a proton in the oxidation process. The plot of peak potential versus pH gave a slope of 63.8 mV pH^{-1}, which is close to the expected value of 59 mV pH^{-1} for participation of equal numbers of protons and electrons in the NVP oxidation process [18–21]. The corresponding equation is shown below:

$E_p = 1.403 - 0.0638\ \mathrm{pH}: r^2 = 0.991.$

Further, the number of electrons transferred, n, was determined by the peak width, $W_{1/2}$ at half height. This was found to be around 83 mV in all the electrolytes, which is close to the theoretical value of 90 mV expected for an electrochemical reaction involving transfer of a single electron [22]. Hence, we propose the electroxidation of NVP involves the transfer of one electron.

Based on the above results, a probable mechanism for electrooxidation of NVP is proposed (Scheme 1). It is proposed that the secondary ring nitrogen of NVP undergoes a single electron oxidation to yield a radical cation, which is further deprotonated to form a radical. Then, the free radical readily combines with another radical to form a dimerized product. This scheme is also in agreement with an earlier report [23].

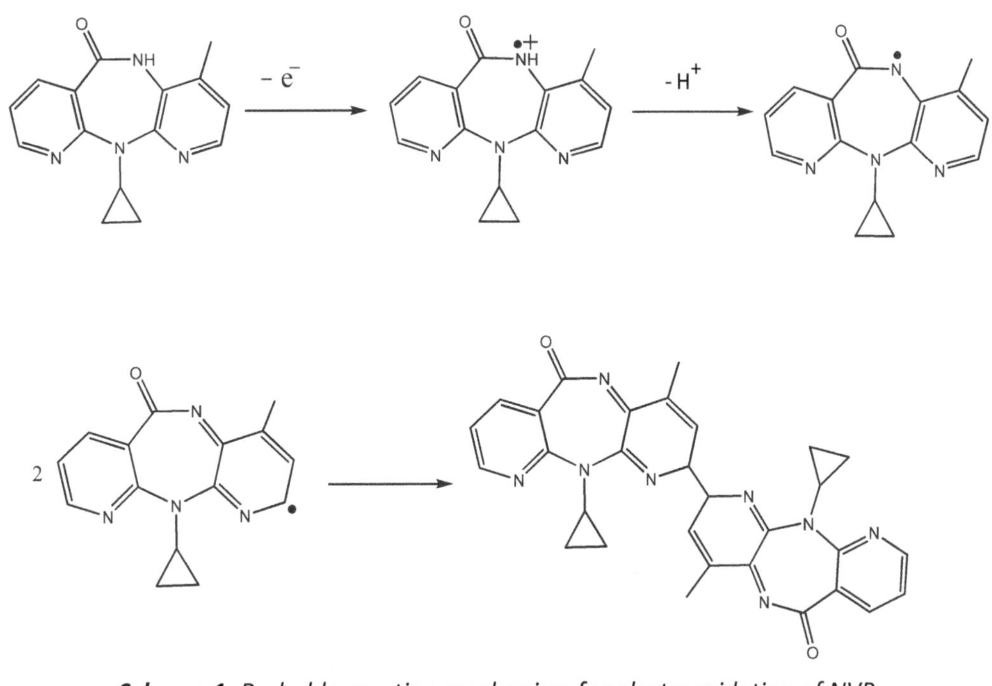

Scheme 1. *Probable reaction mechanism for electrooxidation of NVP.*

Influence of scan rate on electrooxidation of NVP

We examined the influence of the scan rate on the electrochemical behavior of NVP, to understand the nature of the electrode process. For this, we recorded cyclic voltammograms of 50 μM NVP at a GCE at different scan rates (Figure 4). The oxidation peak current of NVP was noted to increase with increasing scan rate, with a positive shift in the peak potential. The plot of values of log i_p versus log v in the scan rate range of 10–350 mV s^{-1} yielded a straight line with a slope of 0.57. This value is close to the theoretical value of 0.5 expected for an ideal reaction condition for a diffusion-controlled electrode process [24].

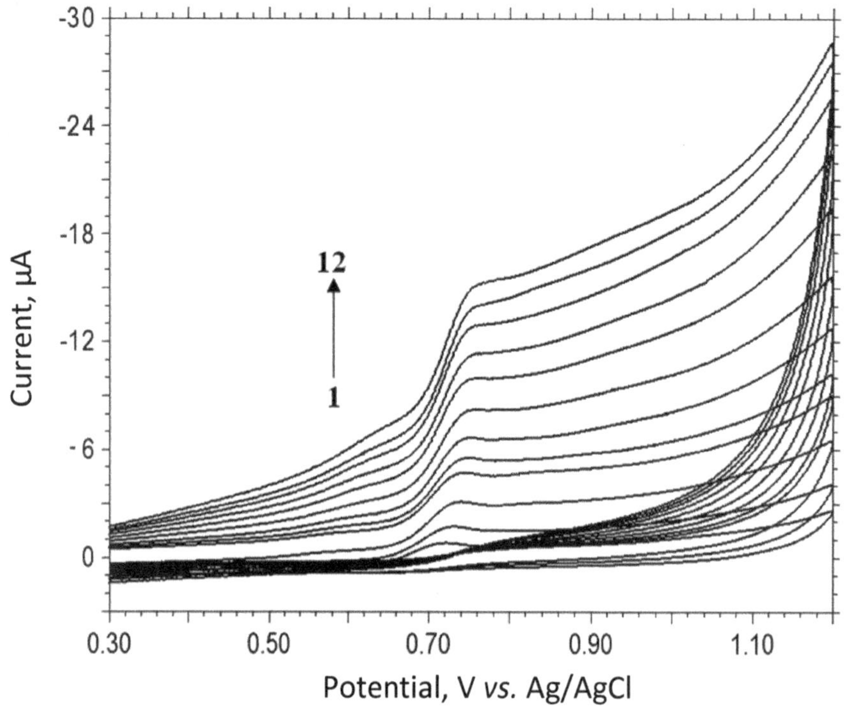

Figure 4. *Cyclic voltammograms for the oxidation of NVP at different scan rates (1–12): 10, 20, 40, 60, 80, 100, 120, 140, 180, 220 and 260, 300, 350 mV s^{-1}.*

Analytical applications

Calibration curve

Considering the electrochemical oxidation of NVP at a GCE, an analytical method was developed using DPV. Differential pulse voltammograms of NVP at different concentrations are shown in Figure 5. Under the optimized conditions, a linear relationship between the peak current and drug concentration was observed in the range of 5.0 – 350 µM (Table 1). Above a concentration of 350 µM, linearity was lost, probably due to adsorption of NVP on the electrode surface. Validation of the optimized procedure for quantitative assay of NVP was examined *via* evaluation of LOD, LOQ, accuracy, precision and recovery values. The LOD and LOQ values were calculated using the equation below [25]:

LOD= 3 s / m,   LOQ= 10 s / m

where s is the standard deviation of the peak currents (five runs), and m is the slope of the calibration curve. The LOD and LOQ values were found to be 1.026 µM and 3.420 µM, respectively (Table 1). Low values of LOD and LOQ confirmed the sensitivity of the proposed method. The inter-day reproducibility of the method was examined by recording the voltammograms of six replicates at 150, 200 and 250 µM. These yielded RSD values of 2.65, 2.50 and 2.24 %, respectively. Further, the RSD values for intra-day assay reproducibility in 150, 200 and 250 µM solutions (n = 6) were found to be 2.33, 2.28 and 2.97 %. The corresponding results are shown in Table 1. The low values of RSD confirm the good precision of the proposed DPV method for assaying NVP. The major advantages of the proposed method are its simplicity, ease of performance, and sufficient sensitivity for NVP.

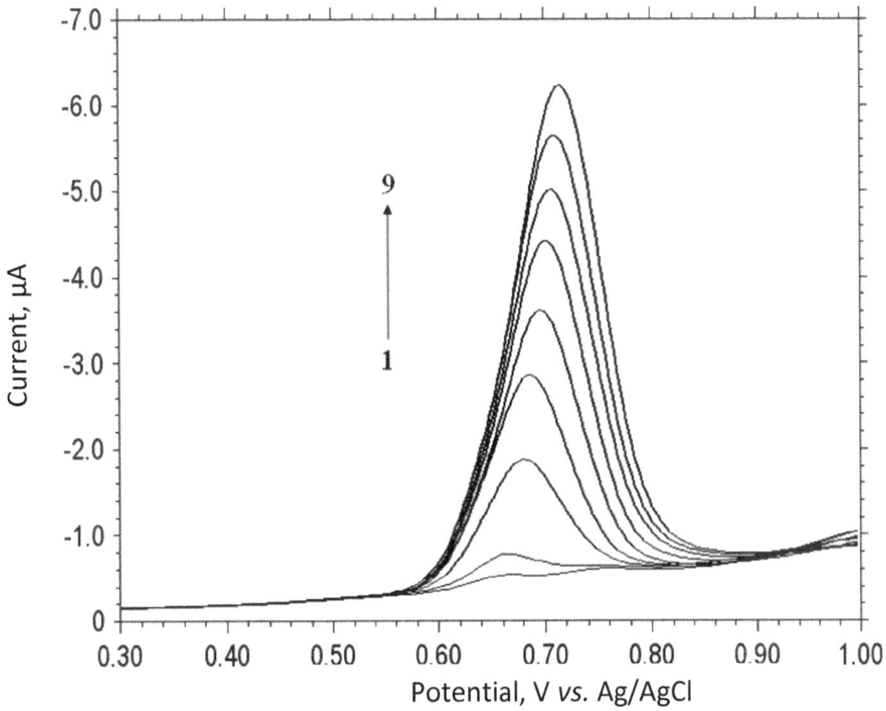

***Figure 5**. Differential pulse voltammograms for increasing concentrations of NVP in phosphate buffer at pH 10 at a GCE. Scan rate: 10 mV s^{-1}; pulse amplitude: 50 mV; and pulse width: 30 ms. NVP concentration was maintained at (1) 5.0, (2) 10, (3) 50, (4) 100, (5) 150, (6) 200, (7) 250, (8) 300, (9) 350 μM.*

***Table 1**. Characteristics of the calibration plot for NVP.*

	DPV
Linearity range, μM	5.0 - 350
LOD, μM	1.026
LOQ, μM	3.420
Inter-day assay RSD* , %	2.65
Intra-day assay RSD*, %	2.97

* For 50 μM NVP

Determination of NVP in urine and blood samples

The proposed method was applied to the determination of NVP in spiked urine samples of healthy volunteers, but not to urine samples of patients treated with NVP. The recoveries from the urine samples were measured by spiking drug free urine with known amounts of NVP, and differential pulse voltammograms were then recorded. The amounts of NVP in the spiked urine samples were then evaluated from the calibration graph. The results of the analysis are listed in Table 2. The average recovery values, higher than 98.94 %, and RSD values less than 2.57 %, indicate the high accuracy and precision of the proposed method.

Further, the proposed method was also applied to the assay of NVP in spiked human serum samples of healthy volunteers, but not to patient serum samples. For this experiment, drug free serum samples were spiked with 150, 200 or 250 μM of NVP; differential pulse voltammograms were then recorded. The amount of NVP in each serum sample was calculated from the calibration plot. The results of the analysis are summarized in Table 2. The percent recovery of NVP was determined by comparing the peak currents of the drug in serum samples with those of pure drug using the calibration curve.

Table 2. *Results of analysis of NVP in spiked human urine and serum samples.*

Urine samples				
NVP added, μM	*n*	Amount found, μM	Average recovery, %	RSD, %
150	5	148.90	99.26	2.57
200	5	197.89	98.94	2.45
250	5	249.78	99.91	2.25
Serum samples				
150	5	149.96	99.97	2.31
200	5	198.02	99.01	2.28
250	5	248.91	99.56	2.07

Analysis of NVP in tablets

The practical analytical application of the DPV method was further established by determining NVP concentrations in tablets. The corresponding results of the analysis are shown in Table 3. The low RSD values again highlighted the reproducibility of the results.

Recovery studies were carried out using a standard addition method. Known quantities of pure NVP were mixed with defined amounts of pre-analyzed formulations; then the mixtures were analyzed as before. The total amount of the drug was then determined, and the amount of drug added was calculated by the difference. The high percentage of recovery indicates that the commonly encountered excipients in the formulation did not interfere with the proposed method.

Table 3. *Determination of NVP in tablets.*

	Nevimune[a]	Neviretro[b]
Labeled amount, mg	200	200
Amount found, mg	198.7	199.8
Recovery, %	99.35	99.90
RSD,[c]%	1. 95	1. 80
Pure NVP added to tablet solution, mg	25	25
Found, mg	25.06	24.92
Recovery, %	100.1	99.7
RSD' %	1.89	2.05

[a]Marketed by Cipla. Ltd., India; [b]Marketed by Alkem (Cytomed) Ltd., India; [c]Average of six determinations

Conclusions

In the present study, the electrochemical behavior of NVP at a GCE was investigated, and various electrochemical parameters are reported for the first time. Electrochemical oxidation of NVP was observed to be irreversible, diffusion controlled and pH dependent. Based on these findings, a simple, rapid and sensitive DPV method was developed for assaying NVP in pharmaceutical formulations and biological samples.

Acknowledgements: *We are grateful to the authorities of Karnataka University, Dharwad, for providing the necessary facilities.*

References

[1] R.N. Rao, D.D. Shinde, *J. Pharm. Biomed. Anal.* **50** (2009) 994-999.

[2] Indian Pharmacopoeia 1996: Addendum 2002. New Delhi: The controller of publications; (2002) 919 - 921.

[3] J.C. Adkins, S. Noble, *Drugs* **56** (1998) 1055-1064.

[4] D. Rey, M. Partisani, G. Hess-Kempf, V. Krantz, M. Priester, C. Cheneau, C. Bernard-Henry, E de Mautort, L. Decroix, J.M. Lang, *J. Acq. Immun. Def. Synd.* **37** (2004) 1454-1456.

[5] C.C.J. Carpenter, M.A. Fischl, S.M. Hammer, M.H. Hirsch, D.M. Jacobsen, D.A. Katzenstein, J.S.G. Montaner, D.D. Richman, M.S. Saag, R.T. Schooley, M.A. Thompson, S. Vella, P.G. Yeni, P.A. Volberding, *J. Amer. Med. Assoc.* **277** (1997) 1962-1969.

[6] R.M. Gulick, J.M. Mellors, D. Havlir, J.J. Eron, C. Gonzalez, D. McMahon, D.D. Richman, F. T. Valentine, L. Jonas,A. Meibohm, E. A. Emini, J. A. Chodakwitz, *New Engl. J. Med.* **337** (1997) 734-739.

[7] W. Cavert, D.W. Notermans, K. Staskus, S.W. Wietgrefe, M. Zupancic, K. Gebhard, K. Henry, Z.Q. Zhang, R. Mills, H. McDade, C.M. Schuwirth, J. Goudsmit, S.A. Danner, A.T. Haase, *Science* **276** (1997) 960.

[8] V.K. Gupta, R. Jain, K. Radhapyari, N. Jadon, S. Agarwal, *Anal. Biochem.* **408** (2011) 179–196.

[9] B. Uslu, S.A. Ozkan, *Anal. Lett.* **40** (2007) 817-853.

[10] B. Uslu, S.A. Ozkan, *Comb. Chem. High T. Scr.* **10** (2007) 495-513.

[11] M.S.M. Ahmed, J.S. Reddy, I.E. Chakravarth, K. Prabhavathi, *J. Chem. Pharm. Res.* **3**(4) (2011) 172-176.

[12] L.A. Zanolli Filho, C.R. Galdez, C.A. Silva, M.F.M. Tavares, D.M. Costa, M.S.A. Prado, *J. Brazil. Chem. Soc.* **22** (2011) 2005-2012.

[13] H.N.Mistri, A.G.Jangid, A. Pudage, N. Gomes, M. Sanyal, P. Shrivastav, *J. Chromatogr. B* **853** (2007) 320-332.

[14] G. Ramachandran, A.K. Hemanthkumar, V. Kumaraswami, S. Swaminathan, *J. Chromatogr. B* **843** (2006) 339-344.

[15] H.N.Mistri, P. Shrivastav, A.G.Jangid, M. Sanyal, *Anal. Lett.* **40** (2007) 1147-1165.

[16] P. Lemmer, S. Schneider, M. Schuman, C. Omes, V. Arendt, J.C. Tayari, L. Fundira, R. Wennig, *Ther. Drug Monit.* **27** (2005) 521-525.

[17] P. Daneshgar, P. Norouzi, M.R. Ganjali, A.O. Seyedler, H. Eshraghi, *Colloid Surface B* **68** (2009) 27-32.

[18] Y. Zhou, J. Zhi, X. Zhang, M. Xu, *Diam. Relat. Mater.* **20** (2011) 18-22.

[19] P. Janeiro, A.M.O. Brett, *Anal. Chim. Acta* **518** (2004) 109-115.

[20] I. Rebelo, J.A.P. Piedade, A.M.O.Brett, *Bioelectrochemistry* **63** (2004) 267-270.

[21] R.N. Goyal, N. Bachheti, A. Tyagi, A.K. Pandey, *Anal. Chim. Acta* **605** (2007) 34-40.

[22] C.M.A. Brett, A.M.O. Brett, Electrochemistry*: Principles, Methods and Aplications,* Oxford Science University Publications, Oxford 1993.

[23] S.S. Kalanur, J. Seetharamappa, *Anal. Lett.* **43** (2010) 618-630.

[24] A.J. Bard, L.R. Faulkner, *Electrochemical Methods: Fundamentals and Applications*, Wiley, New York, 1980.

[25] S. Skrzypek, W. Ciesielski, A. Sokolowski, S. Yilmaz, D.Kazmierczak, *Talanta* **66** (2005) 1146-1151.

5

p-toluene sulfonic acid doped polyaniline carbon nanotube composites: synthesis via different routes and modified properties

ASHOK K. SHARMA✉ and YASHPAL SHARMA*

Department of Materials Science & Nanotechnology, D.C.R University of Science & Technology, Murthal-131 039 (Haryana) India

**Department of Chemistry, G. J. University of Science & Technology, Hisar-125001, India*

✉ Corresponding Author

Abstract

Composites of polyaniline and carbon nanotube (CNT) were prepared by in-situ chemical polymerization method using various aniline concentrations in the initial polymerization solution with p-toluene sulfonic acid (PTS) as secondary dopant and mechanical mixing of the PANI and CNT using different weight ratios of PANI and CNTs. The structural characterizations of the composites were done by Fourier transform infrared (FTIR) and Ultra violet visible spectroscopy (UV-Visible). Scanning electron microscopy (SEM) was used to characterize the surface morphology of the composites. It was found that the composites prepared by in-situ chemical polymerization had smoother surface morphology in comparison to the composites obtained by mechanical mixing. The capacitive studies reveal that the in-situ composite has synergistic effect and the specific capacitance of the composite calculated from cyclic voltammogram (CV) was 385.1 F/g. Thermal studies indicate that the composites are stable as compared to PANI alone showing that the CNT contributes towards thermal stability in the PANI-CNT composites.

Keywords

Cyclic voltammetry; TGA; scanning electron microscopy; composites; conducting polymer; specific capacitance; chemical polymerization; doping, nanomaterials.

Introduction

Due to exceptional electrical, electronic and mechanical properties, conducting polymers are also popular as synthetic metals. Various conducting polymers such as polyaniline, polyacetylene, polypyrrole and polythiophene have been the subject of intensive research because of their application in different fields such as optoelectronic and display devices, and as active electrode materials in primary and secondary batteries, supercapacitors, tissue engineering, drug delivery *etc* [1,2].

Carbon nanotubes after their discovery by Iijima and Ichihashi [3] have been the field of great interest because of their unique electronic, mechanical and magnetic properties [4]. Owing to their unique properties there are varieties of applications of CNT which include supercapacitors and batteries, sensors, biological applications like bone tissue engineering and many more [5]. Recently, composites of the CNT and conducting polymers were a matter of immense research because the composite from these two important materials produce the synergistic effect and produce better applications than the individual materials [6-9]. CNT acts as good filler in the polymer composite compared to other carbon materials because of their high surface area. Various conducting polymers have been investigated to be used in the composite preparation but out of these PANI acts as promising material because of its various properties like air stability, flexibility, light weight, high conductivity, ease of synthesis by chemical and electrochemical method and good processability [10,11]. Recently, PANI was investigated in the synthesis of PANI-CNT composites by using MWNT [7,9], SWNT [8] and carboxylated CNT [6,12].

The composite preparation from the PANI and CNT produces a hybrid material having great mechanical strength, high capacitance properties, electrical properties and good processability [13-16]. There are mainly two methods to prepare such composites: chemical oxidative polymerization method and the electrochemical method. The electrochemical method has limitations in terms of mass production while chemical method gives a high yield and is cheaper [17].

Doping plays an important role in the conducting polymers. Doping is the process which converts the neutral polymer backbone to a charged π conjugated system. This allows electrons to flow through π conjugated system of the conducting polymers due to the formation of conduction bands. Various dopant are used for doping of polyaniline which include *p*-toluene sulphonic acid (PTS), dodecyl benzenesulphonic acid (DBSA), camphor sulphonic acid (CSA), polyvinyl sulfonic acid (PVA) etc. [18]. However, HCl itself acts as a dopant for PANI but using a secondary dopant along with HCl enhance the electronic properties of the material and also the processability.

This study describes the synthesis of some PANI-CNT composites using *p*-toluene sulfonic acid (PTS) as secondary dopant along with HCl by means of chemical oxidative polymerization method and by mechanical mixing of the PTS doped PANI and CNT in the same proportions as used in the *in-situ* chemical oxidative polymerization. The synergistic capacitive effect of the composite was also investigated. The difference in the morphology and thermal behaviour of the composites prepared by both the methods were also investigated.

Experimental

Materials

MWNT (99 %) was purchased from Nanostructured and Amorphous Materials Inc. having outer diameter 20-40 nm and length of 1-2 μm. Aniline (S.D. Fine-Chem. Ltd, 99.5 %) was vacuum distilled prior to use. PVDF, MW Vaco Carbon and *p*-toluene sulfonic acid (PTS) sodium salt (95 %) was purchased from Sigma-Aldrich. All other chemicals like hydrochloric acid (35.4 %), ammonium peroxydisulfate (APS) (98 %), ethanol (99.9 %), sulphuric acid (98 %), activated carbon and N, N-dimethyl formamide (DMF) were obtained from S.D. Fine Chem. Ltd. and used as received without further purification. All chemicals were of analytical grade. Solutions were prepared in deionized water.

Preparation of PANI-CNT composites by in-situ chemical polymerization

For the preparation of PANI-CNT composites by *in-situ* chemical oxidative polymerization method 0.5 g of CNT was dissolved in 1M HCl and then ultrasonicated for 2 hours in order to disperse CNT bundles. Different amounts of aniline monomer (0.5, 1.5, 2.5, 3.5 and 4.5 ml yielding CNT: Aniline ratios of 1:1, 1:3, 1:5, 1:7 and 1:9, wt/v) was dissolved in 1 M HCl and 0.2 M PTS. This content was transferred to the flask containing ultrasonicated CNT. Again the mixture (CNT+Aniline) was put to ultrasonication for 2 hours, to incorporate aniline monomer in the CNT matrix. A freshly prepared pre-cooled ammonium peroxydisulphate (1 M) solution in 1 M HCl was added drop wise to the CNT-aniline mixture. The reaction mixture was stirred for 6 hours at 0-5 °C. Appearance of greenish black color indicated the formation of the PANI-CNT composite. The resulting solution was filtered and washed with distilled water and methanol several times to remove the unreacted oxidant and oligomers if any. The product was dried in vacuum oven at 60 °C overnight. PANI was also prepared as above where 2.5 ml of aniline was dissolved in the initial polymerization solution. Herein, the PANI-CNT composites with different ratios of CNT and aniline monomer will be designated as ratios of CNT: aniline monomer given in parentheses: CPC1 (1:1), CPC2 (1:3), CPC3 (1:5), CPC4 (1:7), CPC5 (1:9). Figure 1 is the schematic representation of the *in-situ* chemical oxidative polymerization method.

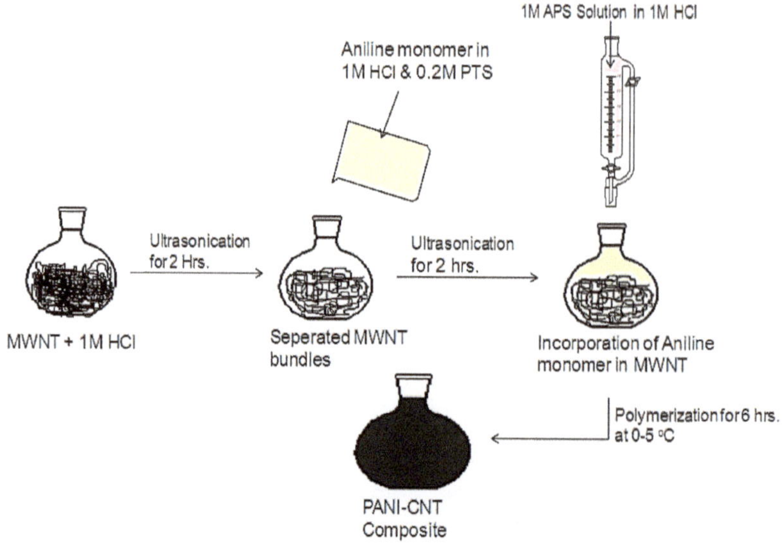

***Figure 1**. Schematic representation of the in-situ chemical oxidative polymerization method*

Preparation of PANI-CNT composites by mechanical mixing method

For the preparation of PANI-CNT composites by mechanical mixing method, PANI was prepared by the *in-situ* chemical route. The as such prepared PTS doped PANI was mixed with CNT in different ratios (CNT:PANI ratios were 1:1, 1:3, 1:5, 1:7 and 1:9) in DMF solvent. The mixture was ultrasonicated for 2-3 hours to disperse the CNT bundles and for better incorporation of the CNT in the polymer matrix. After sonication the mixture was magnetically stirred for 6-7 hours. Finally, the mixture was filtered and washed several times with distilled water and methanol. The PANI-CNT composites were collected by drying the powder in vacuum oven at 60 °C. Herein, the PANI-CNT composites prepared by mechanical mixing method with different ratios of CNT and PANI will be designated as ratios of CNT: PANI given in parentheses CPM1 (1:1), CPM2 (1:3), CPM3 (1:5), CPM4 (1:7), CPM5 (1:9).

Preparation of electrode for cyclic voltammetry

The PANI-CNT composite prepared by *in-situ* chemical oxidative polymerization method was mixed with 10 wt. % of activated carbon and 5 wt. % of MW Vaco Carbon as conductor and 5 wt. % of binder (PVDF) in N,N-Dimethyl formamide (DMF) solvent. The mixture was ultrasonicated for 2 hours and then stirred to form slurry. The slurry was brush coated on 2×2 cm^2 of graphite sheet as a current collector.

Characterization

The chemical structure of the composite was characterized by using Fourier transform infrared spectrophotometer (Shimadzu IR Affinity-I 8000 FT-IR spectrophotometer). Interaction between PANI and CNT was investigated by means of UV-Visible spectroscopic technique (Varian Cary- 5000 spectrophotometer). The scanning electron microscopy (SEM, JEOL-JSM-5600LV) was used to examine the surface morphology and average thickness of deposited PANI over the surface of CNT. The capacitive property of the PANI-CNT composite electrodes was examined by cyclic voltammetry using three-electrode cell system with a reference electrode (saturated calomel electrode) and a counter electrode (Pt) in 1 M H_2SO_4 solutions as electrolyte. Cyclic voltammetry measurements were carried out by potentiostat/galvanostat (CH Instruments 600C series). The geometric surface area of the working electrode was 2×2 cm^2. Cyclic voltammograms were recorded in the voltage windows 0-0.8 V at a scan rate of 2 mV/s. Thermal analyses of the CNT, PANI and PANI-CNT composites was done by using thermogravimetric analyzer (Universal V4.7A TA Instruments).

Results and Discussion

FTIR study

The FTIR spectrum of the PANI-CNT composite is given in Figure 2. The spectrum shows all bands of PANI. The band at 1448.58 and 1581.16 cm^{-1} may be assigned to the benzoid and quinoid ring vibrations, respectively. The band at 1294.11 cm^{-1} is due to the conducting form of the PANI indicating that the PANI exists in the conducting emeraldine salt form. The bands at 1100.79 and 785.68 cm^{-1} may be assigned to the aromatic C-H in plane and aromatic C-H out of plane bending vibrations, respectively [19]. A broad peak at 3402.13 cm^{-1} is due to the N-H stretching vibration of the aromatic amine. The bands at 1068.92 and 674.19 cm^{-1} represent the symmetric stretching of the SO_3^- group [18]. The bands are in agreement with the theoretical predictions and confirm the deposition of PANI over the surface of the CNT *i.e.* formation of PANI-CNT composite using PTS as secondary dopant [20,21].

UV-Visible spectra

UV-Visible spectra of the CNT, PANI and the PANI-CNT composite were recorded in N,N-dimethylformamide (DMF) solvent in the wavelength range from 300 to 800 nm. Figure 3 shows the UV-Visible spectra of CNT, PANI and the PANI-CNT composite. The CNT has no characteristic peak in this region. PANI has its characteristic peak at 540 nm. For the PANI-CNT composite the PANI shows its characteristic peak at 541 nm showing the protonated doped state of the polyaniline indicating the deposition of PANI over the CNT surface and formation of composite [22,23].

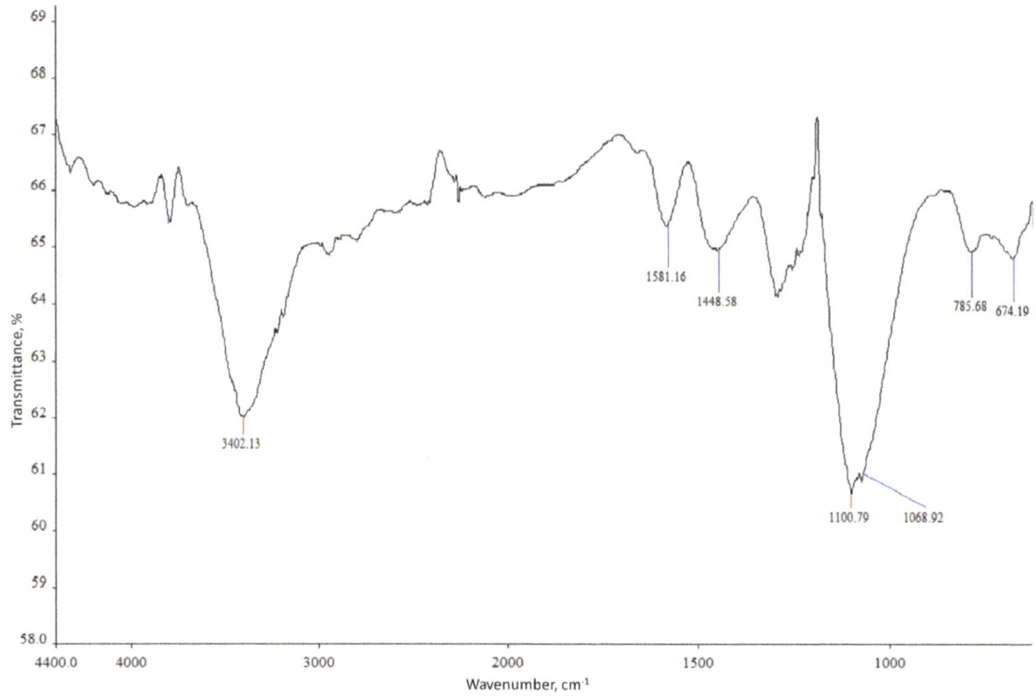

Figure 2. *FTIR spectrum of the PANI-CNT composite*

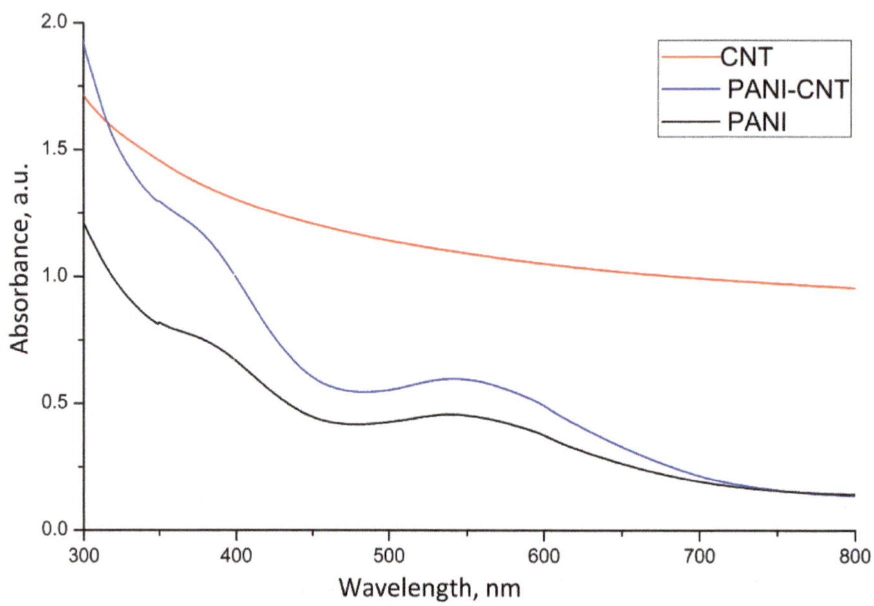

Figure 3. *UV-Visible spectra of the CNT, PANI and PANI-CNT composite*

Surface morphology

Surface morphology of the composites was studied by scanning electron microscopy (SEM). Figure 4a-c shows the SEM micrographs of the CNT, CPC5 and CPM5 respectively. CNT shows smooth surface morphology with average diameter of ≈20-40 nm. Comparing Figure 4b with the SEM micrograph of CNT (Figure 4a) it could be concluded that the polyaniline becomes deposited over the surface of the CNT indicating the formation of the PANI-CNT composite. PANI-CNT composite shows that the average thickness of deposited PANI over the surface of CNT is ≈200-230 nm. Comparing Figure 4c with the SEM micrographs of CNT (Figure 4a) and the

PANI-CNT composite (Figure 4b), it can be said that the composite prepared through mechanical mixing method shows defective and irregular surface morphology which may be caused due to mixing of the two components for long period of time. However, it it is not easy to distinguish the CNT bundles in the composite which is clearly seen in case of PANI-CNT composite prepared by *in-situ* chemical oxidative polymerization method. In the case of the *in-situ* polymerization method, the PANI wrapped the CNT surface in a regular manner using each CNT bundle as can be seen in Figure 4b but in mechanical mixing method the PANI deposition takes place instead of wrapping over the CNT in irregular fashion.

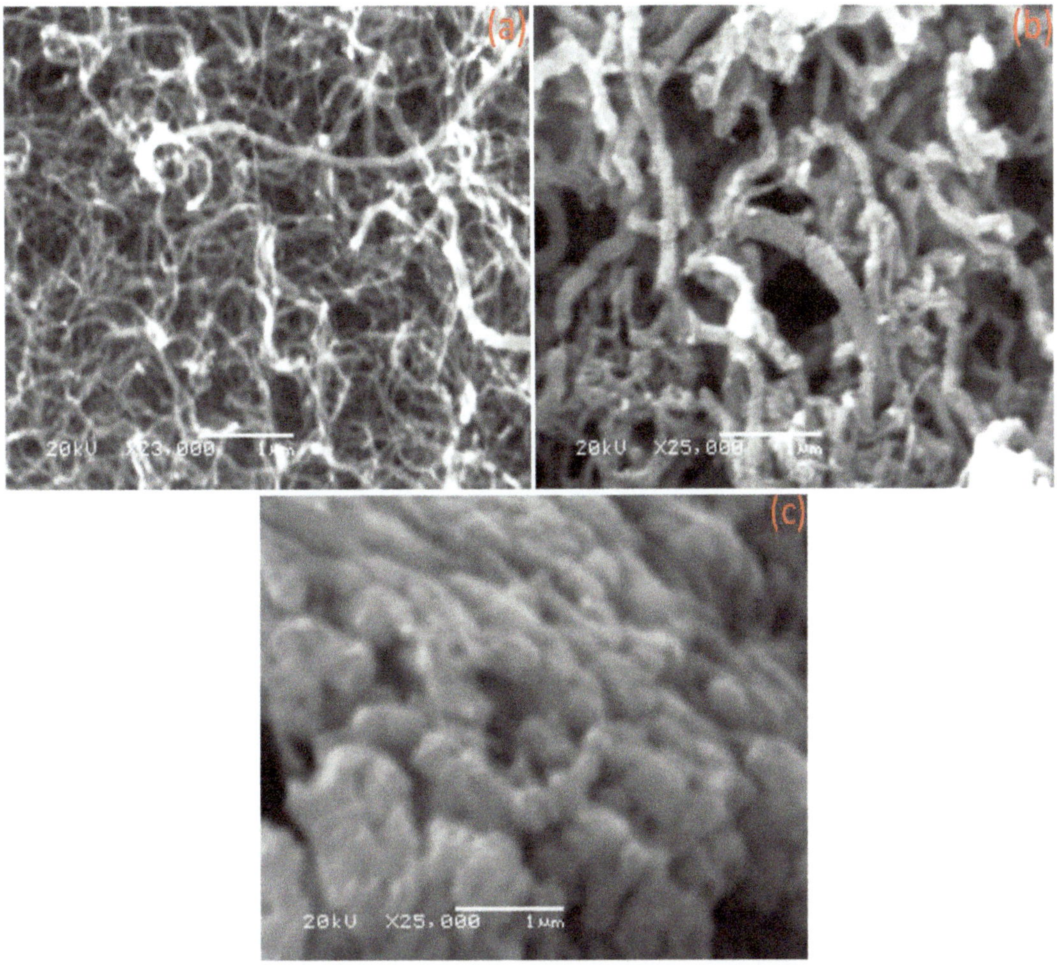

Figure 4. *SEM micrographs of **(a)** CNT, **(b)** CPC5 and **(c)** CPM5 respectively*

Capacitive study

The capacitive characteristic of the CNT, PANI and the PANI-CNT composite prepared by *in-situ* chemical oxidative polymerization was investigated by means of cyclic voltammetry. The capacitance was calculated from the obtained cyclic voltammogram by dividing the average current with scan rate while the specific capacitance was calculated by dividing the capacitance by the weight of active material.

It can be concluded from Figure 5 that CNT behaves like an electric double layer showing almost rectangularly shaped cyclic voltammogram and is almost mirror image with respect to zero current line. For the PANI and PANI-CNT composite the redox peaks are indicative of the higher currents flowing in the PANI-CNT as compared to the PANI alone and can be attributed to the porous structure of the composite in which CNT provides a larger surface area for aniline to polymerize upon. The specific capacitance of CNT, PANI and PANI-CNT composite calculated from cyclic voltammograms was 31.56 F/g, 314.07 F/g and 385.1 F/g, respectively. It can be explained as the synergy contribution from the PANI and the CNT because of better electrolyte access in the active material and improved contact between the current collector and the active material resulting in decreased overall internal resistance in the electrode and better capacitance value [24-27].

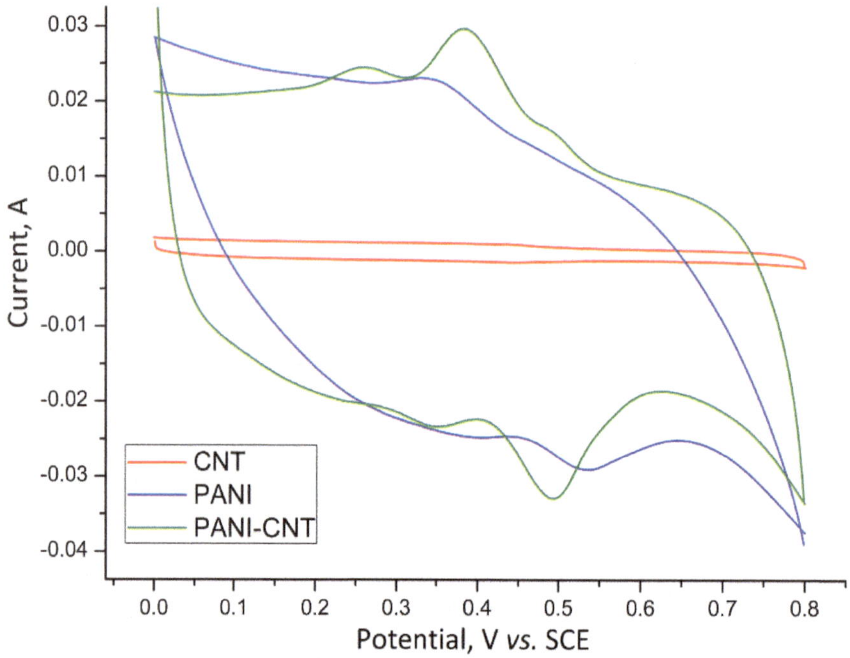

***Figure 5**. Cyclic voltammogram of CNT, PANI and the PANI-CNT composite*

Thermal analysis

Thermogravimetric analysis (TGA) of the CNT, PANI and the CNT-PANI composites were carried out under nitrogen atmosphere in order to investigate the % mass loss with respect to temperature. Figures 6a and 6b show the TGAs of the CNT, PANI and CNT-PANI composites prepared by chemical and mechanical mixing method, respectively. From Figure 6a it can be inferred that the composites show very good thermal stability as compared to the PANI, which can be attributed to the presence of CNT in the original polymer matrix [22]. CNT has highest thermal stability and does not show any decomposition in the temperature range of 30 to 580 °C. The polyaniline on the other hand exhibited a slight decrease in the weight at 114 °C which can be assigned to the loss of water [28]. Up to temperatures of 340 °C approximately 15 % of the weight is lost. Above this temperature PANI showed regular decrease in the weight due to thermal degradation of the polymer.

The order of thermal stability of various PANI-CNT composites, prepared chemically is CPC1 > CPC5 > CPC4 > CPC3 > CPC2 indicating increased thermal stability of the composite as aniline content increases in the initial polymerization solution. Highest thermal stability of CPC1 may be assigned to 1:1 aniline to CNT ratio and hence the lesser amount of PANI. Therefore, CNT in this composite played a dominant role in promoting the thermal stability. In case of higher aniline content the thermal stability is somewhat independent to the CNT content and can be attributed to the layer by layer degradation of the deposited PANI over CNT surface.

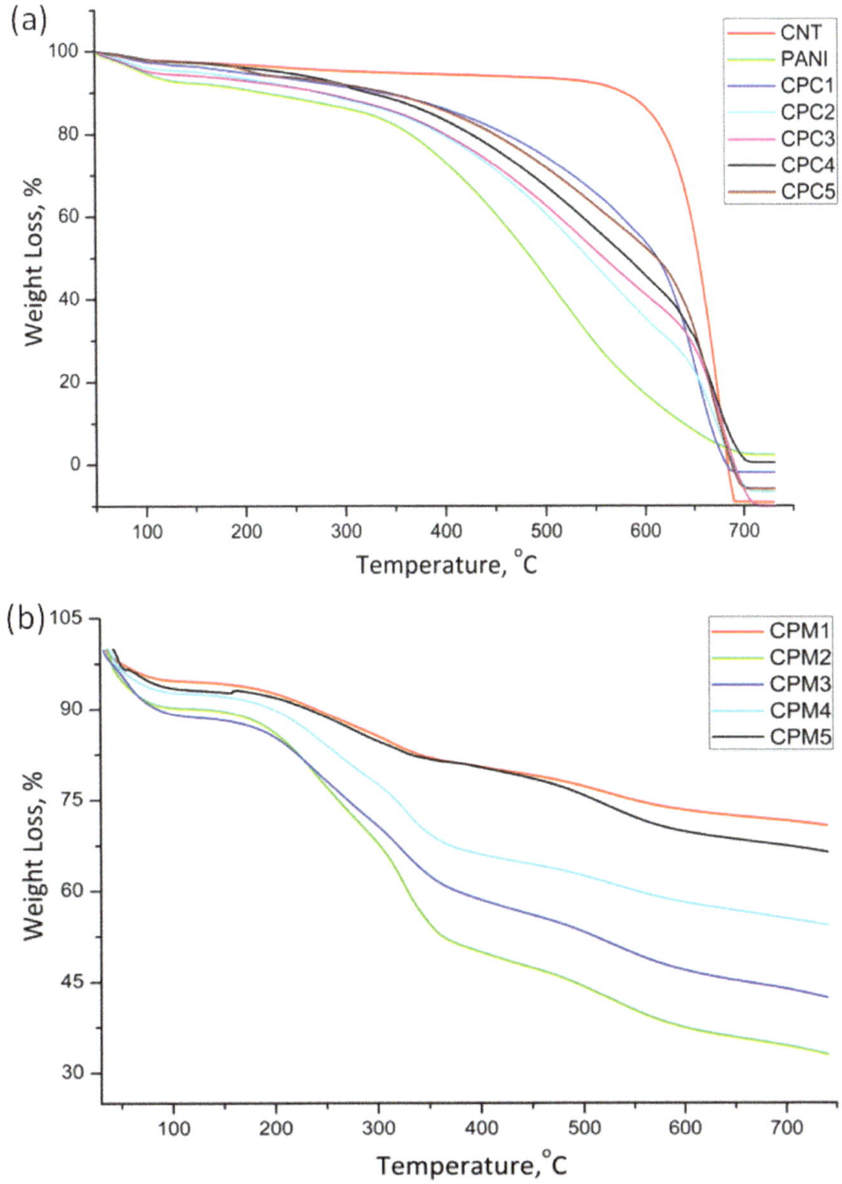

***Figure 6**. TGA Curves of **(a)** CNT, PANI and PANI-CNT composites prepared by in-situ chemical oxidative polymerization method and **(b)** PANI-CNT composites prepared by mechanical mixing method*

***Table 1.** Comparison of % weight loss of two kinds of composites at 380 and 600 °C*

Composite	Weight loss at 380 °C, %	Weight loss at 600 °C, %
CPC1	13	48
CPC2	19	64
CPC3	18	59
CPC4	15	56
CPC5	13	48
CPM1	19	28
CPM2	50	67
CPM3	41	55
CPM4	33	44
CPM5	20	30

A similar trend was obtained for the mechanically mixed composites. For comparison purpose the weight losses of the composites at temperature 380 °C and 600 °C are given in the table I. The results revealed that the composites prepared by chemical method have higher thermal stability compared to composites by mechanical mixing at the temperature range of 380 °C. This can be attributed to the layer by layer deposition of the PANI over the CNT surface in case of chemically prepared composites which can also be seen in the SEM micrograph, whereas no such layered structure is obtained in the mechanically mixed composites. The other reason behind this is the smooth surface morphology of the chemically prepared composites due to separation of the CNT bundles. However, at temperature of 600 °C the reverse order of thermal stability was seen because of presence of higher PANI content in the mechanically mixed composites [29]. Another reason is that in the case of mechanically mixed composites there is a lower chance of separation of CNT bundles and hence the PANI gets deposited over thick CNT bundles resulting in the enhanced stability at higher temperature.

Conclusions

The composites of the PANI-CNT have been successfully synthesized chemically and by mechanical mixing of the PANI and CNT. Composites prepared by chemical polymerization have smooth surface morphology indicative of layer by layer deposition of the PANI over CNT substrate. Whereas the composites prepared by mechanical mixing appear to have irregular deposition of PANI. The thermal stability of the composite is better than the PANI indicating the contribution of CNT towards the thermal stability of the composites. The composite exhibited higher value of specific capacitance than the CNT and PANI alone and is indicative of the synergy. However, CNT did not contributed much to the capacitive value but acts as a good substrate for PANI deposition.

Acknowledgements: *Authors are thankful to the University Grants Commission, New Delhi (INDIA) for providing financial assistance in the form of major research project.*

References

[1] J.W. Schultze, H. Karabulut, *Electrochim. Acta* **50** (2005) 1739–1745.
[2] Y. Sharma, A. Tiwari, S. Hattori, D. Terada, A.K. Sharma, M. Ramalingam, H. Kobayashi, *Int. J. Biol. Macromol.* **51** (2012) 627-631.
[3] S. Iijima, T. Ichihashi, *Nature* **363** (1993) 603–605.
[4] J.W. Mintmire, B.I. Dunlap C.T. White, *Phys. Rev. Lett.* **68** (1992) 631-634.
[5] M. Tasviri, H.A. Rafiee-Pour, H. Ghourchian, M.R. Gholami, *Appl. Nanosci.* **1** (2011) 189–195.
[6] Y.J. Zhang, Y.W. Lin, C.C. Chang, T.M. Wu, *Synth. Met.* **161** (2011) 937–942.
[7] R. Mangu, S. Rajaputra, V.P. Singh, *Nanotechnology* **22** (2011) 215502.
[8] A.A. Mikhaylova, E.K. Tusseeva, N.A. Mayorova, A.Y. Rychagov, Y.M. Volfkovich, A.V. Krestinin, O.A. Khazova, *Electrochim. Acta* **56** (2011) 3656–3665.
[9] L. Zheng, X. Wang, H. An, X. Wang, L. Yi, L. Bai, *J. Solid State Electrochem.* **15** (2011) 675–681.
[10] T.A. Skotheim, R.L. Elsenbaumer, J.R. Reynolds, Handbook of conducting polymers, Marcel Dekker, New York (1997).
[11] P. Ghosh, S.K. Siddhanta, A. Chakrabarti, *Eur. Polym. J.* **35** (1999) 699-710.
[12] T.M. Wu, Y.W. Lin, C.S. Liao, *Carbon* **43** (2005) 734–740.
[13] L.S. Schadler, S.C. Giannaris, P.M. Ajayan, *Appl. Phys. Lett.* **73** (1998) 3842-3844.

[14] H.D. Wagner, O. Lourie, Y. Feldman, R. Tenne, *Appl. Phys. Lett.* **72** (1998) 188-190.
[15] D. Qian, E.C. Dickey, R. Andrews, T. Rantell, *Appl. Phys. Lett.* **76** (2000) 2868-2870.
[16] S.L. Shi, L.Z. Zhang, J.S. Li, *J. Polym. Res.* **16** (2009) 395–399.
[17] E. Frackowiak, V. Khomenko, K. Jurewicz, K. Lota, F. Béguin, *J. Power Sources* **153** (2006) 413-418.
[18] P.D. Gaikwad, D.J. Shirale, V.K. Gade, P.A. Savale, K.P. Kakde, H.J. Kharat, M.D. Shirsat, *Bull. Mat. Sci.* **29** (2006) 417–420.
[19] M.R. Karim, C.J. Lee, Y.T. Park, M.S. Lee, *Synth. Met.* **151** (2005) 131-135.
[20] S.B. Kondawar, S.A. Acharya, S.R. Dhakate, *Adv. Mat. Lett.* **2** (2011) 362-367.
[21] B.S. Kushwah, S.C. Upadhyaya, S.Shukla, A.S. Sikarwar, R.M. S. Sengar, S. Bhaduria, *Adv. Mat. Lett.* **2** (2011) 43-51.
[22] T.M. Wu, Y.W. Lin, *Polymer* **47** (2006) 3576–3582.
[23] A.K. Sharma, Y. Sharma, *Anal. Lett.* **45** (2012) 2075-2085.
[24] S.R. Sivakkumar, W.J. Kim, J.A. Choi, D.R. MacFarlane, M. Forsyth, D.W. Kim, *J. Power Sources* **171** (2007) 1062–1068.
[25] J.H. Kim, Y.S. Lee, A.K. Sharma, C.G. Liu, *Electrochim. Acta* **52** (2006) 1727-1732.
[26] J.H. Kim, A.K. Sharma, Y.S. Lee, *Mat. Lett.* **60** (2006) 1697-1701.
[27] A.K. Sharma, Y. Sharma, R. Malhotra, J.K. Sharma, *Adv. Mat. Lett.* **3** (2012) 82-86.
[28] W. Feng, , X.D. Bai, Y.Q. Lian, J. Liang, X.G. Wang, K. Yoshino, *Carbon* **41** (2003) 1551-1557.
[29] A.M. Showkat, K.P. Lee, A.I. Gopalan, S.H. Kim, S.H. Choi, S.H. Sohn, *J. Appl. Polym. Sci.* **101** (2006) 3721–3729.

6

$Ni\text{-}MoO_2$ cathodes for hydrogen evolution in alkaline solutions. Effect of the conditions of their electrodeposition

GVOZDEN TASIĆ, BORKA JOVIĆ*, UROŠ LAČNJEVAC*, NEDELJKO KRSTAJIĆ** and VLADIMIR JOVIĆ*✉

University of Belgrade, Vinča Institute, Department of Physical Chemistry, 11001 Belgrade, P.O. Box 522, Serbia

**Institute for Multidisciplinary Research University of Belgrade, 11030 Belgrade, P.O. Box 33, Serbia*

***Faculty of Technology and Metallurgy University of Belgrade, 11000 Belgrade, Karnegijeva 4, Serbia*

✉ Corresponding Author

Abstract

The electrodeposition of $Ni\text{-}MoO_2$ composite coatings is a specific process taking place under defined hydrodynamic conditions. In this work the influence of hydrodynamics, current density, and MoO_2 concentration on the electrodeposition of $Ni\text{-}MoO_2$ coatings from a suspension of MoO_2 particles in a Watt's type bath were investigated by comparison of their polarization characteristics for the hydrogen evolution reaction (HER) in 1 M NaOH at room temperature. It was shown that electrolyte and air flow rates influence the process of coating electrodeposition at low concentrations of MoO_2. With increases in MoO_2 concentration the current density became the main parameter influencing the coatings' properties. The best coating, with the lowest over-voltage for the HER, was obtained from the suspension containing 3 g dm^{-3} of MoO_2 particles. The over-voltage of this coating was 57 mV lower for the HER at $j = -0.3$ A cm^{-2} (the current density used in industrial applications) than that for the commercial De Nora's Ni + RuO_2 electrode.

Keywords

$Ni\text{-}MoO_2$; composite coatings; electrodeposition; HER; morphology.

Introduction

One of about 90 hydrogen production routes [1] is alkaline water electrolysis [2]. It is a well-established process that is technologically very simple and produces very clean gases, but it is also energy-inefficient, slow, and expensive. One of the ways to increase the energy efficiency of electrolysers is by developing advanced electrocatalytic materials in order to reduce the electrode over-voltage. In conventional water electrolysis, the cathode is usually made of stainless steel or nickel-based materials (Raney Ni), and operates in 6–9 M KOH solution in a temperature range of 60–90 °C. The overall energy efficiency of electrolysis is related to the hydrogen evolution reaction (HER). In chlor-alkaline electrolysis, HER, as the cathodic reaction in 32 wt.% NaOH at 90 °C, is also responsible for the energy efficiency of electrolysers [3]. Hence, for industrial application, advanced electrocatalytic materials for HER are of great importance to increase energy efficiency. Since porous Ni (Raney Ni) electrodes showed extensive deactivation during electrolysis, several approaches to design and produce new electrocatalytic cathode materials have been presented in the past [4–20]. One of these approaches is based on the Brewer intermetallic bonding theory, which predicts a synergetic activation effect whenever metals of the left half of the d-series are alloyed with metals of the right half of the d-series [4–10]. Accordingly, an increase in electrocatalytic activity has been widely observed for such materials: Ti–Ni [11], Pt–Mo [12], Ni–Mo [9], Ni–Co–$LaNi_5$ [13], Ni–RuO_2 [14], and Ni–transition metals (Fe, Mo, W) [15]).

Recently, particular attention has been given to the replacement of expensive Ni + RuO_2 [14] composite coating with coatings obtained by the electrodeposition of Ni with molybdenum oxide species [16–20]. It was shown that Ni–MoO_3 [16,20] composite coatings, electrodeposited from a suspension containing different concentrations of MoO_3 particles in a Watt's type bath, possess good catalytic activity, but their properties during the "service life" test were not good enough for application in industrial electrolysis [18]. On the other hand, Ni–MoO_2 composite coatings electrodeposited from a suspension containing 3 g dm^{-3} MoO_2 particles in a solution of 0.2 M $NiCl_2$ + 2 M NH_4Cl were found to be a promising replacement [17,18] for commercial Ni + RuO_2 coating obtained by a similar electrodeposition procedure.

Taking into account that the commercial Ni + RuO_2 cathodes were electrodeposited by incorporation of RuO_2 particles in an Ni matrix, in this work an attempt was made to produce Ni–MoO_2 composite coatings from a suspension of MoO_2 particles in a Watt's type bath under different conditions (different concentrations of MoO_2 particles, different current densities, and different hydrodynamic conditions) and to compare their polarization characteristics for HER in 1 M NaOH with that for the commercial Ni + RuO_2 coating. In this way most of the possibilities of obtaining Ni–Mo–oxide coatings by electrochemical deposition would be fulfilled.

Experimental

Electrodeposition of the Ni–MoO_2 composite coatings

The MoO_2 powder was synthesized by a rheological phase reaction route [18]. The average size of MoO_2 powder particles, determined with a Brookhaven Instruments light-scattering system equipped with a BI-200SM goniometer, a BI-9000AT correlator, a temperature controller, and a Coherent INNOVA 70C argon-ion laser, amounted to about 200 nm. Dynamic light scattering measurements were performed using 135 mW laser excitation at a wavelength of 514.5 nm and a detection angle of 90°. All samples were electrodeposited onto Ni 40 mesh from the suspension of MoO_2 powder particles (1–3 g dm^{-3}) in a Watt's type electrolyte containing 210 g dm^{-3} $NiSO_4 \times 7H_2O$, 60 g dm^{-3} $NiCl_2 \times 6H_2O$, 60 g dm^{-3} Na_2SO_4, and 30 g dm^{-3} H_3BO_3 at pH = 3.8 and t = 50 °C in the pilot plant cell (Fig. 1a,b,c). The electrolyte was circulated with the pump, while the flow rate was measured with the flow meter (Fig. 1a). Additional mixing of the electrolyte was provided by the air flow (measured with the flow meter, Fig. 1b) through two pipes with small openings facing the bottom of the cell in order to remove eventually precipitated molybdenum oxide particles from the bottom of the cell and to force particles to float and circulate with the electrolyte (Fig. 1b,c). The temperature of the electrolyte was kept constant by the thermocouple, heater, and control unit (as shown in Fig. 1a). The Ni 40 mesh cathode (dimensions: 5 × 6 cm^2) was connected to a Ni holder (frame, Fig. 1d) and placed between two Ni anode plates (18 × 22 cm, Fig. 1b). A homemade power supply, with a ripple smaller than 1 %, was used to apply the required current/voltage [16,17].

All Ni 40 mesh substrates were first cathodically degreased in a solution of 30 g dm^{-3} NaOH, 30 g dm^{-3} Na_2CO_3, and 40 g dm^{-3} $Na_3PO_4 \times 12H_2O$ for 5 min at j = –100 mA cm^{-2} and t = 60 °C. Neutralization was performed in a solution of 20 g dm^{-3} H_2SO_4 for 5 min. The Ni 40 meshes were then shortly etched in a 1:3 mixture of H_2O:HNO_3 and washed with distilled water before the electrodeposition of the Ni–MoO_2 composite coatings.

HER investigations

All experiments were carried out in 1 M NaOH at room temperature in a three-compartment electrochemical cell with two Pt mesh counter electrodes placed parallel (in separate compartments) to the working electrode and a saturated calomel electrode (SCE) as the reference electrode, connected to the working electrode (in the central compartment) by means of a Luggin capillary. The solution of 1 M NaOH was made from extra pure UV water (Smart2PureUV, TKA) and p.a. chemicals.

All experiments were performed with the potentiostat Reference 600 and PHE 200 or DC 105 Software (Gamry Instruments).

Polarization curve measurements

Electrodes were first submitted to hydrogen evolution at a constant current density of j = –0.3 A cm^{-2} for 800 s (step 1), followed by HER at a constant potential at which the cathodic current density for HER was higher than –0.3 A cm^{-2} (step 2) for 800 s. After this pre-electrolysis, polarization curves were recorded by sweeping the potential at 1 mV s^{-1} from the potential applied in step 2 to the potential value of –1.10 V and recording the current density. The potential values were corrected for the *IR* drop by the current interrupt technique (DC 105).

SEM and EDS analysis of the Ni–MoO_2 composite coatings

The appearance of the surface and the chemical composition of the coatings were investigated by SEM (VEGA TS 5130 MM, Tescan) equipped with an energy-dispersive x-ray spectroscope (EDS; INCAPentaFET-x3, Oxford Instruments).

Results and Discussion

Electrodeposition of Ni–MoO_2 samples

Samples 1–12 were deposited with a charge of 60 C cm^{-2}, while samples 13–21 were deposited with a charge of 120 C cm^{-2}. The conditions of electrodeposition are given in Table 1, while corresponding values of potential taken from the polarization curves for HER in 1 M NaOH at $j = -0.3$ A cm^{-2} are presented in the last column of Table 1.

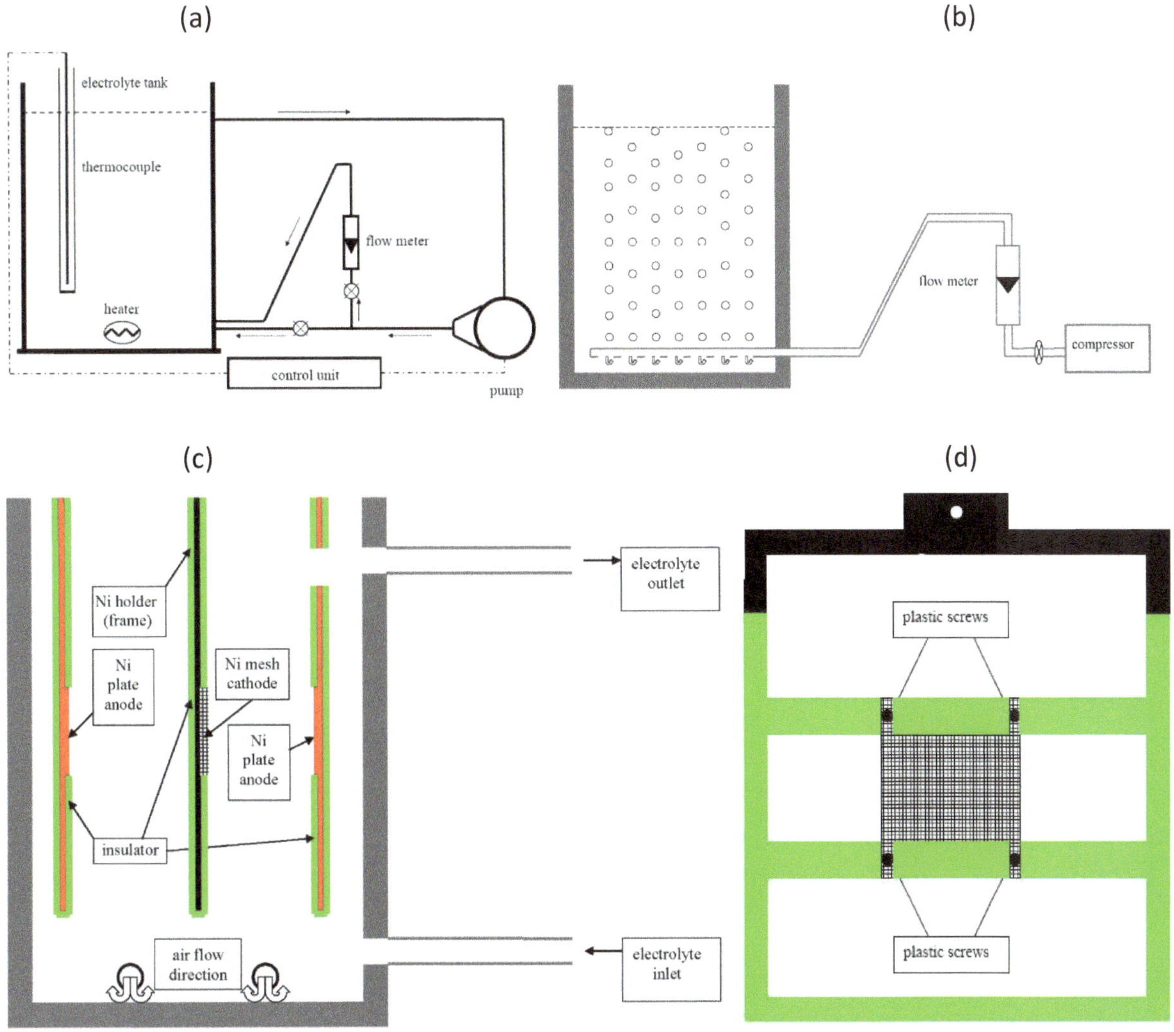

Figure 1. ***(a)*** *Schematic presentation of the system for electrodeposition of coatings.* ***(b)*** *System for additional mixing of electrolyte with air bubbling through two pipes on the bottom of the tank.* ***(c)*** *Position of anodes and cathode.* ***(d)*** *Position of the mesh cathode on the Ni frame.*

Before the addition of MoO_2 powder particles to the electrolyte, electrodeposition of several pure Ni samples in a Watt's type bath was performed at $j = -50$ mA cm^{-2} with different electrolyte and air flow rates. In all cases good quality deposits were obtained with a high current efficiency (> 85 %). When the first amount of MoO_2 powder particles of 1 g dm^{-3} was added to the electrolyte, electrolyte was cycled at the highest rate of 20 dm^3 min^{-1} (with the pump, Fig. 1a) and mixed with the highest air flow of 20 dm^3 min^{-1} (using a compressor, Fig. 1b) for several hours. Electrodeposition of Ni–MoO_2 composite coating started after 24 h of electrolyte aging (following experience from our previous work [17]).

The electrolyte flow rates were set at two values, 10 and 20 dm^3 min^{-1}, while the air flow rates were 1, 5, 10, and 20 dm^3 min^{-1}. Three values of the deposition current densities (j_{dep}) were used: −25 mA cm^{-2}, −50 mA cm^{-2}, and −100 mA cm^{-2}.

Polarization curves for the best samples

Four samples were deposited in the suspension containing 1 g dm^{-3} MoO_2. Their polarization curves for HER were similar, with the best one being that for sample 4. All samples possessed lower over-voltage for HER than Ni 40 mesh with about 50 μm of pure Ni deposited from the Watt's type bath (Ni). The best sample (4) was deposited at the highest current density and flow rates of electrolyte and air (see Table 1). Nine samples were deposited in the suspension containing 2 g dm^{-3} MoO_2. The best polarization characteristics were recorded for sample 13 and the worst ones for sample 11, while all others were placed between these two. Eight samples were deposited in the suspension containing 3 g dm^{-3} MoO_2. The best polarization characteristics were recorded for sample 19, while all others were placed close to the best one. The best polarization characteristics for samples deposited from all three suspensions are presented in Fig. 2, together with polarization characteristics for the deposited Ni coating (Ni) and commercial De Nora's Ni + RuO_2 (DN) coating.

Table 1. *Conditions for electrodeposition of Ni-MoO_2 coatings.*

Sample	MoO_2 g dm^{-3}	$-j_{dep}$ mA cm^{-2}	Electrolyte flow rate, dm^3 min^{-1}	Air flow rate, dm^3 min^{-1}	$E_{(-0.3\ A\ cm^{-2})}$ V
1	1	25	10	1	-1.480
2		25	10	5	-1.446
3		50	10	1	-1.437
4		50	10	5	-1.437
5	2	25	10	1	-1.466
6		25	10	5	-1.469
7		50	10	5	-1.400
8		50	10	1	-1.430
9		50	10	10	-1.400
10		50	10	20	-1.434
11		50	20	10	-1.463
12		50	20	5	-1.420
13		50	20	5	-1.393
14	3	50	20	5	-1.343
15		50	20	10	-1.324
16		25	20	10	-1.358
17		100	20	10	-1.309
18		100	20	10	-1.267
19		50	20	10	-1.270
20		50	10	5	-1.284
21		25	10	5	-1.295
DN					-1.327

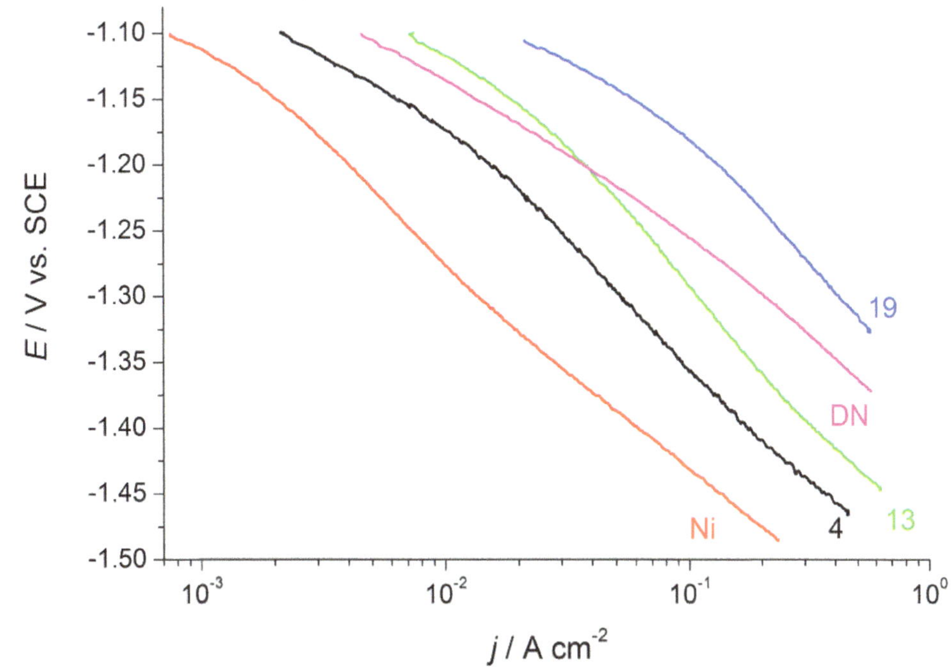

Figure 2. *The best polarization curves recorded in 1 M NaOH at room temperature for samples deposited in the suspensions containing 1 g dm*$^{-3}$ *MoO_2 (sample 4), 2 g dm*$^{-3}$ *MoO_2 (sample 13), and 3 g dm*$^{-3}$ *MoO_2 (sample 19). Polarization curves for deposited Ni (Ni) and commercial De Nora's Ni + RuO_2 electrode (DN).*

Morphology of electrodeposited samples

The typical morphology of the electrodeposited samples is presented in Fig. 3a. Two different types of deposits could be clearly seen: a bright, compact deposit, rich in Ni, around the position where the Ni wires crossed, and a dark deposit characterized by a large number of cracks, rich in Mo, located on the rest of the surfaces of the Ni wires. The Ni-rich and Mo-rich parts of the surfaces were analysed by EDS, as shown in Fig. 3b. The results of the EDS analysis are presented in Table 2.

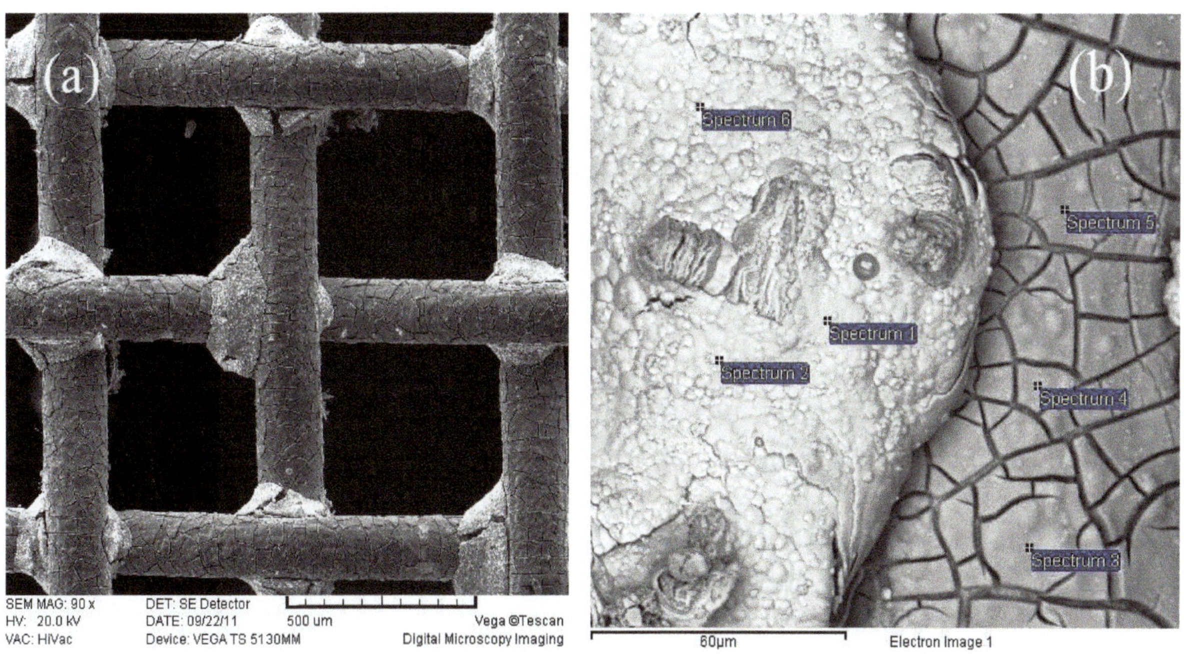

Figure 3. *(a) SEM of the surface of a typical deposit; (b) EDS analysis of Ni-rich and Mo-rich surfaces.*

Table 2. *Chemical compositions of sample surface presented in Fig. 3b.*

Spectrum No.	O content, at. %	Ni content, at. %	Mo content, at. %	Average values		
				O content, at. %	Ni content, at. %	Mo content, at. %
1	36.45	50.57	12.98			
2	30.57	58.24	11.19	37	51	12
6	45.43	44.31	10.26			
3	75.14	4.08	20.77			
4	76.04	4.74	19.22	76	5	19
5	76.10	5.13	18.77			

As can be seen, spectra 1, 2, and 6 correspond to the Ni-rich surface, while spectra 3, 4, and 5 correspond to the Mo-rich surface. Similar morphologies and compositions were obtained for samples electrodeposited from ammoniacal $NiCl_2$ solution [17–19], except that a larger part of the surface of the Ni–MoO_2 composite coatings obtained in our previous work [17–19] was richer in Ni, while a smaller part contained a high percentage of Mo.

In chlor-alkali electrolysis the efficiency of the cathodes results from the combination of a certain activity and stability at the high current densities (3–6 kA m^{-2}) used in technological applications. One of the main reasons for the loss of activity and stability of cathodes during long-term operation is the so-called polarity inversion of the electrodes which takes place during the replacement of old electrodes of an electrolyser with new ones in zero-gap cells. During this operation anodes and cathodes are short-circuited, causing a reverse current flow which may damage the cathodes and negatively affect their activity for HER [3]. Manufacturers can predict how often this operation should be performed during a certain period of time and, in accordance with that, design appropriate accelerated "service-life" tests for cathodes. The service-life test of cathode materials that are promising for use in industrial chlor-alkali electrolysis is the subject of only two papers in the literature [18,21]. The procedure for testing the service life is based on a sequence of galvanostatic polarizations in the HER range and cyclic voltammetries (CVs) in a wide potential range, inducing hydrogen evolution at its negative limit and oxygen evolution at its positive limit (reproducing the conditions of polarity inversion [18,21]). As shown in our previous paper [18] the main reason for the loss of activity of Ni–MoO_2 composite coatings is the dissolution of Mo-rich parts of the coating during the anodic polarization. Although the morphology and composition analysis of Ni–MoO_2 composite coatings obtained in this work indicates the presence of large amount of deposit with Mo-rich parts, their performance during the service-life test will be the subject of our further work.

Conclusions

In this work the influence of hydrodynamics, current density, and MoO_2 concentration on the electrodeposition of Ni-MoO_2 coatings from a suspension of MoO_2 particles in a Watt's type bath was investigated by comparison of their polarization characteristics for HER in 1 M NaOH at room temperature. It was shown that electrolyte and air flow rates influence the process of coatings electrodeposition at low concentrations of MoO_2, while at higher concentrations of MoO_2 the current density became the main parameter influencing the coatings' properties. The best coating was obtained from the suspension containing 3 g dm^{-3} of MoO_2 particles. This coating had a 57 mV lower over-voltage for the HER at $j = -0.3$ A cm^{-2} than a commercial De Nora's Ni + RuO_2 electrode.

Acknowledgements: *The authors are indebted to the Ministry of Education and Science of the Republic of Serbia (Project No. 172054) for the financial support of this work as well as to the Department for the Research and Development of De Nora Industries S.p.A. for providing them with the commercial De Nora's cathode for hydrogen evolution in the chlor-alkali plants.*

References

[1] C.J. Winter, J. Nitsch, *Hydrogen as an energy carrier: technologies, systems, economy*, Springer–Verlag, Berlin, New York, 1988
[2] J.O. Jensen, V. Bandur, N.J. Bjerrum, S.H. Jensen, S. Ebbesen, M. Mogensen, N. Tophoj, L. Yde, *Pre–investigation of water electrolysis*, Energinet.dk, 2008
[3] C. Iwakura, M. Tanaka, S. Nakamatsu, H. Noue, M. Matsuoka, N. Furukawa, *Electrochim. Acta* **40** (1995) 977-982
[4] L. Brewer, in *Alloying*, J.L. Walter, M.R. Jackson, C.T. Sims, Ed(s)., ASM International, New York, USA, 1988, p. 1
[5] L. Brewer L, P.R. Wengert, *Metall. Trans.* **4** (1973) 83–103
[6] L. Brewer, *Science* **161** (1968) 115–122
[7] K.A. Gschneider, in *Solid state physics: advances in research and applications*, F. Seitz, D. Tunold, Ed(s)., Academic Press, New York, USA, 1964, p. 275
[8] M.M. Jakšić, Č.M. Lačnjevac, B.N. Grgur, N.V. Krstajić, *J. New Mat. Electrochem. Systems* **3** (2000) 131-144
[9] J.M. Jakšić, N.V. Krstajić, B.N. Grgur, M.M. Jakšić, *Int. J. Hydrogen Energ.* **23** (1998) 667-681
[10] P. Sabatier, *La Catalyse en Chimie Organique*, Librairie Polytechnique, Paris, France, 1913
[11] N.V. Krstajić, B.N. Grgur, N.S. Mladenović, M.V. Vojnović, M.M. Jakšić, *Electrochim. Acta* **42** (1997) 323-330
[12] B.N. Grgur, N.M. Marković, P.N. Ross, *J. Phys. Chem. B* **102** (1998) 2494-2501
[13] G. Wu, N. Li, C.S. Dai, D.R. Zhou, *Mater. Chem. Phys.* **83** (2004) 307-314
[14] A.C. Tavares, S. Trasatti, *Electrochim. Acta* **45** (2000) 4195-4202
[15] E. Navarro-Flores, Z. Chong, S. Omanovic, *J. Mol. Catal. A* **226** (2005) 179-197
[16] N.V. Krstajić, Lj. Gajić-Krstajić, U. Lačnjevac, B.M. Jović, S. Mora, V.D. Jović, *Int. J. Hydrogen Energ.* 36 (2011) 6441-6449
[17] N.V. Krstajić, Lj. Gajić-Krstajić, U. Lačnjevac, B.M. Jović, S. Mora, V.D. Jović, *Int. J. Hydrogen Energ.* **36** (2011) 6450-6461
[18] V.D. Jović, U. Lačnjevac, B.M. Jović, N.V. Krstajić, *Electrochim. Acta* **63** (2012) 124-130
[19] U.Č. Lačnjevac, B.M. Jović, V.D. Jović, N.V. Krstajić, *J. Electroanal. Chem.* **677–680** (2012) 31-40
[20] Borka M. Jović, Uroš Lačnjevac,Vladimir D. Jović, Ljiljana Gajic-Krstajić, Nedeljko V. Krstajić, *J. Serb. Chem. Soc.*, **77** (2012) 211-224
[21] A.L. Antozzi, C. Bargioni, L. Iacopetti, M. Musiani, L. Vazquez-Gomez, *Electrochim. Acta* **53** (2008) 7410-7416

7

Electrochemical investigations of unexplored anthraquinones and their DNA binding

AFZAL SHAH✉, ABDUR RAUF , ASAD ULLAH, AZEEMA MUNIR, RUMANA QURESHI, IFTIKHAR AHMAD, MUHAMMAD TAHIR SOOMRO, ZIA-UR-REHMAN

Department of Chemistry, Quaid-i-Azam University, 45320, Islamabad, Pakistan and

National Center of Excellence in Analytical Chemistry, University of Sindh, Jamshoro, Pakistan

✉Corresponding Author

Abstract

The redox behaviour of two potential anticancer anthraquinones, 9,10-anthraquinone and 2-chloromethyl-9,10-anthraquinone was investigated in a wide pH range. Cyclic voltammetry based assay was developed for the assessment of the effect of medium, substituents, potential scan rate and number of scans on the voltammetric response of anthraquinones. The electrode reaction mechanism was suggested on the basis of cyclic and differential pulse voltammetric results. The effect of DNA on anthraquinones was also probed at physiological pH which could lead to further investigation of possible citotoxic activity in vitro. The results revealed that anthraquinones interact with DNA more strongly than the clinically used anticancer drug, epirubicin.

Keywords

Anthraquinones; redox mechanism; voltammetry; binding constant.

Introduction

Quinones are known to have a variety of biochemical and physiological functions. Literature survey reveals that these compounds are endowed with antibacterial, antifungal, anti-inflammatory, wound healing, analgesic, antipyretic, antimicrobial, antibiotic and antitumor activities [1,2]. Some derivatives of quinones are common constituents of biologically important molecules such as vitamins K [3]. Others, like mitoxantrone and pixantrone, have been reported to have appreciable antineoplastic activity [4]. It has been reported that the biological action of quinones is due to their electron transfer rates and redox potentials [5]. Consequently, the knowledge of the redox properties of quinones derivatives seems conceivable for a better understanding of their biological action.

Anthraquinones (AQs), an important class of tricyclic compounds, have drawn the utmost attention in medical field as anticancer drugs [6]. The main role of AQs in biological electron transport [7] and industrial processes as redox catalysts [8] have led to the extensive investigations of their electrochemical behavior [9,10]. Quinone-hydroquinone redox complexes are widely used in voltammetric studies due to their robust electrochemistry. The electron transfer process of most of the quinones has been reported to occur through two one-electron steps via a mechanism involving semiquinone free radical [11]. Thus, the study of electron transfer reactions is expected to provide an useful information about molecular structure [12]. Literature survey reveals that in spite of the broad range activities of quinones the electrochemical redox mechanism and DNA binding studies of biologically important AQs, 9,10-anthraquinone (AQ) and 2-chloromethyl-9,10-anthraquinone (CM), is an unexplored matter at physiological pH. To bridge this gap and to offer an explanation for the role of the selected AQs (Scheme 1) in exercising anticancer effect their detailed electrochemistry has been carried out in a wide pH range. Due to the existing resemblance between electrochemical and biological reactions it can be assumed that the redox mechanisms taking place at the electrode and in the body share similar principles [13]. By monitoring the variation in the electrochemical signal of the DNA binding drug it is possible to propose the mechanism of the interaction and the nature of the complex formed. Since AQs participate in several electron transfer processes in the cellular milieu, the present study is expected to provide valuable insights into molecular mechanisms of their action and designing specific DNA-targeted drugs [14].

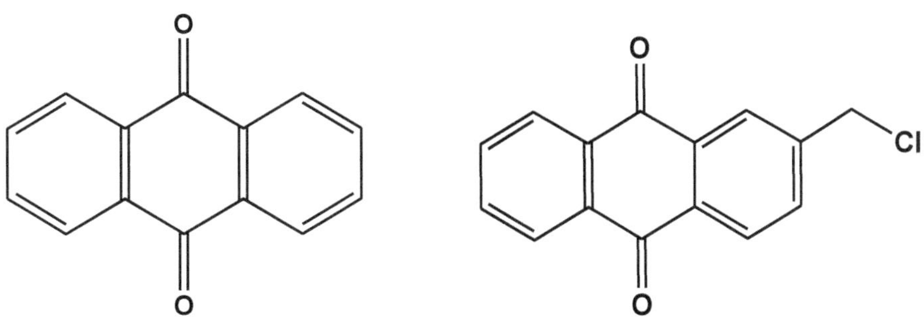

9,10-Anthraquinone (AQ) 2-Chloromethylanthraquinone (CM)

Scheme 1*. Structures of the chemicals used.*

Experimental

Anthraquinones were purchased from Sigma and used without further purification. 0.5 mM working solutions of AQs were prepared in 50 % ethanol and 50% buffer. All supporting electrolyte solutions (Table 1) were prepared using analytical grade reagents. Double stranded DNA was extracted from chicken blood by Falcon method [15,16]. The concentration of the DNA stock solution (2.0 mM) was determined from UV absorbance at 260 nm using a molar extinction coefficient (ε) of 6600 M^{-1} cm^{-1} [17,18]. A ratio of absorbance at 260 and 280 nm of (A_{260}/A_{280}) > 1.8 indicated protein free DNA [19]. Doubly distilled water was used throughout.

Voltammetric experiments were performed using μAutolab running with GPES 4.9 software, Eco-Chemie, The Netherlands. A glassy carbon (GC) (A = 0.07 cm^2) was used as a working electrode, a Pt wire served as a counter electrode and a saturated Ag/AgCl electrode was employed as the reference electrode. Before each experiment the surface of GCE was polished with alumina powder followed by thorough rinsing with distilled water. For reproducible experimental results the clean GC electrode was placed in the supporting electrolyte solution and various cyclic voltammograms were recorded until the achievement of a steady state baseline voltammogram. All the voltammetric experiments were conducted in a high purity argon atmosphere at room temperature (25±1°C). The pH measurements were carried out with a Crison micro pH 2001 pH-meter with an Ingold combined glass electrode.

Table 1. *Supporting electrolytes of 0.1 M ionic strength.*

pH	Composition	pH	Composition
2.0	HCl + KCl	7.0	$NaH_2PO_4 + Na_2HPO_4$
3.0	HAcO + NaAcO	8.0	$NaH_2PO_4 + Na_2HPO_4$
4.0	HAcO + NaAcO	9.0	$NH_3 + NH_4Cl$
5.0	HAcO + NaAcO	10.0	$NH_3 + NH_4Cl$
6.0	NaH_2PO_4+Na_2HPO_4	12.0	NaOH + KCl

Results and Discussion

The cyclic voltammograms of 0.5 mM AQ (Fig. 1) obtained in 0.1 M acetate based supporting electrolyte (pH 3.0) registered a prominent cathodic peak 1_c (more concentration sensitive) at -0.41 V followed by a small reduction peak 2_c at -0.60 V. A single anodic peak 1_a corresponding to oxidation of the reduction product of AQ was recorded in the reverse scan. The E_{p1c}-$E_{p1c/2}$ of 60 mV (peak 1_c) indicated reversible one electron transfer process as reported by the previous investigators [20]. However, the unequal anodic and cathodic peak currents pointed to the quasireversible nature of the overall electrochemical process. The decrease in the cathodic peak current of AQ, as evident from 2nd, 3rd and 4th voltammograms obtained under the same conditions, indicated the adsorption of the reduction product at the electrode surface.

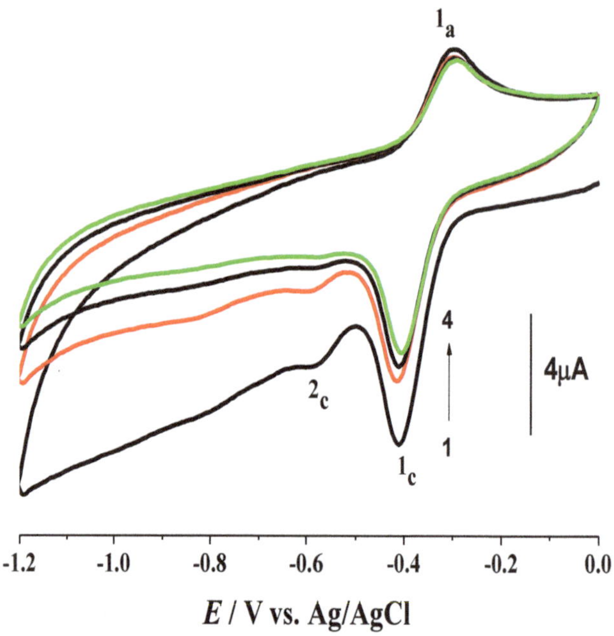

Figure 1*. CVs of 0.5 mM AQ obtained in one experiment in a medium buffered at pH 3.0 using 0.1 M acetate based electrolyte (HAcO+NaAcO): 1st, 2nd, 3rd and 4th scans at 100 mV s^{-1}.*

The appearance of two cathodic peaks in the differential pulse voltammogram -DVP (shown in Fig. 2) confirmed the CV result of two step reduction of AQ. In non-aqueous media quinones undergo two reversible1e processes, but the results of our experiments indicate the appearance of one quasi reversible and another irreversible peak in a 50:50 mixture of ethanol and water. The width at the half peak height ($W_{1/2}$) of 91 and 93 mV (close to the theoretical value of 90.4 mV) indicated the involvement of one electron in each reduction step.

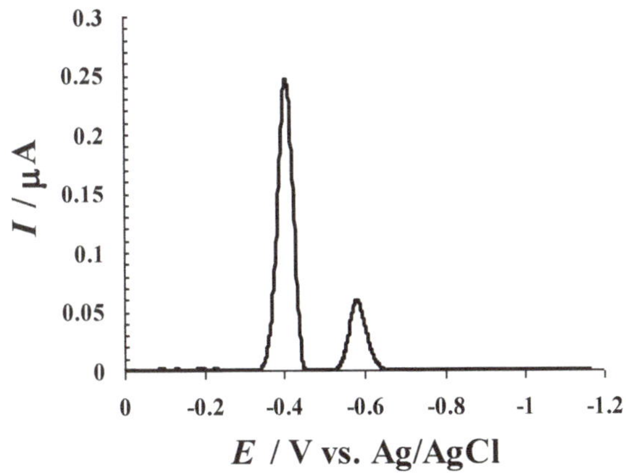

***Figure 2**. First scan DPV of AQ obtained in pH 3.0 at 5 mV s^{-1}; pulse amplitude = 50 mV and pulse period = 200 ms*

For the assessment of the effect of pH, CVs of AQ (Fig. 3) were obtained in acidic, neutral and alkaline media. In contrast to acidic conditions, the reduction of AQ at pH 7.0 was evidenced by a sharp single cathodic peak. The appearance of two broad anodic peaks in the reverse scan showed two step oxidation of the AQ reduction product The cathodic peak width (E_{pc}-$E_{pc/2}$) of 34 mV (close to the theoretical value of 30 mV) indicated the transfer of two-electrons as reported by Bard *et al.* [20]. The equal anodic and cathodic currents at pH 9.0, with peak potential difference of 75 mV can be attributed to the almost reversible nature of the overall redox process under these conditions [21]. The variation in voltammetric response indicated the strong dependency of the AQ redox mechanism on the pH of the medium. With the increase in pH, the potential of 1_c displaces to more negative values. This indicated the involvement of protons in the electro-reduction of AQ. Two peaks were also noticed in the CV of CM at pH 4.5 (Fig. 4) but at positions different from AQ indicating the modulation of the electrophore by the substituent attached at the aromatic ring. The reduction potential of CM was also found to shift cathodically with the increase in pH indicating the involvement of protons during electron transfer processes. On the basis of results obtained from CV and DPV the mechanism shown in Scheme 2 was suggested.

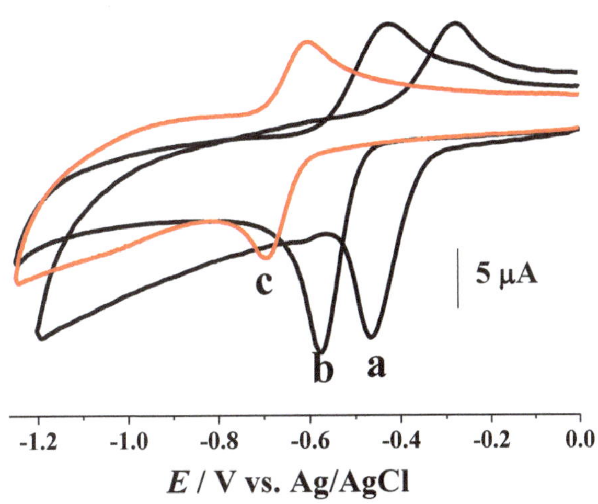

***Figure 3**. 1st scan CVs of 0.5 mM AQ obtained in pH (a) 4.0, (b) 7.0 and (c) 9.0 at 200 mV s^{-1}.*

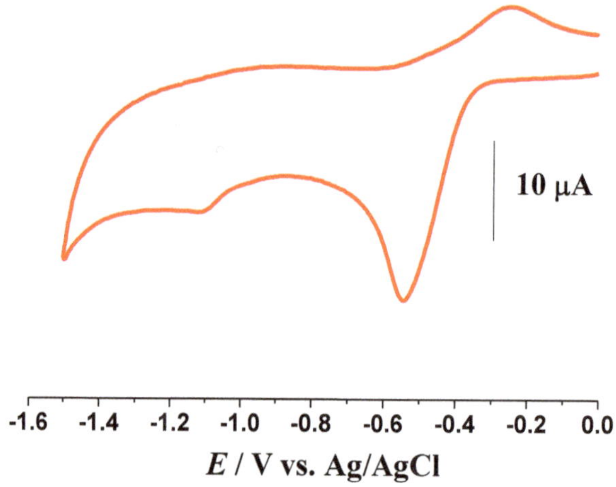

***Figure 4**. 1st scan CV of 0.5 mM CM obtained in pH 4.5 at 200 mV s^{-1}.*

In order to rapidly screen the selected compounds for their anticancer activities, a relatively simple cyclic voltammetric based assay was used without performing other elaborate assays. Although easy to implement, this assay cannot be applied to electroinactive compounds in the potential window of the working electrode. In addition to it, the method is limited to compounds exerting their anticancer effect through direct interaction with DNA. In spite of these obvious limitations the assay can differentiate the binding strength and modes of action of DNA-binding compounds. As the redox behaviour of quinones is sensitive to their covalent or non covalent binding to other molecules, therefore, cyclic voltammetry was used for their DNA binding studies. For CV titrations both the concentration and volume of anthraquinones were kept constant while varying the concentration of DNA in solution. The CVs of anthraquinones in the presence of DNA have been presented at physiological pH 4.5 (pH of lysosomes of the cell) as no report is available in literature on their DNA-binding behavior under such conditions.

$AQ \xrightleftharpoons{e+H^+} AQH \xrightarrow{e+H^+} HAQH$ (Strongly acidic conditions)

$AQ \xrightarrow{2e+2H^+} HAQH \xrightarrow{-e\text{-}H^+} AQH \xrightarrow{-e\text{-}H^+} AQ$ (Neutral conditions)

$AQ \xrightleftharpoons{e} AQ^-$ (Alkaline conditions)

***Scheme 2**. Proposed redox mechanism of AQ*

Typical CV behaviour of 0.5 mM AQ at pH 4.5 and v = 200 mV s^{-1} in the absence and presence of 40, 100 and 120 nM DNA has been shown in Fig. 5. With the gradual addition of DNA, peak 1_c got diminished and peak 2_c increased in height. The positive peak potential shift accompanied with the decrease in current of 1_c is attributed to the disappearance of reactant, AQ and its $1e^-$, $1H^+$ reduction product (HAQ) due to their intercalation into the stacked base pair domain of DNA. The increase in current of peak 2_c can be related to the greater current carrying ability of negatively charged AQ^- in the presence of DNA having anionic phosphate backbone.

The CV of 0.5 mM CM in the presence of 40 nM DNA (shown in Fig. 6) showed quite different behavior from AQ. The cathodic peak potential was shifted (80.3 mV) anodically while the anodic peak potential underwent a drift of 30.2 mV in the cathodic direction accompanied with the decrease in peak currents.

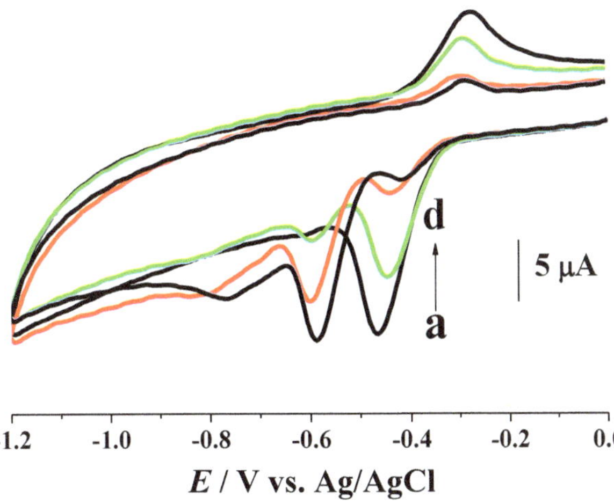

***Figure 5**. CVs of 0.5 mM AQ in the (a) absence and presence of (b) 40, (c) 100 and (d) 120 nM DNA at pH 4.5; v = 200 mV s^{-1}.*

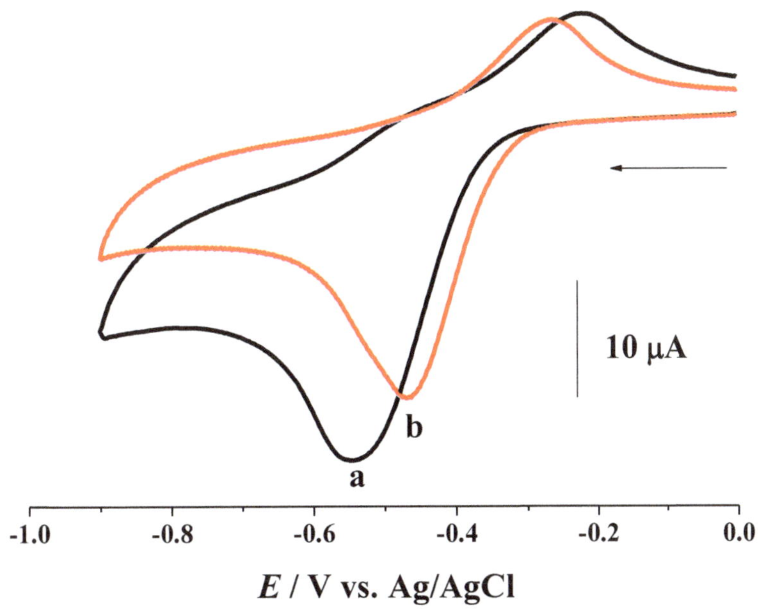

***Figure 6**. CVs of 0.5 mM CM in the (a) absence and presence of (b) 40 nM DNA at pH 4.5; v = 200 mV s^{-1}.*

In general the positive shift in peak potential is suggested for intercalation of the drug into the stacked base pair pockets of DNA [22], while negative shift is related to electrostatic interaction of the drug with the anionic phosphate backbone of DNA [23]. So, the obvious positive peak potential shift in the cathodic peak of CM can be attributed to the intercalation of CM into the base pair pockets of DNA. The shift in the anodic direction further indicates that CM is easier to reduce in the presence of DNA because its reduced form is more strongly bound to DNA than its oxidized form (neutral form). The displacement of the anodic peak in the negative going direction can be linked with the electrostatic interaction of the reduction product of CM with anionic oxygen of DNA. The interaction is expected to alter the DNA replication machinery that may lead to the death of cancerous cells. For such a system, where both forms of the drug interact with DNA, Scheme 3 can be applied [24].

$CM + e + H^+ \leftrightarrow HCM \quad E_f^\circ$

K_{ox} $\qquad$ K_{red}

$CM\text{-}DNA + e + H^+ \leftrightarrow CM\text{-}DNA \quad E_b^\circ$

***Scheme 3**. General redox process of the free and DNA bound AQ.*

Based upon the process discussed in Scheme 3, the following equation is obtained [25]

$$E_b^\circ - E_f^\circ = 0.059 \log (K_{red}/K_{ox}) \tag{1}$$

where $E_f^\circ$and E_b° are the formal potentials of the free and DNA-bound forms of CM respectively.

For a shift of 60 mV caused by the addition of 40 nM DNA into 0.5 mM CM a ratio of K_{ox}/K_{red} was calculated as 0.096, which indicates 10.40 times stronger interaction of the reduced form of the drug with DNA than its oxidized form. The substantial diminution in peak current can be attributed to the formation of slowly diffusing CM-DNA supramolecular complex due to which the concentration of the free drug (mainly responsible for the transfer of current) is lowered.

The dependence of the current function I_p, on the scan rate, v, is an important diagnostic criterion for establishing the type of mechanism. The peak currents of these compounds were found to vary linearly with the square root of the scan rate ($\nu^{1/2}$) as expected for diffusion-limited electrochemical process. The diffusion coefficient was determined by the following form of Randles- Sevcik equation

$$I_{pc} = - 2.99 \times 10^5\, n\, (\alpha_c n)^{1/2} A\, C_\infty D_o^{1/2}\, \nu^{1/2} \tag{2}$$

where n is the number of electrons transferred during the reduction, α_c is the cathodic charge transfer coefficient, A the electrode area in cm^2, D_o the diffusion coefficient in $cm^2\ s^{-1}$, C_∞ the concentration in mol cm^{-3} and ν the scan rate in V s^{-1} [26]. Taking $n = 1$ (as determined from $W_{1/2}$ of DPV) and $\alpha_c = 0.73$ (evaluated from $E_p\text{-}E_{p/2} = 47.7/\alpha n$), the values of D were obtained from the plots of I_{pc} *vs.* $\nu^{1/2}$ at pH 4.5. The electroactive area (0.071 cm^2) of the working electrode was determined as described in literature [27]. From the measured slopes of 1.24×10^{-5} and 2.17×10^{-5} the diffusion coefficients of AQ and CM were determined as 2.49×10^{-6} and 7.58×10^{-6} $cm^2\ s^{-1}$. The general idea about diffusivity is that the molecule having smaller molecular mass should have greater diffusion coefficient but the results of our experiments showing the opposite trend disproved the common concept. The high D value of CM may be due to the presence of electron-withdrawing chlorine substituent group that will lower the electronic density at the electrophore thus increasing its affinity for reduction at the electrode surface. The lower diffusion coefficients of the DNA bound AQ (9.28×10^{-7}) and CM (1.42×10^{-6} cm^2/s) as compared to their free forms is suggestive of slowly diffusing supramolecular drug-DNA adduct formation.

Heterogeneous electron transfer rate constant (k°_{sh}) was evaluated by using Nicholson equation [28] which is based on correlation between ΔE_p and k°_{sh} through a dimensionless parameter ψ by following equation,

$$\psi = \frac{\gamma k^\circ_{sh}}{(\pi a D_o)^{1/2}} \tag{3}$$

where $\gamma = D_O/D_R$, $a = nF\upsilon/RT$ and ν is the scan rate. ψ for different values of ΔE_p can be obtained from the table reported in literature [20]. From the knowledge of ψ , k°_{sh} values of AQ and CM were calculated as 1.02×10^{-3} and 8.51×10^{-4} cm s^{-1} using equation (3). The k°_{sh} values also characterized the reduction processes to be quasi-reversible in nature.

Based upon the decrease in peak current of AQ and CM by the addition of different concentration of DNA, ranging from 20 to 120 nM, the binding constant was calculated from the intercept of the plot of log (1/[DNA]) *versus* log $(I/(I_o-I))$ [29]

$$\log (1/[\text{DNA}]) = \log K + \log (I/(I_o-I)) \quad (4)$$

where K is the binding constant, I_o and I are the peak currents of the drug in the absence and presence of DNA respectively.

The linear fitting of the amperometric data yielded the binding constant of CM and AQ as 2.14×10^6 and 1.50×10^6 M^{-1} indicating their high affinity with DNA. The higher binding constant values than the $K = 4.1 \times 10^5$ M^{-1} of the classical intercalator, proflavin and $K = 3.4 \times 10^4$ M^{-1} of clinically used intercalating anticancer drug, epirubicin [30-32] suggested their preferred anti-cancer drug candidature. However, the clinical use of these drugs demands more research work on their side effects, reactions with Fe(III) in the blood and production of radicals.

Conclusions

The present study shows that anthraquinones can be reduced at the glassy carbon electrode in a pH dependent diffusion controlled manner involving the same number of electrons and protons. The two cathodic peaks, noticed in the CV of CM at positions different from AQ, indicated the modulation of the electrophore by the substituent attached at the aromatic ring. The potential anticancer AQ and CM at physiological pH were found to interact with DNA by intercalative mode as evidenced by the decrease in current intensity and positive peak potential shift. The diffusion coefficients of free anthraquinones were found greater than their DNA-bound forms as expected. The binding constant varied in the sequence: $K_{CM} > K_{AQ}$ with values greater than the clinically used anticancer drugs. The investigation of the electrochemical behaviour of AQ and CM has the potential of providing valuable insights into biological redox reactions of this class of molecules, resulting in a better understanding of the data described for biological systems and increasing the overall knowledge of anthraquinones physiological mechanism of action.

Acknowledgements: *The authors gratefully acknowledge the funds provided by Quaid-i-Azam University and Higher Education Commission Islamabad, Pakistan, for supporting this work.*

References

[1] S. Dettrakul, S. Surerum, S. Rajviroongit, P. Kittakoop, *J. Nat. Prod.* **72** (2009) 861-865.
[2] L.S. Tien, H.L. Chi, L.L. Shih, Y.Y. Ding, *Eur. J. Org. Chem.* **29** (2007) 4831-4836.
[3] X. Gong, R. Gutala, A.K. Jaiswal, *Vitam. Horm.*, **78** (2008) 85-101.
[4] B. Gatto, G. Zagotto, C. Sissi, C. Cera, E. Uriarte, G. Palù, G. Capranico, M. Palumbo, *J. Med. Chem.* **39** (1996) 3114-3122.
[5] A.J. Swallow, In: Function on Quinones in Energy Conserving Systems, Academic Press; New York, 1982.
[6] S. Routier, J.L. Bernier, J.P. Catteau, J.F. Riou, C. Bailly, *Anti-cancer drug des.* **13** (1998) 407-415.
[7] S.B. Mcloughlin, C.R. Lowe, *Enzyme Microb. Tech.* **20** (1997) 2-11.
[8] F. Acramone, *Med. Res. Rev.* **4** (1984) 153-188.

[9] G. Dryhurst, K.M. Kadish, F. Scheller, R. Renneberg, Biological Electrochemistry, Academic Press, London, 1982.
[10] A. Salimi, K. Abdi, G.R. Khayatian, *Microchim. Acta* **144** (2004) 161-169.
[11] A. Bartoszek, *Acta Biochim. Pol.* **49** (2002) 323-331.
[12] M.E. Peover, *J. Chem. Soc.* (1962) 4540-4549.
[13] I. Feliciano, J. Matta, E. Melendez, *J. Biol. Inorg. Chem.* **14** (2009) 1109-1117.
[14] R.K. Gilpin, L.A. Pachla, *Anal. Chem.* **69** (1997) 145-164.
[15] N. Muhammad, A. Shah, Z. Rehman, S. Shuja, S. Ali, R. Qureshi, A. Meetsma, M.N. Tahir, *J. Organomet. Chem.* **694** (2009) 3431-3437.
[16] Z. Rehman, A. Shah, N. Muhammad, S. Ali, R. Qureshi, I.S. Butler, *Eur. J. Med. Chem.* **44** (2009) 3986-3993.
[17] A. Shah, A.M. Khan, R. Qureshi, F.L. Ansari, M.F. Nazar, S.S. Shah, *Int. J. Mol. Sci.* **9** (2008) 1424-1434.
[18] M.J. Han, Z.M. Duan, Q. Hao, S.Z. Zheng, K.Z. Wang, *J. Phys. Chem. C.*, **111** (2007) 16577-16585.
[19] G.D. Liu, X. Yang, Z.P. Chen, G.L. Shen, R.Q. Yu, *Anal. Sci.* **16** (2000) 1255-1259.
[20] A.J. Bard, L.R. Faulkner, Electrochemical Method: Fundamentals and Application J.Wiley and Sons, New York, 1980.
[21] C. Fernandez-Sanchez, T. Tzanov, G.M. Gubitz, A. Cavaco-Paulo, *Bioelectrochemistry* **58** (2002) 149-156.
[22] M. Aslanoglu, *Anal. Sci.* **22** (2006) 439-445.
[23] N. Li, Y. Ma, C. Yang, L. Guo, X. Yan, *Biophys. Chem.* **116** (2005) 199-205.
[24] X. Chu, G.L. Shen, J.H. Jiang, T.F. Kang, B. Xiong, R.Q. Yu, *Anal. Chim. Acta* **373** (1998) 29-38.
[25] T.W. Welch, H.H. Thorp, *J. Phys. Chem.* **100** (1996) 13829-13836.
[26] C.M.A. Brett, A.M. Oliveira Brett, Electrochemistry. Principles, methods and applications, Oxford University Press, UK, 1993.
[27] R.S. Nicholson, I. Shain, *Anal. Chem.* **36** (1964) 706-723.
[28] V.C. Diculescu, T.A. Enache, P.J. Oliveira, A.M. Oliveira Brett, *Electroanal.* **21** (2009) 1027-1034.
[29] A. Shah, R. Qureshi, A.M. Khan, R.A. Khera., F.L. Ansari, *J. Brazil. Chem. Soc.* **21** (2010) 447-451.
[30] M.T. Carter, M. Rodriguez, A.J. Bard, *J. Am. Chem. Soc.* **111** (1989) 8901-8911.
[31] S. Neidle, Z. Abraham, *CRC Cr. Rev. Biochem.* **17** (1984) 73-121.
[32] S. Charak, D.K. Jangir, G. Tyagi, R. Mehrotra, *J. Mol. Struct.* **1000** (2011) 150-154.

8

Ir-Ni oxide as a promising material for nerve and brain stimulating electrodes

Joan Stilling[✉], Nehar Ullah and Sasha Omanovic

Department of Chemical Engineering, McGill University, 3610 University Street, Montreal, Quebec, Canada, H3A 0C5

[✉]*Corresponding author*

Abstract

Tremendous potential for successful medical device development lies in both electrical stimulation therapies and neuronal prosthetic devices, which can be utilized in an extensive number of neurological disorders. These technologies rely on the successful electrical stimulation of biological tissue (i.e. neurons) through the use of electrodes. However, this technology faces the principal problem of poor stimulus selectivity due to the currently available electrode's large size relative to its targeted population of neurons. Irreversible damage to both the stimulated tissue and electrode are limiting factors in miniaturization of this technology, as charge density increases with decreasing electrode size. In an attempt to find an equilibrium between these two opposing constraints (electrode size and charge density), the objective of this work was to develop a novel iridium-nickel oxide ($Ir_{0.2}$-$Ni_{0.8}$-oxide) coating that could intrinsically offer high charge storage capacity. Thermal decomposition was used to fabricate titanium oxide, iridium oxide, nickel oxide, and bimetallic iridium-nickel oxide coatings on titanium electrode substrates. The $Ir_{0.2}$-$Ni_{0.8}$-oxide coating yielded the highest intrinsic (material property) and extrinsic (material property + surface area) charge storage capacity (CSC) among the investigated materials, exceeding the performance of the current state-of-the-art neural stimulating electrode, Ir-oxide. This indicates that the $Ir_{0.2}$-$Ni_{0.8}$-oxide material is a promising alternative to currently used Ir-oxide, Pt, Au and carbon-based stimulating electrodes.

Keywords

Neural stimulation electrodes; metal oxides; iridium; nickel; charge storage capacity

Introduction

With a recently emerging focus on the development of neural prostheses (technology which is able to interact with the body's nervous system) to the utilization of electrical stimulation as a therapy for various neurological disorders, many efforts have been directed towards overcoming the extensive challenges related to electrical stimulation electrode design. Controlled and selective stimulation of the human's central and/or peripheral nervous system is the key to success for all neural stimulation treatment techniques, including deep brain stimulation (DBS), functional electrical stimulation (FES), bladder, intraspinal and epidural stimulation. DBS is used as a treatment for Parkinson's disease, epilepsy, essential tremor, dystonia, and medication resistant psychiatric diseases such as obsessive compulsive disorder (OCD) and depression, while FES, bladder, and spinal stimulation are utilized in stroke, multiple sclerosis, and spinal cord injury rehabilitation therapy[1-6]. The immensely widespread applicability of electrical stimulation in treating numerous pathologies demonstrates its incredible utility in successful medical device development.

In all such applications, a neural electrode works as a bridge - transferring information between the external electronic control device and the human biological system (neurons) [7]. Neural prostheses based on **stimulation** and **recording** of neurons (*i.e.* the cells that transmit electrical information in the body) involve the use of electrodes, which are chronically interfaced to the nervous system. Recording electrodes typically pick up information from sensory systems whereas stimulating electrodes, which are the focus of this article, often communicate with motor systems. Metallic biomaterials have been the main material of choice for neural electrodes. For stimulation; titanium (Ti), titanium nitride (TiN), stainless steel (SS), platinum (Pt), Pt alloys, iridium (Ir), ruthenium (Ru), and rhodium (Rh), which support charge injection by capacitive and faradic mechanisms are available, while materials used for recording electrodes include stainless steel, Pt-Ir alloys, Ir oxide and Ti-nitride [7, 8].

Implantation of neural stimulators in each clinical application requires a safe protocol of operation involving their size, charge injection, and service time [9-11]. However, there are many challenges related to the development of such devices. One of the key problems with current electrodes is their lack of stimulus selectivity, which occurs as a result of large electrode size relative to the targeted neuronal population. For this reason, desirable electrodes are characterized by having a small enough geometric size for targeted and selective neuronal activation, while still possessing the ability to inject a sufficient charge density that does not induce harmful effects on neural function, cell structure, or the electrode itself.

To address the issues mentioned above related to the neural stimulation electrodes, the biocompatibility properties of the material in response to stimulation can be investigated. Metallic biomaterials offer many advantages over the other materials, such as high mechanical strength, wear resistance, inertness, and potential to produce an oxide layer on the surface of the material. However, metallic biomaterials are vulnerable to irreversible faradaic reactions such as water electrolysis, corrosion and the release of toxic ions through gas evolution or metal dissolution [10,12,13]. The performance of the electrode depletes with time as a result of the constantly changing and aggressive environment prone to corrosion in which the electrode is implanted. All of these susceptibilities may induce an immune or necrotic response in adjacent tissue, which can lead to fouling of the electrode surface and loss of functionality [12-14]. Hence, a material with a relatively high corrosion resistance is a desirable feature for a stimulating electrode. It is also essential that the material has a high charge storage capacity; one that effectively stimulates an action potential in a nerve [15]. Various noble metals such as Pt and Pt/Ir alloys have a long history of use as neural electrodes. However, the maximum charge injection (Q_{inj}) limit of Pt (100-300 μC cm^{-2} geometrical area) is insufficient in most cases for nerve stimulation [7,16,17]. Titanium is also used extensively in the biomedical field and neuroscience research due to its high corrosion resistance, biocompatibility, low cost and non-toxicity [18-20]. However, its low charge injection capacity limits its use as a neural stimulation electrode material. Thus, faradaic electrode coatings based on films of Ir oxide have been developed in response to the need for microelectrodes with higher charge injection capacities. Activated iridium oxide film (AIROF) microelectrodes, prepared electrochemically, are widely used for neural applications as they offer a significant improvement in the charge injection limit (2-3 mC cm^{-2} geometrical area) with a reversible faradic reaction ($Ir^{3+} \leftrightarrow Ir^{4+} + e^-$) when compared to other materials. However, activated iridium oxide delaminates under high current pulsing and deposits particles into the surrounding tissues. Additionally, the brittle mechanical properties of iridium oxide make it very difficult to fabricate microelectrodes [7,10,21-24].

New methods have been adopted to address this problem by combining other precursors like ruthenium, titanium, tantalum and tin with iridium to form mixed metal oxide films. However, the long-term stability and higher charge storage/injection of these coatings still needs to be optimized for better performance [8,21,25-28]. Some lower cost transition metal oxides such as nickel, manganese, cobalt and their composites have also been investigated as an alternative electrode material for neural stimulating electrode applications. Nickel for example, has comparable electrochemical behavior to iridium, is low in cost, and readily available. Yet, its non-biocompatible structure, low corrosion resistance and low specific capacitance, limit the feasibility of using pure Ni as a neural electrode coating material [12,29,30]. Despite the drawbacks of pure nickel, it may be combined with other precursors (metals), such as iridium, to utilize its beneficial properties. The potential of bimetallic coatings such as this may help to solve some of the problems encountered with current electrodes.

As outlined above, there is an evident need for the investigation into new materials for neural stimulating electrodes. The aim of this study was to alter electrode surface characteristics through the development of a novel nickel-iridium bimetallic coating on a titanium substrate. In an effort to reduce the aforementioned problems, electrodes were thermally prepared, electrochemically tested, and results were compared with the current industry standard (state-of-the-art), iridium oxide coating (control).

Experimental

Selection of electrode coating materials

Various metal oxide coatings were deposited on a titanium substrate and then characterized using a number of different electrochemical and surface characterization techniques. A titanium plate (99.2 %, metal basis) was used as the substrate. Titanium was chosen due to its excellent biocompatibility characteristics, chemical stability, mechanical strength, and biocompatibility [31]. During experimentation, pure iridium oxide coatings were produced as a control sample; this is currently one of the most common materials applied to neural electrodes due to its high charge storage capacity when compared to other pure metals. However, this stimulating electrode suffers from a number of drawbacks, including loss of the charge-injection capability with time, de-lamination of the oxide coating, fouling, and high cost [7,32].

To address these issues and develop a better stimulating neural electrode, alternative metals were investigated. Nickel was chosen on an empirical basis due to its stability and low cost [12]. In addition, other laboratories have previously demonstrated that this metal shows adequate electrochemical properties in terms of its redox and capacitive behavior, which are similar to that of iridium. Both a one hundred percent nickel oxide coating and an uncoated titanium electrode substrate were created as additional controls for experimentation.

In an attempt to obtain a coating that performs superior to both the pure nickel and iridium oxide in terms of coating stability and charge injection capability, an $Ir_{0.2}$-$Ni_{0.8}$-oxide coating was produced. This bimetallic coating combination of iridium with nickel is a novel idea for use in the application of neural electrodes.

Electrode preparation

Metal oxide coatings were formed through a thermal decomposition method on a flat titanium substrate (purity 99.2 % metals basis, Alfa Aesar 10398). To coat the titanium plates, a 0.15 M precursor solution was prepared by dissolving the corresponding metal precursor salts into iso-propanol (purity 99.9%, Fisher Scientific A416-1). $NiSO_4 \times 6H_2O$ (Sigma Aldrich 227676) was the precursor used to prepare the nickel oxide coating and $IrCl_3 \times 3H_2O$ (Acros Organics 195500050) was used for the iridium oxide coating. It was assumed that the molar ratio of the metals in the precursor solution gave the same molar ratio in the electrode coating.

Titanium substrates used in all experiments were 1 cm-diameter discs with a thickness of 0.2 cm. In order to create a surface to which the precursor solution could adhere, the substrate was polished and etched. The titanium substrate plate was first wet-polished using 600-grit SiC sandpaper [33]. Then, the polished plate was rinsed thoroughly with abundant deionized water and sonicated for 30 minutes in a water bath to remove polishing residue. Next, the polished plate was etched in a boiling solution of hydrochloric acid (33 wt. %, Fisher Scientific) and deionized water (1:1 by volume) for 30 minutes [34,35]. After etching, the plate was again thoroughly rinsed with deionized water and then dried in argon.

The metal precursor coating solution was applied uniformly on the freshly prepared titanium substrate with a paint brush. After applying the first coating, the sample was placed in an oven at 383 K for 5 minutes (in order to vaporize the solvent), followed by annealing of the sample at 737 K in a furnace for 15 minutes. The sample was then removed from the furnace and allowed to cool for 5 minutes before another coating was applied. The same procedure was repeated six times in order to form six coatings on the titanium substrate. Finally, the sample was annealed in the furnace at 737 K for a period of one hour to oxidize the coating [26,28,36-38].

Electrochemical measurements

A standard three-electrode electrochemical cell was employed in all electrochemical experiments. Both a graphite counter electrode and saturated calomel reference electrode (SCE) were utilized. All potentials in this paper are expressed with respect to SCE. A titanium plate coated by Ir-, Ni- or $Ir_{0.2}$-$Ni_{0.8}$-oxide was applied as the working electrode (only one side of the titanium plate was coated with the metal–oxide coating). To utilize the coated side of the working electrode samples during experimentation, a custom-made electrode holder was used to expose a 0.785 cm^2 geometrical area of the working electrode to the supporting electrolyte. Electro-chemical experiments were employed in an aqueous 0.16 M NaCl phosphate buffered solution at pH 7.4. The buffer solution, which is commonly used to simulate human-body fluids, was prepared by mixing appropriate amounts of sodium chloride (purity ≥99.5 %, Fluka Chemika 71381), sodium dihydrogen phosphate anhydrous (purity 99 %, Fluka Chemika 71496) and sodium phosphate dibasic (purity 99.5 %, Fisher Scientific S374). A 5 M NaOH solution (Fisher Scientific SS256) was used to adjust the pH of the buffer solution. In order to maintain an oxygen-free electrolyte, argon (99.998 % pure) was purged, both through the electrolyte 30 minutes prior to electrochemical measurements and during the electrochemical experiments. Electrochemical measurements were performed using an AUTOLAB potentiostat/galvanostat/FRA PGSTAT 30 controlled by FRA2 and GPES v. 4.9 software.

The general electrochemical behaviour and charge storage capacity of the uncoated Ti (control) electrode and the Ir-, Ni- and $Ir_{0.2}$-$Ni_{0.8}$-oxide electrodes were determined using cyclic voltammetry (CV).

To ensure complete characterization of the surface processes occurring at the electrode-solution interface and in the oxide phase, electrochemical impedance spectroscopy (EIS) measurements were made in 0.16 M NaCl phosphate buffer solution over seven frequency decades, from 100 kHz to 50 mHz. The alternating current (AC) voltage amplitude of ±10 mV was employed. The corresponding spectra were modeled employing the two equivalent electronic circuits (EEC) presented in Figure 1.

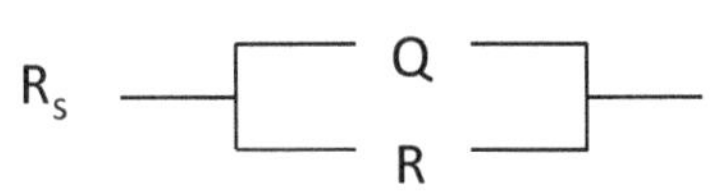

Figure 1. (a) *Equivalent circuit model #1*

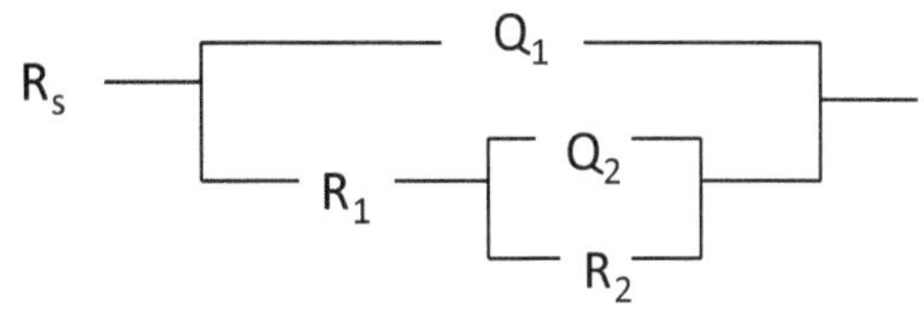

Figure 1. (b) *Equivalent circuit model #2*

The values of solution resistance (R_s), polarization resistance or total resistance to charge transfer ($R = R_1 + R_2$ for circuit in Fig. 1(b)), and capacitance were determined by fitting the experimental data to either of the two models. A constant phase element (CPE), Q, was utilized instead of pure capacitance; this was due to the distribution of the relaxation times as a result of heterogeneities present at the micro level, such as surface roughness [39]. In Fig. 1(a), Q represents the double-layer capacitance, while R represents the polarization resistance. In Fig. 2(b), Q_1 and Q_2 are the double-layer and pseudo-capacitance, respectively (the latter related to the redox transitions in the oxide phase). R_1 and R_2 are the electron-transfer and mass-transport resistance, respectively (the latter related to transport of protons in the oxide phase to balance the metal oxide charge change). The $Ir_{0.2}$-$Ni_{0.8}$-oxide coating was best modeled using the equivalent circuit element shown in Fig. 1(a), while the remaining coatings were modeled using the equivalent circuit model shown in Fig. 1(b).

Surface morphology/topography

Surface morphology/topography of the various metal coatings was investigated using a Phillips XL-30 field emission scanning electron microscope (FE-SEM).

Results and discussion

Surface characterization by scanning electron microscopy (SEM)

In Figure 2, scanning electron micrographs of the three different coatings applied to the titanium substrate are shown. In Figure 2(a, b), a highly roughened $Ir_{0.2}$-$Ni_{0.8}$-oxide coating can be observed. This is shown by its highly tortuous, inhomogeneous surface architecture. The iridium oxide coating shown in Figure 2(c, d) does not demonstrate as rough a surface as in Figure 2(a, b). The surface has deep cracks uniformly dispersed throughout the coating, as indicated by the thin black regions present throughout the image. This surface topography/morphology is less favorable than that in Figure 2(a, b), as less *true* surface area is available per *geometric* surface area to deliver charge. As a result, wider potential windows may be required to deliver a charge sufficient for neuronal stimulation. The nickel oxide coating shown in Figure 2(e, f) demonstrates slight cracks throughout surface. However, the troughs do not appear as deep and wide as those present in the iridium oxide coating shown in Figure 2(c, d). Further, the nickel oxide coating's small roughness value (lower true surface area) can be observed when compared to the $Ir_{0.2}$-$Ni_{0.8}$ oxide coating shown in Figure 2(a, b). Surface roughness has a strong effect on an electrode's charge storage capacity, which will be demonstrated and discussed later in the article.

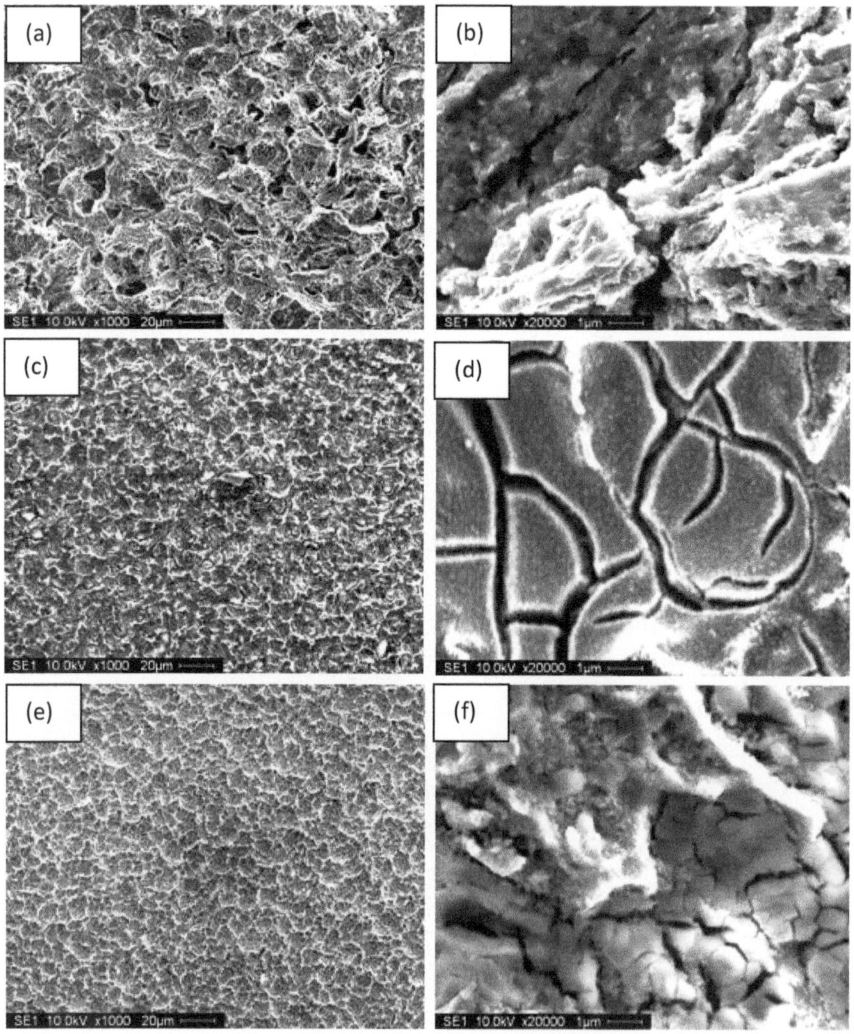

Figure 2. *SEM micrographs of the metal oxide coatings formed on a Ti substrate after CV experiments in 0.16 M saline phosphate buffer solution at pH = 7.4,* ***(a, b)*** *$Ir_{0.2}$-$Ni_{0.8}$-oxide* ***(c, d)*** *iridium oxide* ***(e, f)*** *nickel oxide.*

Electrochemical Characterization by Cyclic Voltammetry

Cyclic voltammetry (CV) was used to assess the electrochemical redox behavior of the electrode coatings within the region between hydrogen and oxygen evolution; -1.0 V to 1.0 V. These measurements allowed for the characterization of the stimulation electrodes by their charge storage capacity (CSC). CSC is defined as the total amount of charge per unit geometric surface area (GSA), that may be stored reversibly in the electrode [13] and is sufficient enough to stimulate an action potential in a nerve without causing any damage to human tissues [7]. Hence, in order to investigate whether the developed coatings are capable of delivering sufficient amount of charge during stimulation, CSC of various coatings were determined using CV data presented in Figure 3. The presented behavior is typical of that displayed by purely capacitive and pseudo-capacitive materials [7,17,40-43]. A few conclusions can be made by analyzing Figure 3. First, there is a large difference in current response among the investigated materials; at a fixed potential, the recorded current increases in the order of Ti- < Ni = Ir < $Ir_{0.2}$-$Ni_{0.8}$-oxide, indicating that the charge storage capacity behaves in the same manner. Second, when analyzing the $Ir_{0.2}$-$Ni_{0.8}$-oxide coating response, one can see that there are two redox peaks; an anodic peak at ca. 0.2 V and a cathodic peak at ca. -0.1 V. These peaks correspond to reversible redox transitions in the oxide phase, namely the Ir(III)/Ir(IV) and Ni(II)/Ni(IV) transitions [41-44]. These results demonstrate that the various electrode coatings behave quite differently when a charge is applied to the electrode.

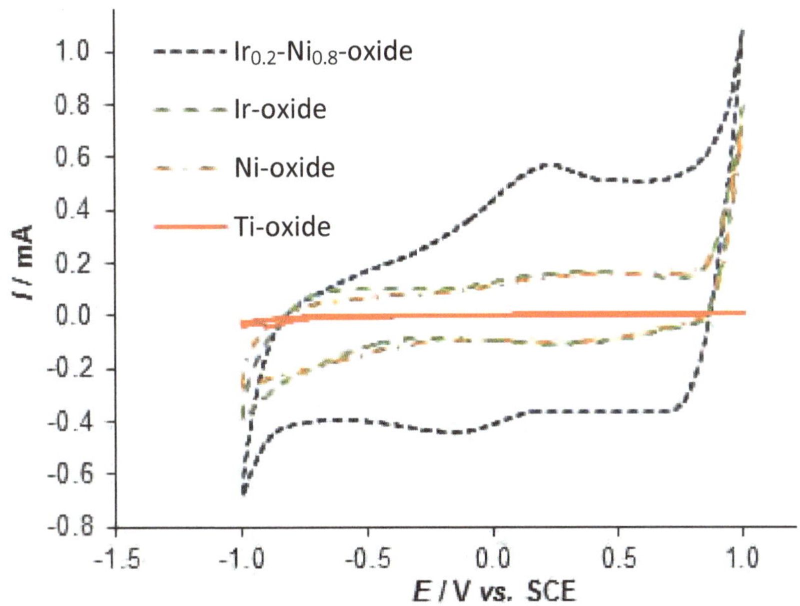

Figure 3. *Cyclic voltammograms of Ti-, Ni-, Ir- and $Ir_{0.2}$-$Ni_{0.8}$-oxide coatings recorded in 0.16 M saline phosphate buffer pH = 7.4 at a scan rate, sr = 100 mV s^{-1}.*

To evaluate the amount of CSC that can be delivered during the anodic and cathodic polarization of the electrode, the corresponding anodic and cathodic areas under the curves were integrated and then normalized with respect to geometric surface area (GSA)in order to determine the **extrinsic** charge storage capacity ($CSC_{Extrinsic}$). Figure 4, presents $CSC_{Extrinsic}$ of the various metal oxide coatings investigated. The $Ir_{0.2}$-$Ni_{0.8}$-oxide coating has the highest $CSC_{Extrinsic}$ of all the materials tested, with a value of 30.3±4.8 mC cm^{-2}. Ir-oxide, which is the current state-of-the-art neural stimulation material, demonstrated a significantly lower $CSC_{Extrinsic}$ than the bimetallic composite, yielding a value of 9.8±1.0 mC cm^{-2}. Pure Ni-oxide also failed to display high a CSC and restrained to a value of 2.8±1.1 mC cm^{-2}, while the titanium (substrate) fell short of all other metal oxide coatings, exhibiting a CSC of only 0.23 ±0.07 mC cm^{-2}. The fact that the titanium oxide sample has a very low CSC value further indicates that adding a coating effectively increases the extrinsic CSC of the electrode. Additionally, it can be concluded that the data obtained from the samples with coatings are a result of the properties of the coating itself and not the titanium substrate. The highest extrinsic CSC of $Ir_{0.2}$-$Ni_{0.8}$-oxide coating is most likely due to its large surface roughness value (determined in current studies, but not shown here). The fairly low percent relative standard deviations for all of the samples in Figure 4 indicate that the coatings were able to store charge in a reproducible manner. To the best of authors' knowledge, a higher $CSC_{Extrinsic}$

value has not been reported in the literature for Ir-Ni-oxide coatings under the experimental condition applied.

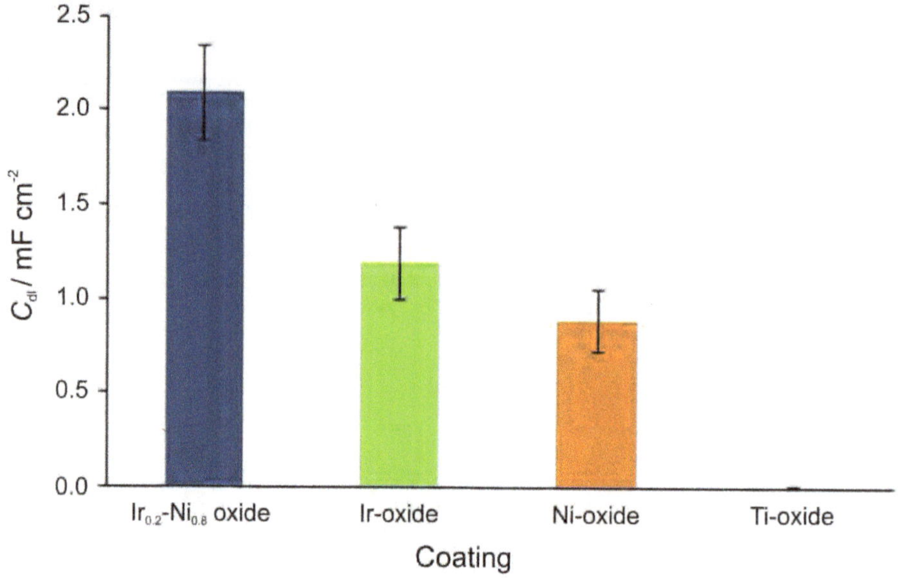

Figure 4. *Total extrinsic charge storage capacity ($CSC_{Extrinsic}$) of various metal oxide coatings determined from CVs in Figure 3, for at least three sample (coating) replicates. The error bars represent the corresponding standard deviation.*

In order to investigate the charge storage capacity of the material alone (without considering the effect of surface area), the *intrinsic* charge storage capacity was further examined. For this purpose, a true electrode surface area was first determined form electrochemical impedance spectroscopy results.

Characterization by Electrochemical Impedance Spectroscopy (EIS)

Figure 5 represents an example of the EIS spectra obtained for $Ir_{0.2}$-$Ni_{0.8}$-oxide coating at the -0.5 V. Similar EIS responses were obtained for all other oxide coatings, however they are not included in this analysis. To quantify the EIS result, the experimental spectrum presented in Figure 5 was modeled using non-linear least-squares fit analysis (NLLS) software [45] and the electrical equivalent circuit (EEC) presented in Figure 1(a). The modeled data are represented by the dashed line in Figure 5. It is evident that there is a substantial agreement between the experimental data (circles) and modeled data (dashed line), confirming the validity of the proposed EEC in describing the impedance behavior of the investigated system under the given experimental conditions. The same EIS analysis of the other developed metal oxide coatings was carried out, using the EEC in Figure 1(b). In order to determine the true electrode surface area of the metal oxide coatings, EEC parameters were analyzed for the contribution of double-layer capacitance.

Low impedance is required for the stimulation of neural networks. Since capacitance is inversely proportional to impedance, a large capacitance is desired [9]. To investigate the capacitive behavior of the developed coatings, a true value of the double-layer capacitance (C_{dl}) was calculated using the equation proposed by Brug *et al.* [46]:

$$C_{dl} = [Q(R_s^{-1} + R^{-1})^{n-1}]^{1/n} \quad (1)$$

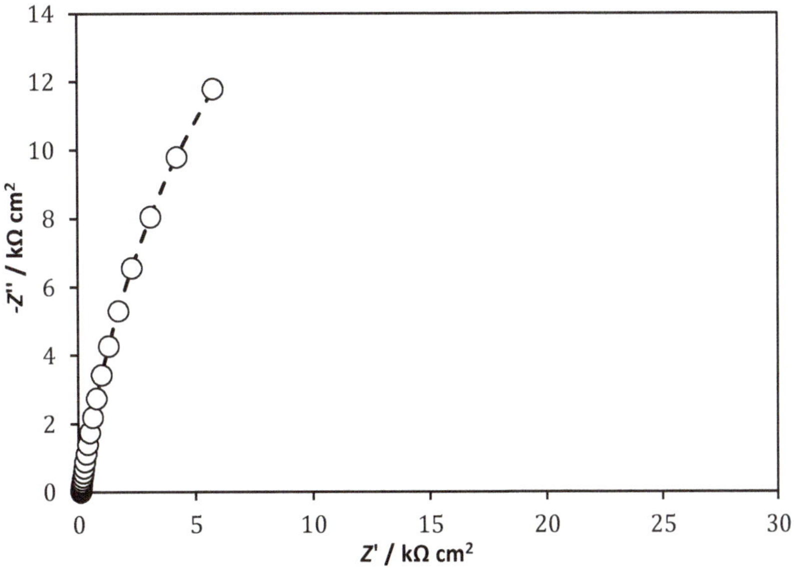

Figure 5. *Nyquist plot of the $Ir_{0.2}$-$Ni_{0.8}$-oxide coating recorded in 0.16 M saline phosphate buffer at electrode potential of -0.5 V. The dashed line represents the simulated spectra obtained using the equivalent electrical circuit model in Figure 1(a).*

C_{dl} / mF cm^{-2} values for all coatings were calculated at -0.5 V and the resulting values are presented in Figure 6. Analysis of the figure demonstrates that the $Ir_{0.2}$-Ni_{08}-oxide coating has a capacitance value of 2.09±0.25 mF cm^{-2}. In contrast, iridium oxide (the state-of-the-art material for neural stimulation electrodes) has the value of 1.18±0.18 mF cm^{-2}, almost half of the value obtained for the $Ir_{0.2}$-Ni_{08}-oxide coating. Furthermore, the pure Ni-oxide coating yielded a value of 0.88±0.16 mF cm^{-2}, while the naked Ti substrate exhibited the lowest value of all substrates tested, at a value of 0.005±0.002 mF cm^{-2}. This reflects the superiority of the $Ir_{0.2}$-$Ni_{0.8}$-oxide coating in terms of double layer capacitance and thus, high charge storage capacity.

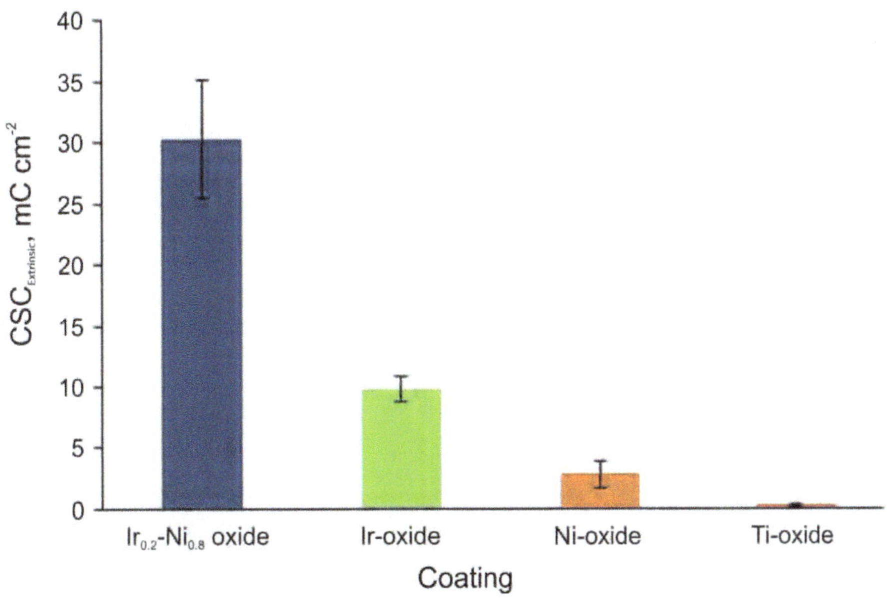

Figure 6. *Dependence of the double-layer capacitance on the applied dc potential of various metal oxide coatings obtained by fitting EIS spectra.*

The next step in our analysis was to determine the **intrinsic** charge storage capacity ($CSC_{Intrinsic}$) of the coatings. Corresponding $CSC_{Intrinsic}$ values were calculated by normalizing the extrinsic charge storage capacity ($CSC_{Extrinsic}$) values plotted in Figure 4 with respect to true electrochemical active surface (TEAS) area of the electrode coating. This was done by employing the following equation [47,48]:

$$CSC_{Intrinsic} = \frac{CSC_{Extrinsic} C_{Theoritical}}{C_{dl}} \quad (2)$$

where $C_{Theoritical}$ /25 μF cm^{-2} represents the theoretical value of the double layer capacitance [13]. Calculated values of $CSC_{Intrinsic}$ for all coatings are plotted in Figure 7.

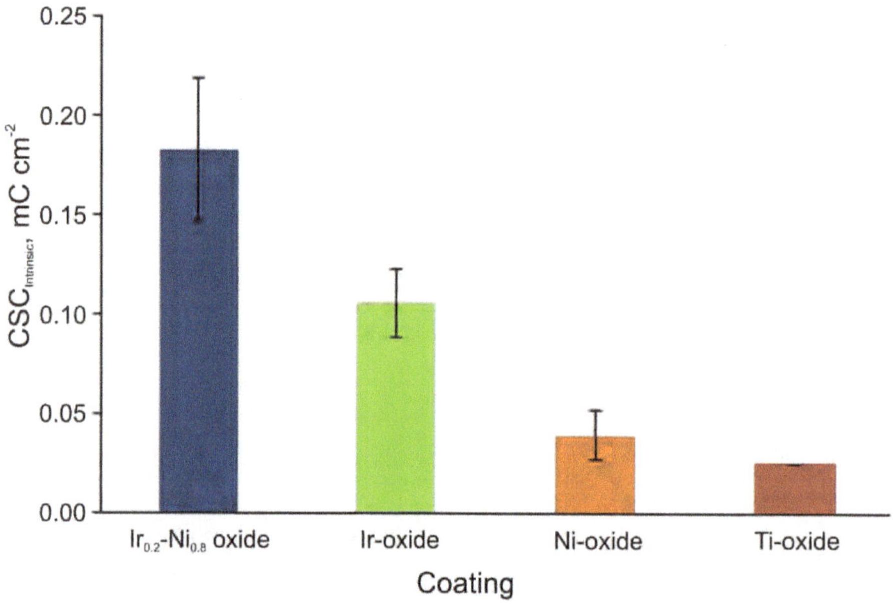

Figure 7. *Total intrinsic charge storage capacity ($CSC_{Intrinsic}$) of various metal oxide coatings obtained from CVs recorded in 0.16 M saline phosphate buffer pH = 7.4, at a scan rate sr = 100 mV s^{-1}.*

It is evident from Figure 7 that the $Ir_{0.2}$-$Ni_{0.8}$ oxide bimetallic coating has the highest intrinsic charge storage capacity as compared to other oxide coatings. Lower $CSC_{Intrinsic}$ values indicate that large electrode sizes would be required in neural applications to deliver the required amount of charge. The high intrinsic CSC ability of the Ir-Ni composite proves that it has potential to be a useful material in the effort of neural electrode miniaturization, while still providing the charge necessary for successful stimulation.

Conclusions

In this study, the electrochemical properties of Ir-, Ni- and $Ir_{0.2}$-$Ni_{0.8}$-oxide coatings formed on a Ti substrate were investigated in 0.16 M NaCl phosphate buffered solution, at pH of 7.4, for the purpose of possible application of the materials in the technology of stimulating neural electrodes.

SEM results showed contrast between the stable, rough surface of the $Ir_{0.2}$-$Ni_{0.8}$ oxide coating and the Ni- and Ir-oxide coatings, which displayed the undesirable characteristic of slight cracking. In addition, it was also determined that the $Ir_{0.2}$-$Ni_{0.8}$-oxide coating had the highest charge storage capacity (CSC) both intrinsically and extrinsically. The $Ir_{0.2}$-$Ni_{0.8}$-oxide coating's extrinsic CSC of 30 mC cm^{-2} proved to be higher than the CSC offered by the current state-of-the-art neural stimulating electrode, Ir-oxide. EIS results also demonstrated that the $Ir_{0.2}$-$Ni_{0.8}$-oxide coating was optimal, achieving the largest capacitance (lowest impedance) value.

This novel bimetallic Ir-Ni-oxide coating serves as an extremely promising alternative to currently used pure Ir-oxide, Pt, Au and carbon-based stimulating electrodes. *In-vivo* investigations are now solicited to further determine biocompatibility and feasibility for use as a neural stimulating electrode in the various biomedical applications.

Acknowledgements: *The authors would like to acknowledge the Natural Science and Engineering Research Council of Canada for providing the support for this research.*

References

[1] S. Miocinovic, S. Somayajula, S. Chitnis, J.L. Vitek, *JAMA Neurology* **70 (2)** (2013) 163-171

[2] J. L. Collinger, S. Foldes, T.M. Bruns, B. Wodlinger, R. Gaunt, D. J. Weber, *Journal of Spinal Cord Medicine* **36 (4)** (2013) 258-172

[3] W. D. Memberg, K. H. Polasek, R. L. Hart, A. M. Bryden, K. L. Kilgore, G. A. Nemunaitis, H. A. Hoyen, M. W. Keith, R. F. Kirsch, *Archives of Physical Medicine and Rehabiition* **95 (6)** (2014) 1201-1211

[4] B. M. Doucet, A. Lam, L. Griffin, *Yale* Journal *of Bioogy & Medicine* **85 (2)** (2012) 201-215

[5] L. R. Sheffler, J. Chae, Physical Medicines & Rehabilition Clinics of North America **24 (2)** (2013) 305-323

[6] D. Terzic, A. Abosch, *Journal of* Neurosurgical *Sciences* **56 (4)** (2012) 267-277

[7] S.F Cogan, Annual Review of Biomedical Engineering **10 (1)** (2008) 275-309

[8] R. Fröhlich, A. Rzany, J. Riedmüller, A. Bolz, M. Schaldach, *Journal of Materials Science: Materials in Medicine* **7 (7)** (1996) 393-397

[9] Y. Lu, T. Wang, Z. Cai, Y. Cao, H. Yang, Y. Y. Duan, *Sensors and Actuators B: Chemical* **137 (1)** (2009) 334-339

[10] T.M. Silva, J. Rito, M.G.S. Ferreira, I. Fonseca, K. Watkins, *Journal of Materials Science: Materials in Medicine* **7 (5)** (1996) 261-264

[11] P.R. Troyk, S.F. Cogan, *Neural* Engineering, B. He, Ed (s)., Springer US, Chicago, USA, 2005, p. 1-48

[12] Y. Xie, L. Zhou, C. Huang, H. Huang, J. Lu, *Electrochimica Acta* **53 (10)** (2008) 3643-3649

[13] D. R. Merrill, M. Bikson, J. G. R. Jefferys, Journal *of Neuroscience Methods* **141 (2)** (2005) 171-198

[14] Biomaterials, http://wikisites.mcgill.ca/bme501/images/a/a3/Class_Juncker_part1_071010.pdf (October 21, 2010)

[15] L. S. Robblee, M. Randolph, (EIC Laboratories, Inc.), US4677989, (1987)

[16] T. L. Rose, L. S. Robblee, *IEEE* Transactions *on Biomedical Engineering* **37 (11)** (1990) 1118-1120

[17] Y. Lu, Z. Cai, Y. Cao, H. Yang, Y.Y. Duan, *Electrochemistry Communications* **10 (5)** (2008) 778-782

[18] Y. Xie, D. Fu, *Materials Chemistry and* Physics **122 (1)** (2010) 23-29

[19] M. Vuković, D. Marijan, D. Čukman, P. Pervan, M. Milun, *Journal of Materials Science* **34 (4)** (1999) 869-874

[20] M.P. Neupane, I.S. Park, S.J. Lee, K.A. Kim, M.H. Lee, T.S. Bae, *International Journal of Electrochemical Science* **4 (2)** (2009) 197-207

[21] I.-S. Lee, C.-N. Whang, J.-C. Park, D.-H. Lee, W.-S. Seo, *Biomaterials* **24 (13)** (2003) 2225-2231

[22] K. Wang, H.A. Fishman, H. Dai, J.S. Harris, *Nano Letters* **6 (9)** (2006) 2043-2048

[23] S. F. Cogan, A. A. Guzelian, W. F. Agnew, T. G. H. Yuen, D. B. McCreery, *Journal of Neuroscience Methods* **137 (2)** (2004) 141-150

[24] T. M. Silva, J. E. Rito, A. M. P. Simões, M. G. S. Ferreira, M. da Cunha Belo, K. G. Watkins, *Electrochimica Acta* **43(1-2)** (1998) 203-211

[25] T. Arikawa, Y. Takasu, Y. Murakami, K. Asakura, Y. Iwasawa, *The Journal of Physical Chemistry B* **102 (19)** (1998) 3736-3741

[26] V. V. Gorodetskii, V. A. Neburchilov, Russian *Journal of Electrochemistry* **39 (10)** (2003) 1111-1115

[27] J. Kristóf, J. Liszi, P. Szabó, A. Barbieri, A. de Battisti, *Journal of Applied Electrochemistry* **23 (6)** (1993) 615-624

[28] L. K. Xu, J. D. Scantlebury, *Corrosion* Science **45 (12)** (2003) 2729-2740

[29] I.-H. Kim, J.-H. Kim, K.-B. Kim, Electrochemical *and Solid-State Letters* **8 (7)** (2005) A369-A372

[30] V. Srinivasan, J. W. Weidner, *Journal of the Electrochemical Society* **147 (3)** (2000) 880-885

[31] I. S. P. Madhav Prasad Neupane, Sook Jeong Lee, Kyoung A Kim, Min Ho Lee, Tae Sung Bae, *International Journal of Electrochemical Science* **(4)** (2009) 197 -207

[32] L. S. Robblee, (EIC Laboratories, Inc.), US4717581, (1988)

[33] S. H. Bonilla, C.F. Zinola, *Electrochimica Acta* **43 (3-4)** (1998) 423-426

[34] P. Fernandez-Ibañez, S. Malato, *O. Enea, Catalysis Today* **54 (2-3)** (1999) 329-339

[35] J. M. Hu, *Acta Metallurgica Sinica* **13 (3)** (2000) 849

[36] S. Barison, S. Daolio, M. Fabrizio, A. De Battisti, *Rapid Communications in Mass Spectrometry* **18 (3)** (2004) 278-284

[37] A. De Battisti, G. Lodi, L. Nanni, G. Battaglin, A. Benedetti, *Canadian Journal of Chemistry* **75 (11)** (1997) 1759-1765

[38] S. Barison, S. Daolio, M. Fabrizio, A. D. Battisti, *Rapid Communications in Mass Spectrometry* **18 (3)** (2004) 278-284

[39] G. J. Brug, A. L. G. van den Eeden, M. Sluyters-Rehbach, J. H. Sluyters, *Journal of Electroanalytical Chemistry and Interfacial Electrochemistry* **176 (1-2)** (1984) 275-295

[40] I. S. Lee, C. N. Whang, K. Choi, M. S. Choo, Y. H. Lee, *Biomaterials* **23 (11)** (2002) 2375-2380

[41] A. Armutlulu, J.K. Kim, M. Kim, S. A. B. Allen, M. G. Allen, *Transducers and Eurosensors XXVII: The 17th International Conference on Solid-State Sensors, Actuators and Microsystems,* Nickel-oxide-based supercapacitors with high aspect ratio concentric cylindrical electrodes, Barcelona, Spain, 2013, p. 1480

[42] K. R. Prasad, N. Miura, *Applied Physics* Letters **85 (18)** (2004) 4199-4201

[43] A. K. K. Kyaw, H. Tantang, T. Wu, L. Ke, C. Peh, Z. H. Huang, X. T. Zeng, H. V. Demir, Q. Zhang, X. W. Sun, *Applied Physics Letters* **99 (2)** (2011) 1107-1-1107-3

[44] N. Ullah, S. Omanovic, Journal of Surface and Coatings Technology (2014) p. submitted

[45] B. A. Boukamp, in *Equivalent Circuit Users manual, Report CT88/265/128*, University of Twente, Department of Chemical Technology, The Netherlands, 1989

[46] G. J. Brug, A. L. G. Van Den Eeden, M. Sluyters-Rehbach, J. H. Sluyters, *Journal of Electroanalytical Chemistry* **176 (1-2)** (1984) 275-295

[47] D. Profeti, T. A. F. Lassali, P. Olivi, Journal *of Applied Electrochemistry* **36 (8)** (2006) 83-888

[48] S. Trasatti, O. A. Petrii, *Journal of* Electroanalytical *Chemistry* **327 (1-2)** (1992) 353-376

9

Novel Mannich bases bearing pyrazolone moiety Synthesis, characterization and electrochemical studies

KRISHNA NAIK✉, ALURU RAGHAVENDRA GURU PRASAD*, YADATI NARASIMHA SPOORTHY and LAKSHMANA RAO KRISHNA RAO RAVINDRANATH

Sri Krishnadevaraya University, Anantapur, A.P., India

**ICFAI Foundation for Higher Education, Hyderabad, A.P., India*

✉ Corresponding Author

Abstract

The present investigation describes a series of new {4-[3-Methyl-5-oxo-4-(4^{l}-substituted phenyl hydrazono)-4,5-dihydro-pyrazol-1-yl]-phenoxy}-acetic acid (2-oxo-1-piperidine-1--ylmethyl-1,2-dihydro–indol-3-ylidene)-hydrazides synthesized by the Mannich reaction of {4-[3-Methyl-5-oxo-4-(4^{l}-substituted phenyl hydrazono)-4,5-dihydro-pyrazol-1-yl]--phenoxy}-acetic acid (2-oxo-1,2-dihydro-indol-3-ylidene)-hydrazide with aqueous formaldehyde and a solution of piperidine in dimethylformamide. These novel Mannich bases were characterized by elemental analysis, IR, ^{1}H NMR and mass spectral data. Electrochemical behavior of these compounds were studied by two techniques namely polarography and cyclic voltammetry. The results from both the techniques were compared and the reduction mechanism in acidic as well as basic medium was proposed.

Keywords

Mannich bases, Synthesis, Polarography, Cyclic voltammetry, Comparison, Reduction mechanism.

Introduction

Pyrazolones [1-5] and related heterocycles are widely used in medicinal chemistry as they possess wide range of biological and pharmacological properties. Further, Mannich bases exhibit pronounced biological activities. The key feature of Mannich reaction is that the amino carbonyl products are valuable synthons for synthesis [6-8] and can be readily converted to derivatives that possess useful applications in paint industry, polymer industry and particularly in medicine and the pharmaceutical industry [9-14]. Due to this reason, Mannich bases have engrossed a great deal of attention of pharmaceutical chemists. In view of the above mentioned facts, we propose to synthesize certain novel Mannich bases containing a pyrazolone moiety.

The pharmacological properties of many compounds have been quantitatively related to their reduction process [15-18]. Structure-activity relationship studies [19,20] demonstrated that the reduction potential correlates with the antimicrobial activity of certain compounds.

Extensive work on electrochemical behavior of medicinally important compounds has been already reported from these laboratories [21,22]. In this article, the cyclic voltammetric behavior of Mannich bases on a hanging mercury drop electrode and modified carbon paste electrode is reported. A comparison of polarographic behavior of the compounds with their cyclic voltammetric behavior is also presented.

Experimental

All the chemicals and reagents used in the studies were analytical reagent grade obtained from Merck. Britton-Robinson buffer solutions were prepared from the appropriate volumes of acetic acid (0.04 M), phosphoric acid (0.04 M), boric acid (0.04 M) and sodium hydroxide (0.2M). pH meter, Model LI – 10 manufactured by ELICO Private Limited, Hyderabad, India was used for pH measurements.

A CL-25 Pen Recording Polarograph manufactured by ELICO Private Limited, Hyderabad, India was used to record current-voltage curves. The capillary having the characteristics 1.80 $mg^{2/3}\ s^{-1/2}$ at h = 80 cm was employed in the studies.

The cyclic voltammeter used consists of an X-Y recorder (Model RE 0074), a PAR 175 Potentiostat and an PAR 175 Universal Programmer. A single compartment cell Model 303 SMDE with silver wire as reference electrode and platinum wire as counter electrode was used for cyclic voltammetric studies. A stationary mercury drop electrode (SMDE 303) with a drop area 0.0096 cm^2 was used as the working electrode.

General polarographic procedures (Scheme 1)

8.0 mL of the buffer solution of desired pH, 2.0 mL of Mannich base solution (1.0×10^{-2} M) in dimethylformamide, 6.0 mL of dimethylformamide (DMF) and 4.0 mL of distilled water were mixed thoroughly in the polarographic cell. The polarograms were recorded after the expulsion of dissolved oxygen with nitrogen gas. Geltain was used as the maximum suppressor in all the investigations except in the experiments where the effect of surfactants was studied.

Synthesis of Mannich bases

{4-[3-Methyl-5-oxo-4-(4^{l}-substituted phenyl hydrazono)-4,5-dihydro-pyrazol-1-yl]-phenoxy}-acetic acid hydrazide I were synthesized and characterized by the procedures reported in the literature [23].

Synthesis of {4-[3-Methyl-5-oxo-4--(4^{l}-substituted phenyl hydrazono)-4,5-dihydro-pyrazol-1-yl]--phenoxy}-acetic acid (2-oxo-1,2-dihydro-indol-3-ylidene)- hydrazide IV.

The required isatins were prepared by the procedure described by Marvel and Heirs [24].

Synthesis of isonitrosoacetanilide (II)

22 g of chloral hydrate and 300 mL of water was taken into a one litre round bottom flask. 3.25 g of sodium sulphate, 12 g of aniline (in 75 mL of water), 12 mL of concentrated hydrochloric acid and 27g of hydroxyl amine hydrochloride (in 25 mL of water) were added and heated over a

wire gauge for about 45 minutes. The reaction mixture was cooled, the solid separated was filtered and dried.

a. Synthesis of Isatin 46 (R = H). (III)

A miture of 8 mL of concentrated H_2SO_4, 18 g of dry isonitrosoacetanilide taken in a round bottom flask was heated to 80 °C for about 10 minutes. The reaction mixture was cooled to room temperature and poured onto crushed ice. The precipitated isatin was filtered, washed several times with cold water and dried.

c. Synthesis of {4-[3-Methyl-5-oxo-4-(4'-substituted phenyl hydrazono)-4,5-dihydro-pyrazol-1-yl]-phenoxy}-acetic acid(2-oxo-1,2-dihydro-indol-3-ylidene)-hydrazide IV.

A mixture of I (R = H) and III in 1 : 1 molar ratio when heated in DMF-water bath for 45 minutes, resulted in a compound with melting point 214 °C. Based on spectral data, the compound was assigned structure as {4-[3-Methyl-5-oxo-4-(phenyl - hydrazono)-4,5-dihydro-pyrazol-1-yl]-phenoxy}-acetic acid (2-oxo-1,2-dihydro – indol-3-ylidene)- hydrazide IVa (R = H).

Other members of the series were prepared by similar procedures and their characterization data are given in the Table 1.

Table 1. *Characterization of {4-[3-Methyl-5-oxo-4-(4'-substituted phenyl hydrazono)-4,5-dihydro-pyrazol-1-yl]-phenoxy}-acetic acid (2-oxo-1,2-dihydro-indol-3-ylidene)-hydrazide.*

Compound	R	M.P. °C	Yield, %	Molecular formula	Mass fraction found, % (Calc., %)					
					C	H	N	O	Cl	Br
IVa	H	214	70	$C_{26}H_{25}N_7O_4$	62.52 (62.48)	5.01 (4.89)	19.63 (19.50)	12.82 (12.70)		
IVb	CH_3	241	70	$C_{27}H_{27}N_7O_4$	63.15 (63.02)	5.26 (5.15)	19.10 (18.95)	12.47 (12.32)		
IVc	OCH_3	234	70	$C_{27}H_{27}N_7O_4$	61.24 (61.12)	5.10 (4.95)	18.52 (18.38)	15.12 (15.00)		
IVd	OC_2H_5	224	75	$C_{28}H_{29}N_7O_5$	61.87 (61.75)	5.34 (5.25)	18.04 (17.88)	14.73 (14.62)		
IVe	Cl	225	75	$C_{26}H_{24}ClN_7O_4$	58.48 (58.35)	4.49 (4.32)	18.38 (18.22)	11.99 (11.75)	6.65 (6.52)	
IVf	Br	243	80	$C_{26}H_{24}BrN_7O_4$	53.98 (53.82)	4.15 (3.98)	16.95 (16.80)	11.07 (10.95)		13.82 (13.65)

IR Spectral details

The IR (KBr) spectrum (Figure 1) of {4-[3-Methyl-5-oxo-4-(phenyl hydrazono)-4,5-dihydro-pyrazol-1-yl]-phenoxy}-acetic acid (2-oxo-1,2-dihydro–indol-3-ylidene)- hydrazide (hydrazone) IVa showed characteristic strong absorption bands at 3205 (NH), 3170 (Indole NH), 1602 (C = N), 1656 (pyrazoline C = O), 1700 (Indole C = O) and 1618 (CONH). The spectral data and the respective assignments of IV are given in the Table 2.

^{1}H NMR Spectral details

The ^{1}H NMR (200 MHz) spectrum (Figure 2) of {4-[3-Methyl-5-oxo-4-(phenyl hydrazono)-4,5--dihydro-pyrazol-1-yl]-phenoxy}-acetic acid (2-oxo-1,2-dihydro–indol-3-ylidene)-hydrazide (hydrazone) IVa in DMSO – d_6 showed the signals at δ1.1 (s, 3H, CH_3), δ10.93 (s, 1H, Ar – NH), δ5.78 (s, 2H, N-CH_2 – CO), δ12.75 (s, 1H, Indole NH) and δ7.1 – 7.3 (m, 9H, Ar – H) (Table 3).

Table 2: *IR (KBr) Spectral data of {4-[3-Methyl-5-oxo-4-(4$^{/}$-substituted phenyl hydrazono)-4,5-dihydro-pyrazol-1-yl]-phenoxy}-acetic acid (2-oxo-1,2-dihydro-indol-3-ylidene)-hydrazide.*

Compound	ν_{max} / cm^{-1}					
	>NH	Indole NH	>C=N-	Pyrazoloine C=O	Indole>C=O	>CO-NH-
IVa	3205	3170	1602	1656	1700	1618
IVb	3180	3140	1600	1654	1700	1622
IVc	3100	3150	1505	1654	1701	1625
IVd	3195	3155	1604	1654	1701	1624
IVe	3175	3140	1605	1654	1701	1624
IVf	3190	3150	1604	1654	1701	1624

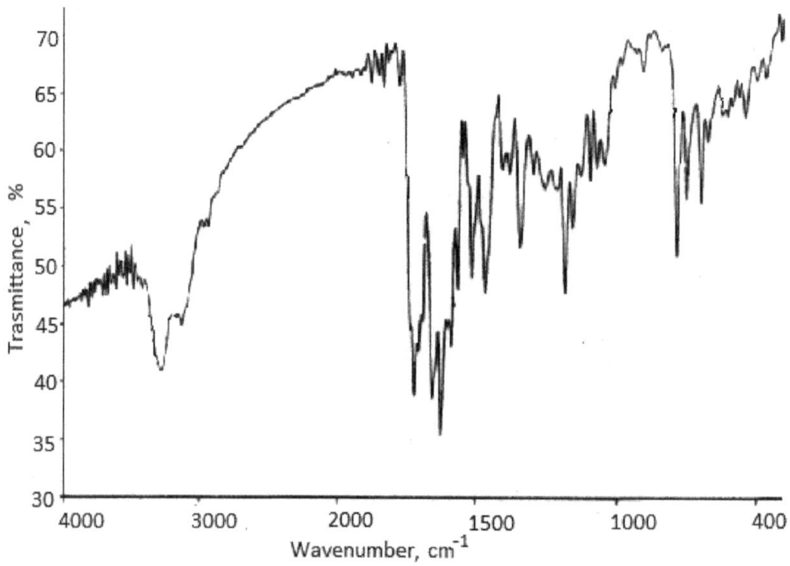

Figure 1. *IR Spectrum of {4-[3-methyl-5-oxo-4-(phenyl hydrazono)-4,5-dihydro-pyrazol-1-yl]--phenoxy}-acetic acid (2-oxo-1,2-dihydro-indol-3-ylidene)-hydrazide*

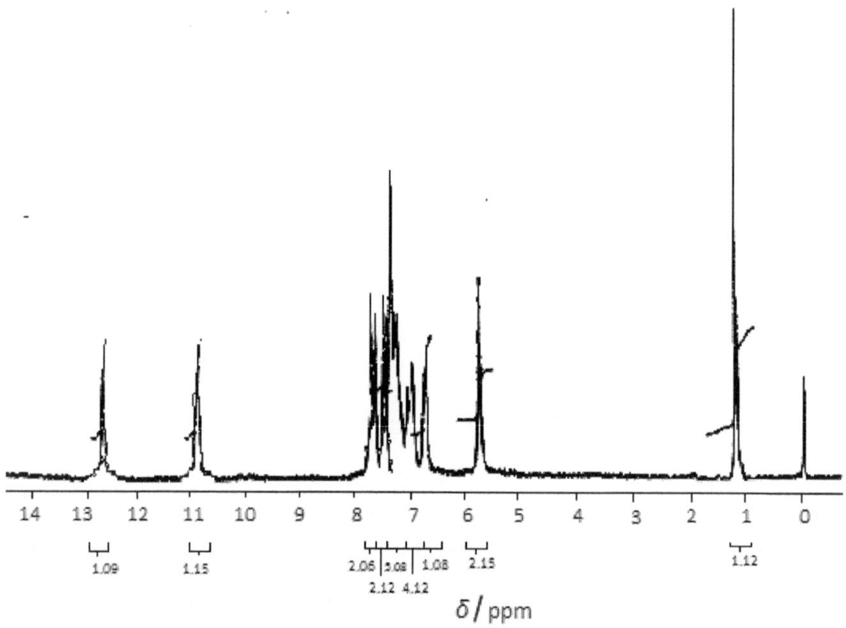

Figure 2. *^{1}H NMR spectrum of {4-[3-methyl-5-oxo-4-(phenyl hydrazono)-4,5-dihydro-pyrazol-1--yl]--phenoxy}-acetic acid (2-oxo-1,2-dihydro-indol-3-ylidene)-hydrazide*

Table 3. *^{1}H NMR Spectral data of {4-[3-Methyl-5-oxo-4-(4'-substituted phenyl hydrazono)-4,5-dihydro-pyrazol-1-yl]-phenoxy}-acetic acid (2-oxo-1,2-dihydro-indol-3-ylidene)-hydrazide.*

Compound	δ / ppm
IVa	1.1 (s, 3H, CH_3), 5.78 (s, 2H, N-CH_2-CO), 6.8 (s, 1H, Ar–NH), 6.9-7.0 (m, 4H, C_6H_4), 7.1–7.3 (m, 5H, Ar-H), 7.4 (d, 2H, C_6H_4), 7.7 (d, 2H, C_6H_4)' 10.93 (s, 1H, CONH), 12.75 (s, 1H, Indole NH)
IVb	0.95 (s, 3H, CH_3), 1.14 (s, 3H, CH_3), 5.82 (s, 2H, N-CH_2-CO), 6.8 (s, H, Ar-NH), 6.9-7.0 (m, 4H, C_6H_4), 7.1-7.3 (m, 4H, Ar–H), 7.4 (d, 2H, C_6H_4), 7.7 (d, 2H, C_6H_4), 10.95 (s, 1H, CONH), 12.75 (s, 1H, Indole NH)
IVc	1.12 (s, 3H, CH_3), 3.24 (s, 3H, OCH_3), 5.8 (s, 2H, N-CH_2CO), 6.8 (s, 1H, Ar-NH), 6.9-7.0 (m, 4H, C_6H_4), 7.1-7.3 (m, 4H, Ar-H), 7.4 (d, 2H, C_6H_4) and 7.7 (d, 2H, C_6H_4), 10.95 (s, 1H, CONH), 12.75 (s, 1H, Indole NH)
IVd	0.95 (s, 3H, CH_3), 1.15 (t, 3H, CH_3), 3.16 (q, 2H, O-CH_2), 5.76 (s, 2H, N-CH_2CO), 6.8 (s, 1H, Ar-NH), 6.9-7.0 (m, 4H, C_6H_4), 7.1-7.3 (m, 4H, Ar-H), 7.4 (d, 2H, C_6H_4) and 7.7 (d, 2H, C_6H_4), 10.91 (s, H, CO-NH), 12.72 (s, 1H, Indole NH)
IVe	1.05 (s, 3H, CH_3), 5.8 (s, 2H, N-CH_2CO), 6.8 (s, H, Ar-NH), 6.9-7.0 (m, 4H, C_6H_4), 7.1-7.3 (m, 4H, Ar-H), 7.4 (d, 2H,C_6H_4) and 7.7 (d, 2H, C_6H_4), 10.93 (s, H, CO-NH), 12.7 (s, 1H, Indole NH)
IVf	1.02 (s, 3H, CH_3), 5.78 (s, 2H, N-CH_2CO), 6.8 (s, H, Ar-NH), 6.9-7.0 (m, 4H, C_6H_4), 7.1-7.3 (m, 4H, Ar-H), 7.4 (d, 2H, C_6H_4) and 7.7 (d, 2H, C_6H_4), 10.93 (s, H, CO-NH), 12.75 (s, H, Indole NH)

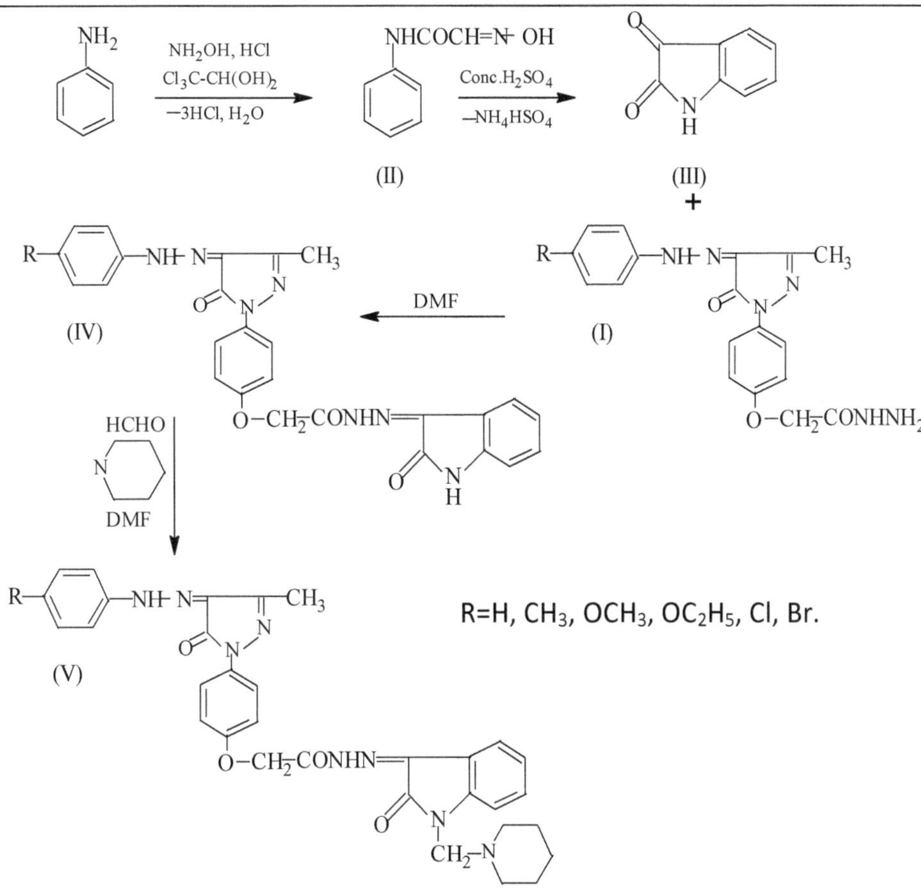

Scheme 1. *Synthesis of {4-[3-Methyl-5-oxo-4-(4'-{4-[3-Methyl-5-oxo-4-(4'-substituted phenyl hydrazono)-4,5-dihydro-pyrazol-1-yl]-phenoxy}-acetic acid (2-oxo-1- piperidine-1-ylmethyl-1,2 –dihydro – indol-5-ylidene)- hydrazide (V)*

d. Synthesis of {4-[3-Methyl-5-oxo-4-(4^{l}-substituted phenyl hydrazono)-4,5-dihydro-pyrazol-1-yl]-phenoxy}-acetic acid (2-oxo-1-piperidine-1-ylmethyl-1,2-dihydro–indol-3-ylidene)--hydrazide (V).

A mixture of hydrazone IVa (R = H), aqueous formaldehyde and a solution of piperidine in DMF were stirred for about six hours at room temperature to give {4-[3-Methyl-5-oxo-4-(phenyl hydrazono)-4,5-dihydro-pyrazol-1-yl]-phenoxy}-acetic acid (2-oxo-1-piperidine-1-ylmethyl-1,2-di-hydro–indol-3-ylidene)- hydrazide V a (R = H).

Similar treatment of hydrazones IV b-f with piperidine in presence of formaldehyde in DMF at room temperature yielded respective {4-[3-Methyl-5-oxo-4-(4^{l}-substituted phenyl hydrazono)-4,5--dihydro-pyrazol-1-yl]-phenoxy}-acetic acid-(2-oxo-1-piperidine-1-ylmethyl-1,2-dihydro–indol-3-ylidene)- hydrazide V b-f (R = CH_3, OC_2H_5, -Cl, -Br).

The structures of the compounds V a- f were established on the basis of elemental analysis and spectral data (Table 4) and spectral data (Table 5 and Table 6).

IR Spectral details

The IR (KBr) spectrum (Figure 3) of {4-[3-Methyl-5-oxo-4-(phenyl hydrazono)-4,5-dihydro-pyrazol-1--yl]-phenoxy}-acetic acid (2-oxo-1-piperidine-1-ylmethyl-1,2-dihydro–indol-3-ylidene)- hydrazide V a exhibited characteristic bands at around 3195 (NH), 1610 (C = N), 1676 (pyrazoline C = O), 1720 (Indole C = O), 1654 (C – NH) and 2933 cm^{-1} (CH_2) (Table 5).

The spectral data and the respective assignments of V are detailed below.

Table 4. *Characterization data of {4-[3-Methyl-5-oxo-4-(4^{l}-substituted phenyl hydrazono)-4,5-dihydro-pyrazol-1-yl]-phenoxy}-acetic acid (2-oxo-1-piperidine-1-ylmethyl-1,2-dihydro-indol-3-ylidene) - hydrazide V*

Compd	R	M.P.	Yield, %	Molecular formula	Mass fraction Found, % (Calc., %)					
					C	H	N	O	Cl	Br
Va	H	158	70	$C_{32}H_{32}N_8O_4$	64.86 (64.72)	5.40 (5.28)	19.91 (19.86)	10.81 (10.75)		
Vb	CH_3	164	70	$C_{33}H_{34}N_8O_4$	65.34 (65.21)	5.61 (5.45)	18.48 (18.34)	10.56 (10.38)		
Vc	OCH_3	167	70	$C_{33}H_{34}N_8O_5$	63.66 (63.56)	5.46 (5.32)	18.00 (17.85)	12.86 (12.64)		
Vd	OC_2H_5	159	75	$C_{34}H_{36}N_8O_5$	64.15 (64.01)	5.66 (5.45)	17.61 (17.50)	12.56 (12.42)		
Ve	Cl	161	75	$C_{32}H_{31}ClN_8O_4$	61.29 (61.15)	4.94 (4.78)	17.87 (17.73)	10.21 (10.8)	5.66 (5.45)	
Vf	Br	160	80	$C_{32}H_{31}BrN_8O$	57.23 (57.08)	4.62 (4.42)	16.69 (16.54)	9.53 (9.38)		11.90 (11.70)

Table 5. *IR (KBr) spectral data of {4-[3-Methyl-5-oxo-4-(4^{l}-substituted phenyl hydrazono)-4,5-dihydro-pyrazol-1-yl]-phenoxy}-acetic acid (2-oxo-1-piperidine-1-ylmethyl-1,2-dihydro-indol-3-ylidene)- hydrazide V.*

Compound	ν_{max} / cm^{-1}					
	>NH	>C=N-	Pyrazoloine C=O	Indole C=O	>CO-NH-	>CH_2-
Va	3195	1610	1676	1720	1654	2933
Vb	3170	1616	1674	1715	1658	2920
Vc	3120	1610	1680	1712	1654	2625
Vd	3175	1614	1674	1711	1656	2915
Ve	3155	1616	1674	1714	1658	2920
Vf	3170	1614	1674	1716	1626	2625

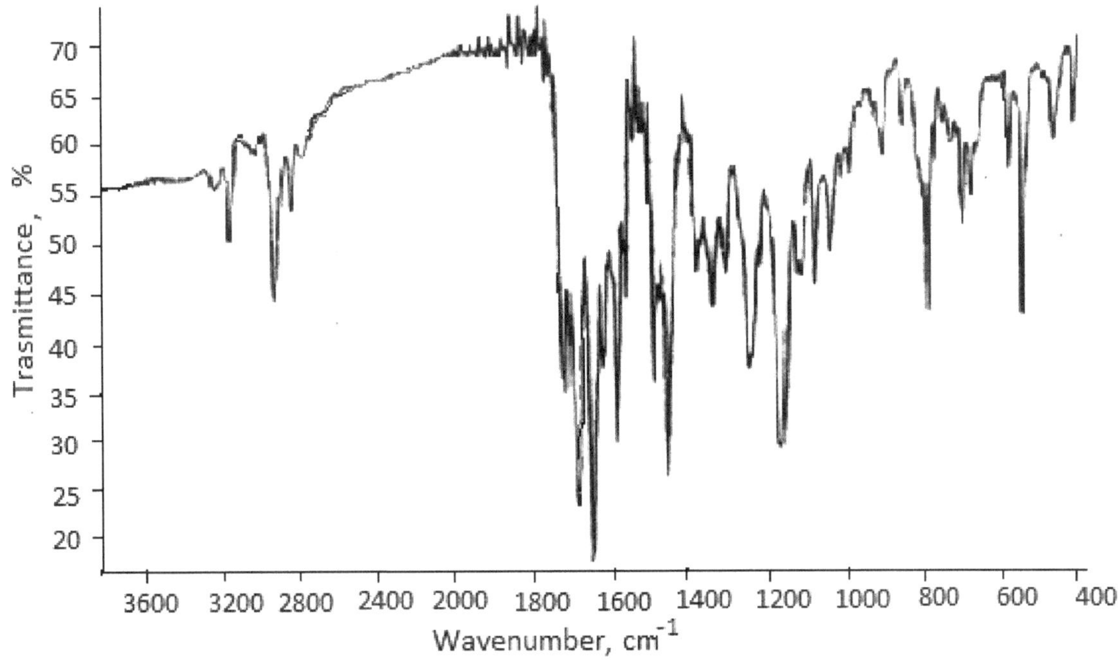

Figure 3. *IR Spectrum of {4-[3-methyl-5-oxo-4-(phenyl hydrazono)-4,5-dihydro pyrazol-1-yl]-phenoxy} –acetic acid (2-oxo-1-piperidine-1-ylmethyl-1,2-dihydro-indol-3-ylidene)-hydrazide*

^{1}H NMR Spectral details

The ^{1}H NMR (200 MHz) spectrum (Figure 4) of 3-[methyl-5-oxo-4-(4^l-*substituted* phenyl hydrazono)-4,5-dihydro-pyrazol-1-yl]-phenoxy}-acetic acid (2-oxo-1-piperidine-1-yl-methyl-1,2-dihydro–indol-3-ylidene)- hydrazide V was recorded in DMSO-d_6 and the data is given below. The appearance of a signal at δ 4.5 due to N–CH_2-N, confirmed the formation of Mannich bases (Table 6).

Table 6. *^{1}H NMR Spectral data of {4-[3-Methyl-5-oxo-4-(4^l-substituted phenyl hydrazono)-4,5-dihydro-pyrazol-1-yl]-phenoxy}-acetic acid (2-oxo-1-piperidine-1-ylmethyl-1,2-dihydro-indol-3-ylidene)- hydrazide*

Compound	δ / ppm
Va	1.14 (s, 3H, CH_3), 1.45 (m, 6H, $(CH_2)_3$), 2.56 (t, 4H, CH_2-N-CH_2), 4.45 (s, 2H, N-CH_2-N), 5.9 (s, 2H, N-CH_2CO), 6.8 (s, 1H, Ar-NH), 6.9-7.0 (m, 4H, C_6H_4), 7.1-7.3 (m, 5H, Ar -H), 7.4 (d, 2H, C_6H_4), 7.7 (d, 2H, C_6H_4), 9.5 (s, H, -CONH).
Vb	0.97 (s, 3H, CH_3), 1.12 (s, 3H, CH_3), 5.90 (s, 2H, N-CH_2-CO), 6.81 (s, H, Ar-NH), 6.9-7.0 (m, 4H, C_6H_4), 7.1-7.3 (m, 4H, Ar–H), 7.4 (d, 2H, C_6H_4), 7.7 (d, 2H, C_6H_4), 9.6 (s, 1H, CONH), 1.45 (m, 6H, $(CH_2)_3$), 2.54 (t, 4H, CH_2-N-CH_2), 4.48 (s, 2H, N-CH_2-N).
Vc	1.15 (s, 3H, CH_3), 3.12(s, 3H, OCH_3), 5.91 (s, 2H, N-CH_2CO), 6.8 (s, 1H, Ar-NH), 6.9-7.0 (m, 4H, C_6H_4), 7.1-7.3 (m, 4H, Ar-H), 7.4 (d, 2H, C_6H_4) and 7.7 (d, 2H, C_6H_4), 2.51 (t, 4H, CH_2-N-CH_2), 1.41 (m, 6H, $(CH_2)_3$), 4.41(s, 2H, N-CH_2-N), 9.51 (s, 1H, CONH).
Vd	0.98 (s, 3H, CH_3), 1.11 (t, 3H, CH_3), 3.11 (q, 2H, O-CH_2), 2.58 (t, 4H, CH_2-N-CH_2), 4.42 (s, 2H, N-CH_2-N), 5.81 (s, 2H, N-CH_2CO), 6.82(s, 1H, Ar-NH), 6.9-7.0 (m, 4H, C_6H_4), 7.1-7.3 (m, 4H, Ar-H), 1.45 (m, 6H, $(CH_2)_3$), 7.4 (d, 2H, C_6H_4) and 7.7 (d, 2H, C_6H_4), 9.41 (s, H, CO-NH).
Ve	1.07 (s, 3H, CH_3), 2.57(t, 4H, CH_2-N-CH_2), 4.47 (s, 2H, N-CH_2-N), 5.92 (s, 2H, N-CH_2CO), 6.8 (s, H, Ar-NH), 6.9-7.0 (m, 4H, C_6H_4), 7.1-7.3 (m, 4H, Ar-H), 7.4 (d, 2H,C_6H_4) and 7.7 (d, 2H, C_6H_4), 1.47 (m, 6H, $(CH_2)_3$), 9.38 (s, H, CO-NH).
Vf	1.06 (s, 3H, CH_3), 1.43 (m, 6H, $(CH_2)_3$), 5.91 (s, 2H, N-CH_2CO), 6.79 (s, H, Ar-NH), 6.9-7.0 (m, 4H, C_6H_4), 7.1-7.3 (m, 4H, Ar-H), 7.4 (d, 2H, C_6H_4) and 7.7 (d, 2H, C_6H_4), 9.51 (s, H, CO-NH), 2.52 (t, 4H, CH_2-N-CH_2), 4.49 (s, 2H, N-CH_2-N).

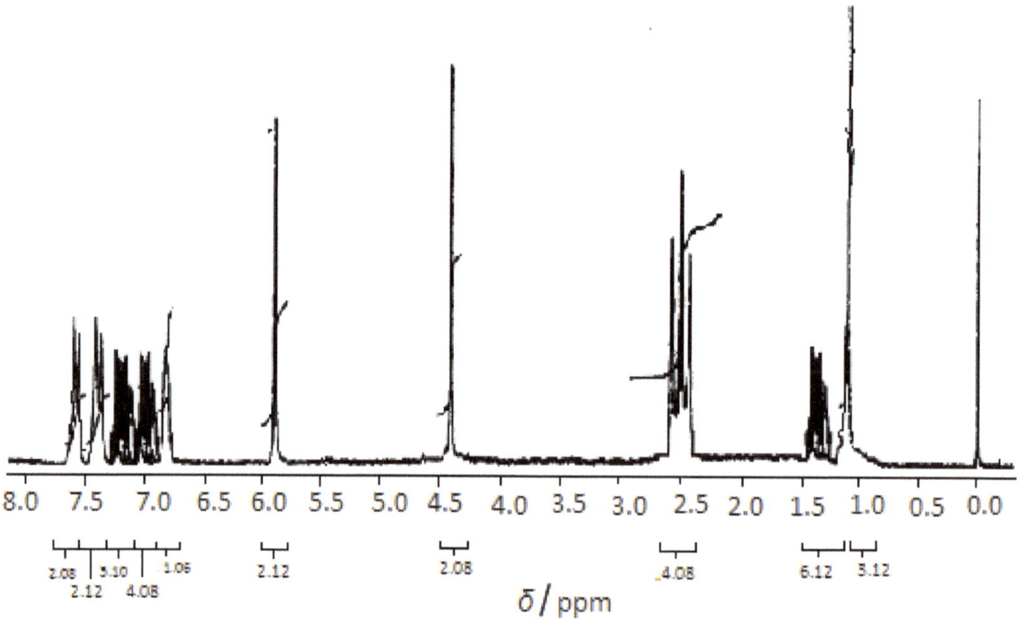

***Figure 4.** ^{1}HNMR spectrum of {4-[3-methyl-5-oxo-4-(phenyl hydrazono)-4,5-dihydro pyrazol-1-yl]-phenoxy} – acetic acid (2-oxo-1-piperidine-1-ylmethyl-1,2-dihydro-indol-3-ylidene)-hydrazide*

Mass spectral details

The mass spectrum of {4-[3-Methyl-5-oxo-4-(phenyl hydrazono)-4,5-dihydro-pyrazol-1-yl]-phenoxy}-acetic acid (2-oxo-1-piperidine-1-yl methyl-1,2-dihydro-indol-3-ylidene)-hydrazide Va (R = H, X = CH_2) exhibited the molecular (M^+) ion peak at m/z 592.

The fragmentation pattern noticed in the mass spectrum of {4-[3-Methyl-5-oxo-4-(4^{I}-phenyl hydrazono)-4,5-dihydro-pyrazol-1-yl]-phenoxy}-acetic acid (2-oxo-1-piperidine-1-yl methyl-1,2-dihydro-indol-3-ylidene)-hydrazide Va (R = H, X = CH_2) is presented in Scheme 2.

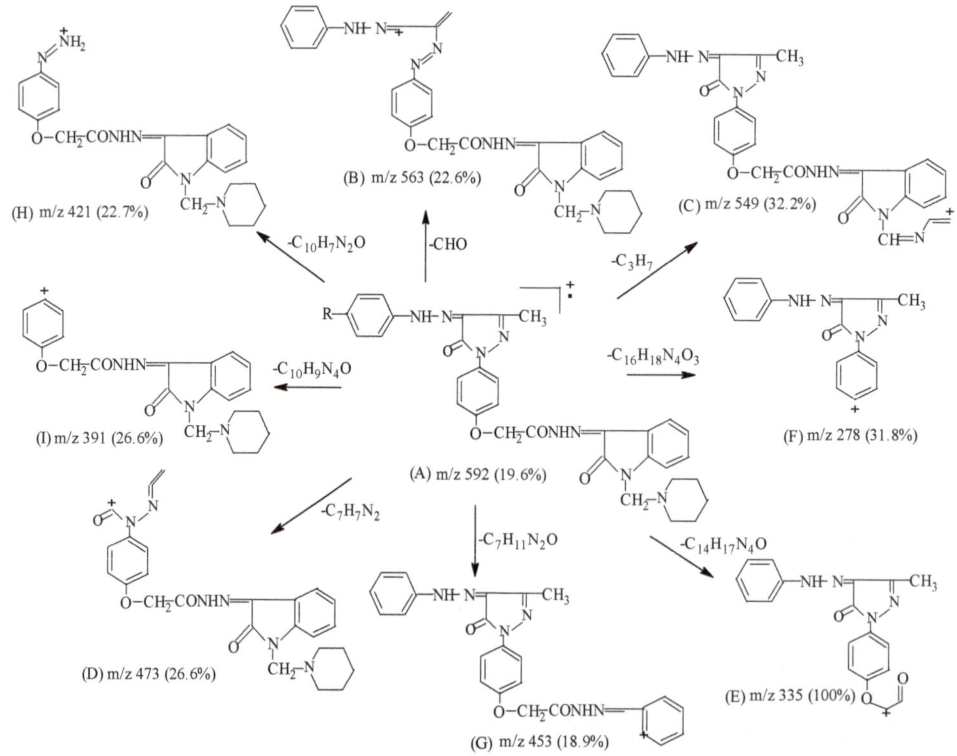

***Scheme 2.** Fragmentation details of {4-[3-Methyl-5-oxo-4-(phenyl hydrazono)-4,5-dihydro-pyrazol-1- -yl]-phenoxy}-acetic acid (2-oxo-1-piperidine-1-ylmethyl-1,2-dihydro-indol-3-ylidene)-hydrazide*

The molecular ion A was observed at m/z 592 (19.6 %). Disintegration of molecular ion A yielded the cation B at m/z 563 (22.6 %). Elimination of C_3H_7 molecule from molecular ion A resulted in the formation of cation C at m/z 549 (32.2 %). Loss of $C_7H_7N_2$ radical from A resulted in the formation of cation D at m/z 473 (26.6 %). Expulsion of $C_{14}H_{17}N_4O$ radical from molecular ion yielded cation E at m/z 335 (100 %). The other important fragments noticed were 278 (31.8 %, F), 453 (18.9 %, G), 421 (22.7 %, H) and 391 (26.6 %, I).

Results and Discussion

Polarographic behaviour of Mannich bases

General polarographic behaviour

{4-[3-methyl-5-oxo-4-(4^{l}-substituted phenyl hydrazono)-4,5-dihydro-pyrazol-1-yl]-phenoxy}-acetic acid (2-oxo-1-piperidine-1-ylmethyl-1,2-dihydro–indol-3-ylidene)-hydrazides Va-f (Table 7) gave two cathodic waves in the pH range 1.1-7.1 and three cathodic waves in the pH range 8.1-10.1. The Polarograms of Va is shown in the Figure 5. An inspection of structure of above compounds showed that the sites susceptible for reduction at the dropping mercury electrode were exocylic >C=N, exocyclic >C=O and cyclic >C=N.

Table 7. Details of Mannich bases synthesized

Compound	Substituent (-R)
Va	H
Vb	4^{l}-CH_3
Vc	4^{l}-OCH_3
Vd	4^{l}-OC_2H_5
Ve	4^{l}-Cl
Vf	4^{l}- Br

Among these groups exocyclic azomethine group was more susceptible for reduction than other groups. However [3-methyl-4,5-dioxo-4,5-dihydro-pyrazol-1-yl]-acetic acid hydrazide does not give reduction wave under experimental conditions. This was probably due to stabilization [25] of the pyrazoline-5-one by ketoenol tautomerism. These observations unambiguously suggest that the waves observed in the present studies were due to reduction of two exocyclic azomethine groups.

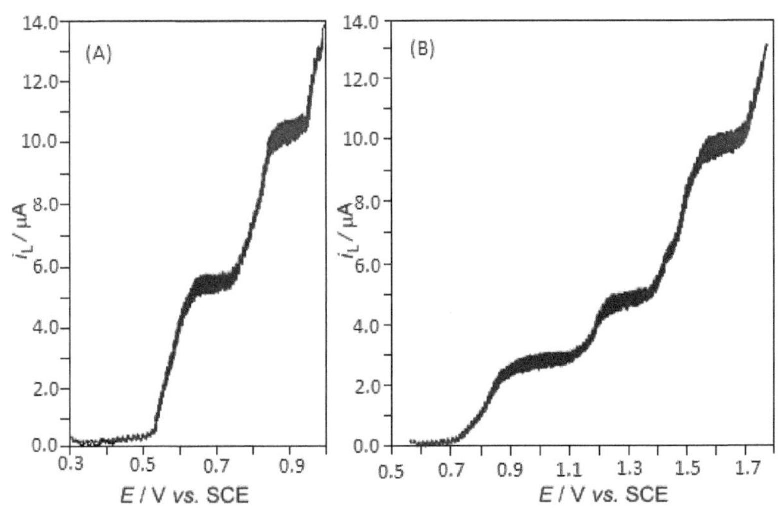

***Figure 5**. Polarograms of {4-[3-Methyl-5-oxo-4-(phenyl hydrazono)-4,5-dihydro-pyrazol-1- -yl]--phenoxy}-acetic acid (2-oxo-1- -piperidine-1-ylmethyl-1,2-dihydro- indol-3-ylidene)-hydrazide ; c = 1 mM; Medium : Aqueous dimethylformamide (40 % v/v). A represents the polarogram at pH 4.1 and B that at indicate 8.1 respectively.*

Half wave potential-pH relation

Half wave potentials of first and second waves shifted to more negative potentials with increase in pH of the medium in the range 1.1-7.1 (Table 1 in Supplementary material). The typical $E_{1/2}$– pH graphs are shown in Figure 6. The half wave potentials of the waves were not altered in alkaline medium. The value of $\Delta E_{½}/\Delta$pH for both waves lie in the range of 0.089-0.094. The $E_{1/2}$-pH plots observed in the pH range 1.1-10.1 suggest that both the protonated form (acidic) and the deprotonated form (basic) of the depolariser were electroactive.

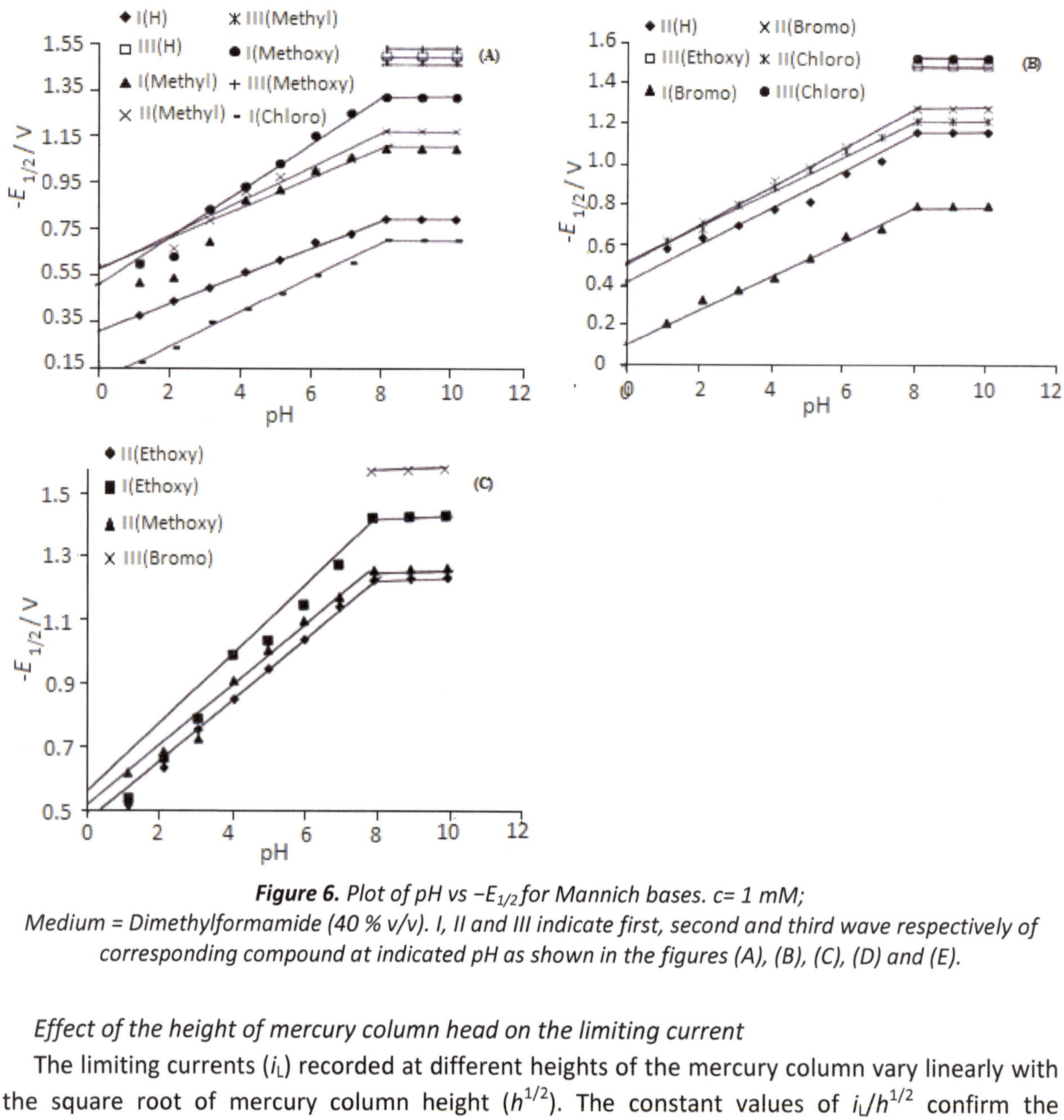

Figure 6. *Plot of pH vs* $-E_{1/2}$ *for Mannich bases. c= 1 mM; Medium = Dimethylformamide (40 % v/v). I, II and III indicate first, second and third wave respectively of corresponding compound at indicated pH as shown in the figures (A), (B), (C), (D) and (E).*

Effect of the height of mercury column head on the limiting current

The limiting currents (i_L) recorded at different heights of the mercury column vary linearly with the square root of mercury column height ($h^{1/2}$). The constant values of $i_L/h^{1/2}$ confirm the diffusion controlled nature of first and second waves. $i_L - h^{1/2}$ plots are shown in Figure 7.

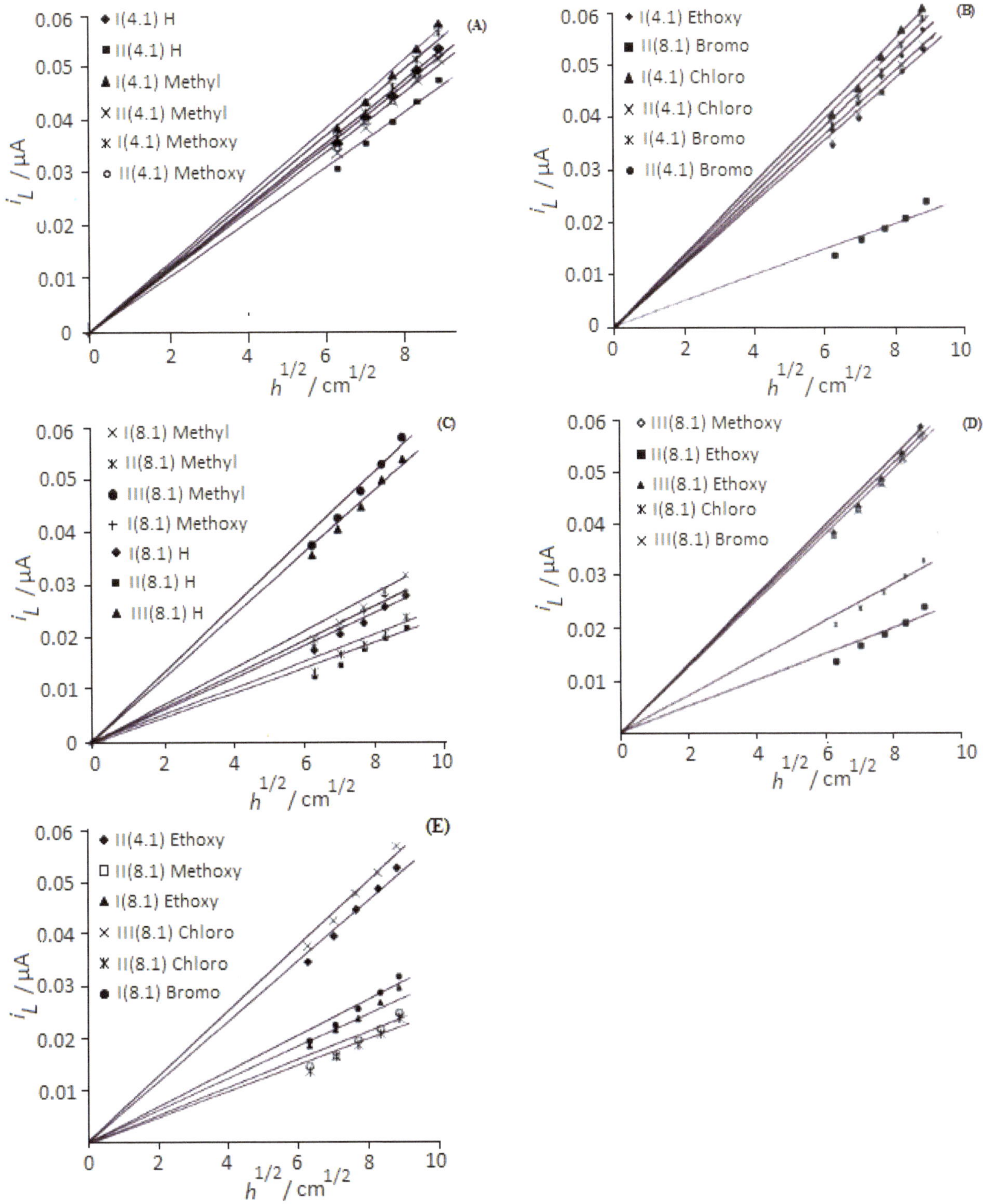

Figure 7. *Effect of mercury column height (h) on limiting current (i_L). c= 1 mM Medium = Dimethylformamide (40 % v/v). I, II and III indicate first, second and third wave resp.*

Effect of concentration on the diffusion current

The influence of concentration on the limiting current in the range 0.5-3.5 mM was recorded in solutions of pH 4.1. The limiting current-concentration plots are presented in Figure 8. The plots were linear and passing through the origin. This observation further confirms the diffusion controlled nature of two waves. (First wave and second wave).

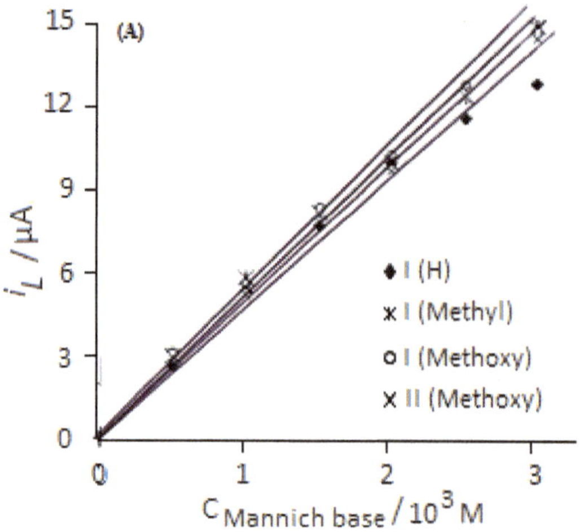

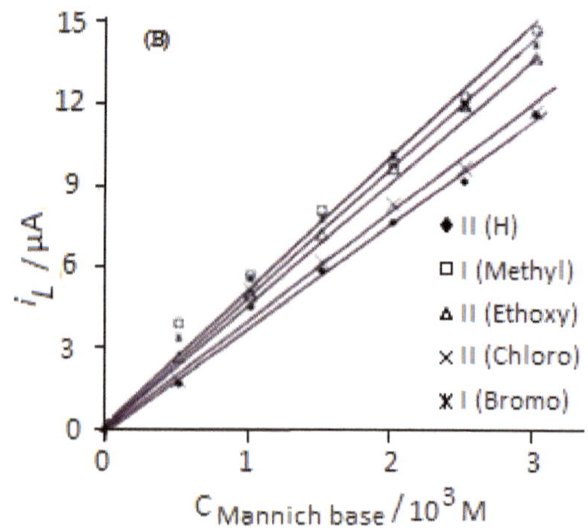

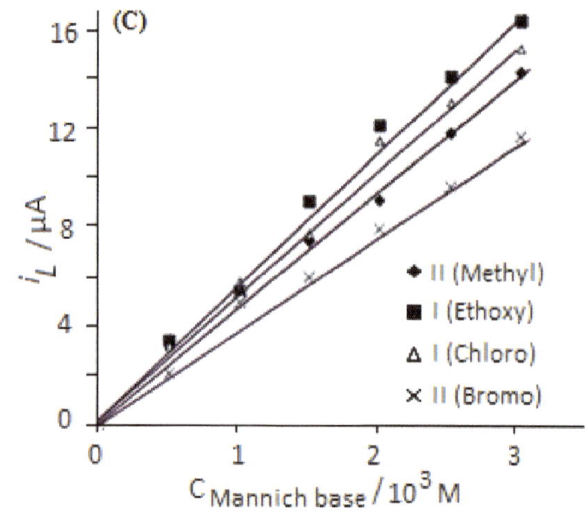

Figure 8. *Effect of concentration on limiting current (i_l). Medium = Dimethylformamide (40% v/v).I and II indicate first and second wave respectively of corresponding compound as indicated in the figures (A), (B) and (C).*

Nature of the electrode process

The linear E_{dme} versus log (i/i_d-i) plots at typical pH 4.1 are shown in Figure 9. The slopes were linear in the range 0.066-0.10 V and were not in agreement with the theoretical values 0.030 V and 0.015 V expected for two electron and four electron reversible reduction process respectively. This indicates the irreversible nature of reduction process. The irreversible nature of the polarographic waves was further confirmed by employing Tome's criteria [26]. The α_{na} values are presented in Table 1 in Supplementary material. The irreversible nature of the two waves was attributed to the bulky group present at the end of >C=N-NH-linkage [27].

Nature of the electrode process

The linear E_{dme} versus log (i/i_d-i) plots at typical pH 4.1 are shown in Figure 9. The slopes were linear in the range 0.066-0.10 V and were not in agreement with the theoretical values 0.030 V and 0.015 V expected for two electron and four electron reversible reduction process respectively. This indicates the irreversible nature of reduction process. The irreversible nature of the polarographic waves was further confirmed by employing Tome's criteria [26]. The α_{na} values are presented in Table 1 in Supplementary material. The irreversible nature of the two waves was attributed to the bulky group present at the end of >C=N-NH-linkage [27].

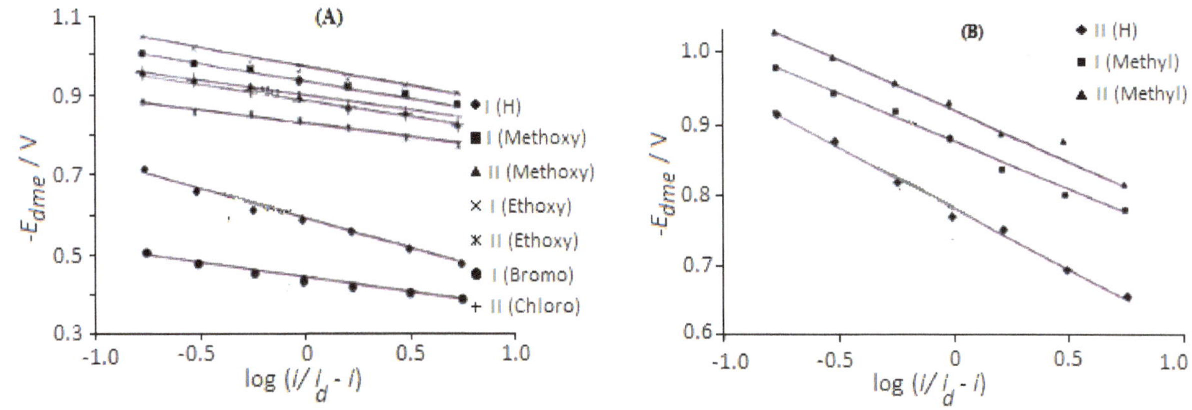

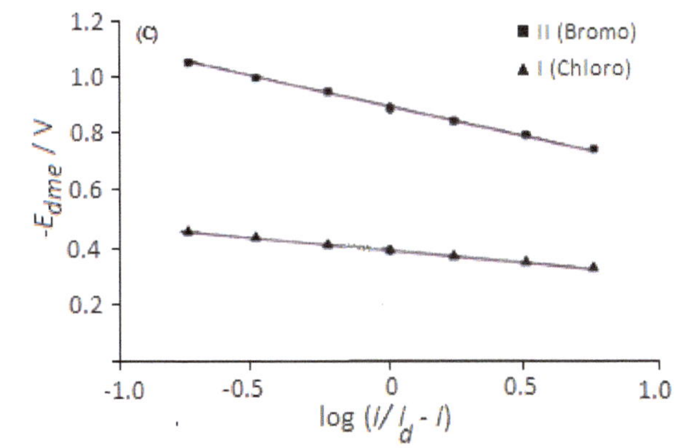

Figure 9. *Semi log plots of Mannich bases. pH 4.1; c = 1 mM; Medium = Dimethylform-amide (40 % v/v).I and II indicate first and second wave wave respectively of corresponding compound as indicated in the figures (A), (B) and (C).*

Kinetic parameters of the electrode reaction

The kinetic parameters k^{o}_{fh} and ΔG^* of the electrode reaction at typical pH values evaluated for first and second waves are presented in Table 1 in Supplementary material. The magnitude of k^{o}_{fh} decreases and ΔG^* increases with the increase in pH of the medium. This indicates irreversible nature of the electrode process [28].

Controlled potential electrolysis

The electrochemical reduction of {4-[3-methyl-5-oxo-4-(4'-substituted phenyl hydrazono)-4,5--dihydro-pyrazol-1-yl]-phenoxy}-acetic acid (2-oxo-1-piperidine-1-ylmethyl-1,2-dihydro–indol-3--ylidene)-hydrazide (V) has been studied by the method of controlled potential electrolysis at pH 4.1.

Controlled potential electrolysis was carried out in a Lingane H-type cell. A large pool of mercury at the bottom of the large compartment was used as cathode and a similar pool of mercury at the bottom of smaller compartment served as anode. The cathode compartment contains 10 mL of V (0.01 M), 30 mL of DMF, 20 mL of KCl (1.0 M) and 40 mL of the buffer solution of required pH (4.1). A potential of -1.2 V was applied and maintained at constant potential by the manual control of output from the battery. The electrolysis was followed by recording the decrease in current with time. The number of electrons per molecule was calculated from i-t curves and was found to be 8 [29]. 1 mL of solution was withdrawn from the electrolysis cell to confirm the presence of aniline by standard spot test [30]. Part of the experimental solution was partially evaporated on water bath to half of its volume, allowed to cool to room temperature and was extracted with ether. The ether layer was evaporated under reduced pressure. The yellow crystalline solid obtained was identified to be 2-(5-thioxo- 4,5- dihydro-[1,3,4]-oxadiazole-2--ylmethyl)-2,4-dihydro-pyrazol-3-one 'A' by elemental analysis, TLC, IR, and ^{1}H NMR spectral data.

Remaining experimental solution was made slightly alkaline and extracted with ethylacetate. The ethylacetate extract was subjected to column chromatography and three compounds were separated by using benzene-methanol 4:1 (v/v) solvent as an eluent. The compounds B, C and D were found to be aniline, (4-Amino-3-methyl-5-oxo-4,5-dihydro-pyrazol-1-yl)-acetic acid hydrazide and 3-Amino-1--piperidin-1-ylmethyl-1,3-dihydro-indol-2-one respectively.

Table 8. *Characterization data of 2-(5-thioxo-4,5-dihydro-[1,3,4]-oxadiazole-2-ylmethyl)-2,4-di-hydro-pyrazol-3-one*

Mass fraction of element found, % (Calc., %)	C 36.34 (36.36), N 28.31 (28.27), O 16.18 (16.14), S 16.11 (16.18)
IR (KBr) Spectral data (ν_{max} / cm^{-1})	3126 (oxadiazole NH); 3180 (NH); 1603 (C = N); 1670 (C = O); 1134 (C = S)
^1H NMR Spectral data (δ /ppm)	2.3(s, H, CH_3); 5.45 (s, 2H, N – CH_2), 14.7 (s, H, thiol-thione tautomeric proton NH), 3.9 & 4.1 (NH_2)
Mass spectral data: m/e (Relative abundance, %)	172(77); 171(100); 156(10); 143(33.3); 140(40); 127(47.7); 112(57.7); 100(8.8); 84(13.3); 59(53.3); 57(34.4); 51(28.8); 29(38.8); 27(12.2)

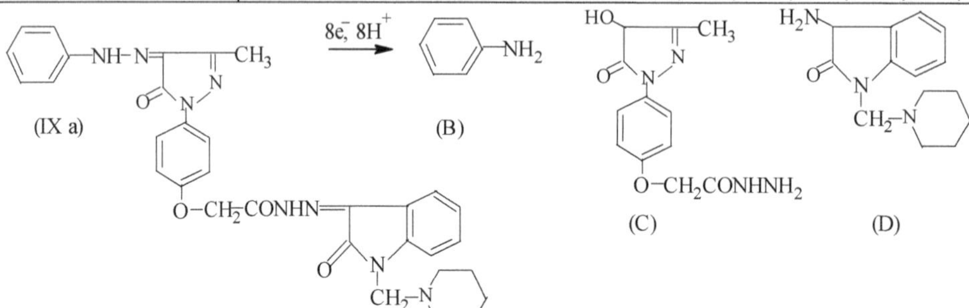

The compounds were characterized by IR and ^1H NMR data. The compound B was further confirmed as aniline by azodye test.

Table 9. *Elemental analysis and spectral data of B , C and D.*

Comp.	Mass fraction of element found, % (Calc., %)	IR Spectral data (ν_{max} / cm^{-1})	^1H NMR Spectral data (δ/ ppm)
B	C 77. 31 (77.33), N 15. 13 (15.09)	3180 (NH), 760 (ArH).	4.0 (s, 2H, ArNH), 6.46 – 7.01 (m, 5H, ArH)
C	C 38. 90 (38.96), N 37.68 (37.78), O 17.45 (17.34)	1676 (C=O), 1654 (CONH), 2933 (CH_2).	0.9 (s, 3H, CH_3), 4.09 (s, 2H, N – CH_2), 3.9 & 4.2 (s, 2H, NH_2), 6.0 (s, 2H, $CONH_2$).
D	C 68.59 (68.51), N 17.21 (17.17), O 6.45 (6.54)	3180 (NH), 1666 (C = O).	4.52 (s, 2H, NCH_2N), 2.0 (s, 2H, NH_2), 2.70 (t, 4H, NCH_2), 6.44 (m, 4H, ArH).

Reduction mechanism in acidic medium

Based on the experimental results (Table 10), it was proposed that two azomethine groups in the compounds Va-f were reduced separately involving four electrons. The reduction steps appear as two waves in solutions of pH 1.1-7.1.

It was clear that compounds Va-f were reduced at the dropping mercury electrode through a mechanism which involves the azomethine, imine intermediate and amine via usual sequence.

Hydrazone → Imine → Amine

V was protonated at azomethine group to yield protonated form II. The weak >C=N-NH undergo cleavage at >C=N-NH single bond [31,32] with the uptake of four electrons and two protons to form the unstable imine intermediates IV and V which subsequently undergo two electron reduction to form VI and VII.

It was reported [33,34] that the above mentioned steps of the reduction (formation of VI & VII) occur at the same potential. The mechanism (Scheme 3) proposed was also supported by the results obtained in cyclic voltammetry.

Reduction mechanism in basic medium

In alkaline medium (pH>pK_a), V exists in the azomethine anionic form (I). The latter in alkaline solutions, was susceptible to chemical cleavage partially into the corresponding carbonyl compounds III, IV and V as shown in the Scheme 4. The first and second waves noticed in alkaline solutions were attributed to the two 4 electron reduction processes of azomethine anionic form to the species containing amino group. The third wave was attributed to the two 2 electron reduction process of heterocyclic carbonyl compounds (IV and V) obtained during the chemical cleavage of dianion I. The heterocyclic carbonyl compounds were reduced to carbinols by two electron reduction process. The decrease in the height of the first and second waves with increase in alkali concentration was attributed to the partial chemical cleavage of dianion.

Effect of substituents on polarographic behaviour

It is known that the polarographic reduction of compound depends on the nature of the compound, the position of the substituent and the reaction medium. Heyrovsky [35,36] correlated the polarographic behaviour of a representative number of compounds with their structure. He opined that the the reducibility of given compound is influenced by conjugated double bonds, triple bonds and aromatic rings present in the substrate.

$E_{1/2}$-σ plots for the compounds under investigation are presented in Figure 10. The values of specific reaction constant (ρ) [38] are presented in Table 11.

Table 10. *Millicoulometric [37] data of {4-[3-methyl-5-oxo-4(4^{l}-substituted phenyl hydrazono)-4,5-di -hidro-pyrazole-1-yl)-phenoxy}-acetic acid (2-oxo-1-piperidine-1-ylmethyl-1,2-dihdro-indol-3-ylidene)- -hydrazide at pH 4.1 and 8.1.*

pH	Current, µA			Time, s			n value		
	First wave	Second wave	Third wave	First wave	Second wave	Third wave	First wave	Second wave	Third wave
4.1	5.4	4.8	-	-	-	-	-	-	-
	4.4	4.0	-	7200	7200	-	3.2	3.2	-
8.1	3.4	3.2	-	10800	10800	-	3.0	3.2	-
	2.7	2.1	5.3	-	-	-	-	-	-
	2.4	1.9	3.5	7200	7200	7200	2.8	2.8	2.1
	2.1	1.7	2.5	10800	10800	10800	3.5	3.4	1.6

It was apparent from Table 1 in Supplementary material that the values of *D*, $E_{1/2}$/pH, α_{na} and *I* (diffusion current constant) were practically in the same range for entire reaction series under study. Thus it was possible to discuss the effect of substituents quantitatively in terms of the Hammett equation [39]. The values of the Hammett substituent constants were taken from the literature [39]. $E_{1/2}$–σ plots in media of pH 4.1 and 9.1 for all the compounds under study are shown in Figure 10. The values of specific reaction constant (ρ) calculated from the graphs were found to be in the range of 0.15-0.85. The specific reaction constant values are all positive [40,41] (Table 11) and low indicating that the nucleophilic reaction was taking place. This confirms that the electron uptake was the potential rate determining step [42].

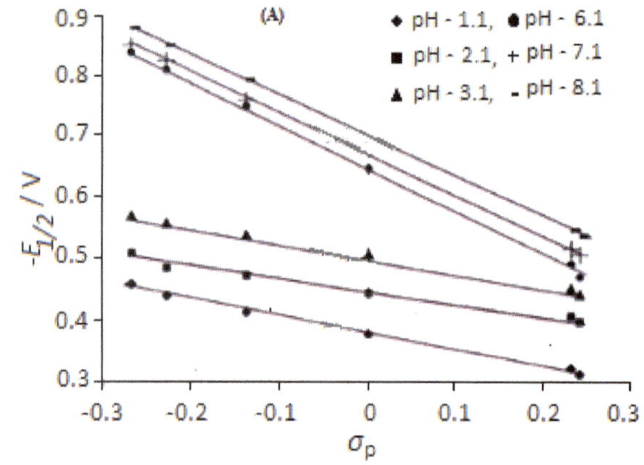

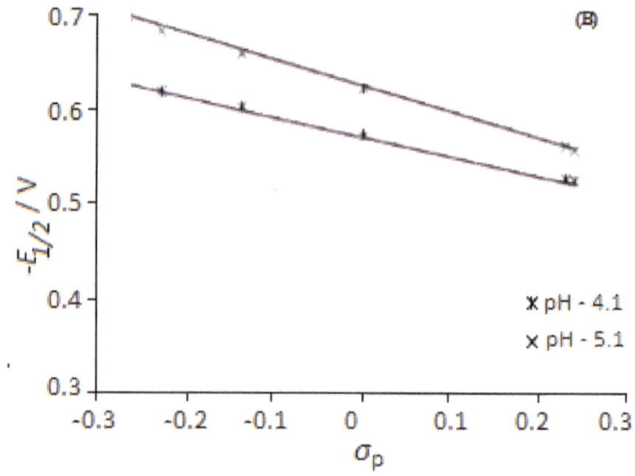

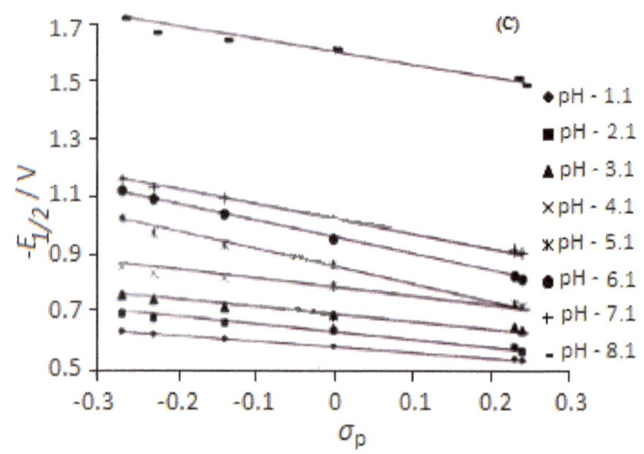

Figure 10. *$-E_{1/2}$–σ_P plots of Mannich bases (A, B: First wave, C: Second wave at indicated pH).* c= *1 mM; Medium = Dimethylformamide (40 % v/v).*

Cyclic voltammetric studies of {4-[3-methyl-5-oxo-4(4$^{\prime}$-substituted phenyl hyarazono)-4,5-dihydro-pyrazol-1-yl]phenoxy}-acetic acid (2-oxo-1-piperidine-1-ylmethyl-1,2-dihydro-indol-3-ylidene)-hydrazide (V a-f) at hanging mercury drop electrode

The cyclic voltammetric data of compounds Va-f at pH 2.1, 4.1, 6.1, 8.1 and 10.1 are given in Tables 2-7 in Supplementary material. The voltammograms contain two cathodic peaks in acidic solutions and three cathodic peaks in basic solutions at all sweep rates. The cyclicvoltammograms of Va is shown in the Figure 11.

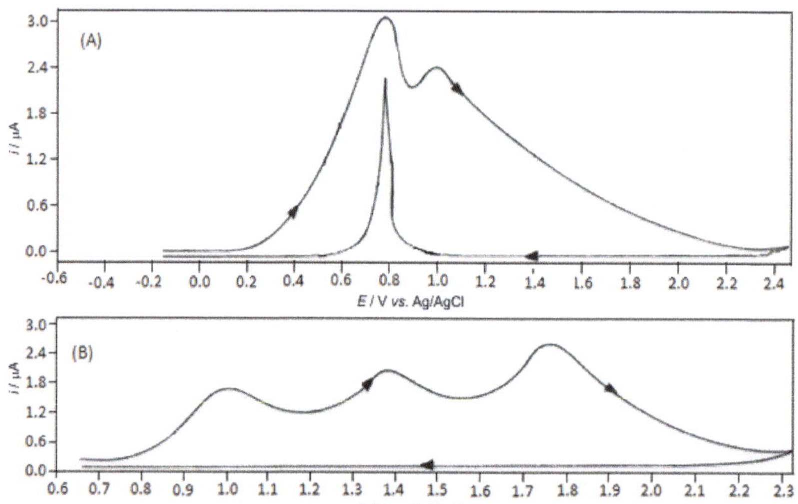

Figure 11. *Cyclic Voltammograms of {4-[3-Methyl-5-oxo-4-(phenyl hydrazono)-4,5-dihydro-pyrazol-1-yl]--phenoxy}-acetic acid (2-oxo-1-piperidine-1-ylmethyl-1,2-dihydro-indol-3-ylidene)-hydrazide; A represents the cyclicvoltammogram at pH 4.1 and B that at pH 8.1.* c = *1 mM; Medium: Aqueous dimethylformamide (40 % v/v); sweep rate 100 mv/s.*

Influence of scan rate on peak potential and peak current

The cathodic peak potentials were shifted to more negative values with the increase in scan rate. The cathodic peak currents increase with the increase in scan rate. The results are shown in Tables 2-7 in Supplementary material.

Nature of the electrode process

i_{pc} *vs.* $\gamma^{1/2}$ was a linear plot passing through the origin for the cathodic peaks. This suggests the diffusion controlled nature of the reduction process. This fact was further confirmed by the increase of peak currents with the increase in the concentration of the depolarizer.

The irreversible nature of the electrode process was indicated by following observations.

1. The anodic peak was absent in the reverse scan.
2. The value of ($E_{pc/2}$ – E_{pc}) was greater than 56.5/*n* mV, where *n* represents number of electronics involved in the corresponding electrode process.
3. The E_{pc} shifted towards more negative potentials with increase in the concentration of the depolarizer.
4. The current function ($i_{pc}/\gamma^{1/2}C$) was independent of scan rate and
5. ($i_{pc}/\gamma^{1/2}$) *vs.* υ graph was similar to the one expected for case II of Nicholson-Shain Criteria [43].

Table 11. *Effect of pH (Britton-Robinson buffers) on the reaction constant for the reduction of {4-(3-methyl-5-oxo-4-(4^{l}-substituted phenyl hydrozono)-4,5 dihydro-1-yl]-phenoxy}-acetic acid (2-oxo-1 piperidine – 1- ylmethyl-1,2-dihydro-indol-3-ylidene)-hydrazide.*

pH	ρ value	
	First wave	Second wave
1.1	0.29	0.20
2.1	0.25	0.25
3.1	0.25	0.25
4.1	0.20	0.20
5.1	0.33	0.58
6.1	0.66	0.66
7.1	0.58	0.58
8.1	0.66	0.50
9.1	0.66	0.50
10.1	0.66	0.50

Effect of pH on the peak potential and peak current

The results presented in ***Tables 2-7*** in Supplementary material show that the cathodic peak potential shifts towards more negative values with the increase in pH of the solution.

The first and the second cathodic peak currents decrease with increase in pH of solutions from 2.1 to 10.1, while peak current corresponding to the third wave noticed in alkaline solutions remains constant. The results were similar to those observed in polarographic studies. Hence the reduction mechanism at HMDE was assumed to be same as that noticed at DME.

Voltammograms recorded in media of pH 4.1 under repeated cycles showed that the peak height decreases with increase of repetition of cycles. However, no significant change was noticed in the shape of the voltammogram. This may be ascribed to the adsorption of the depolarizer on the mercury solution interface.

Reduction pattern in acidic media

From the position of the peak on potential axis, it was proposed that first and second cathodic peaks obtained in acidic solutions were due to the reduction of azomethine group to amines. The reduction mechanism is shown in Scheme 3.

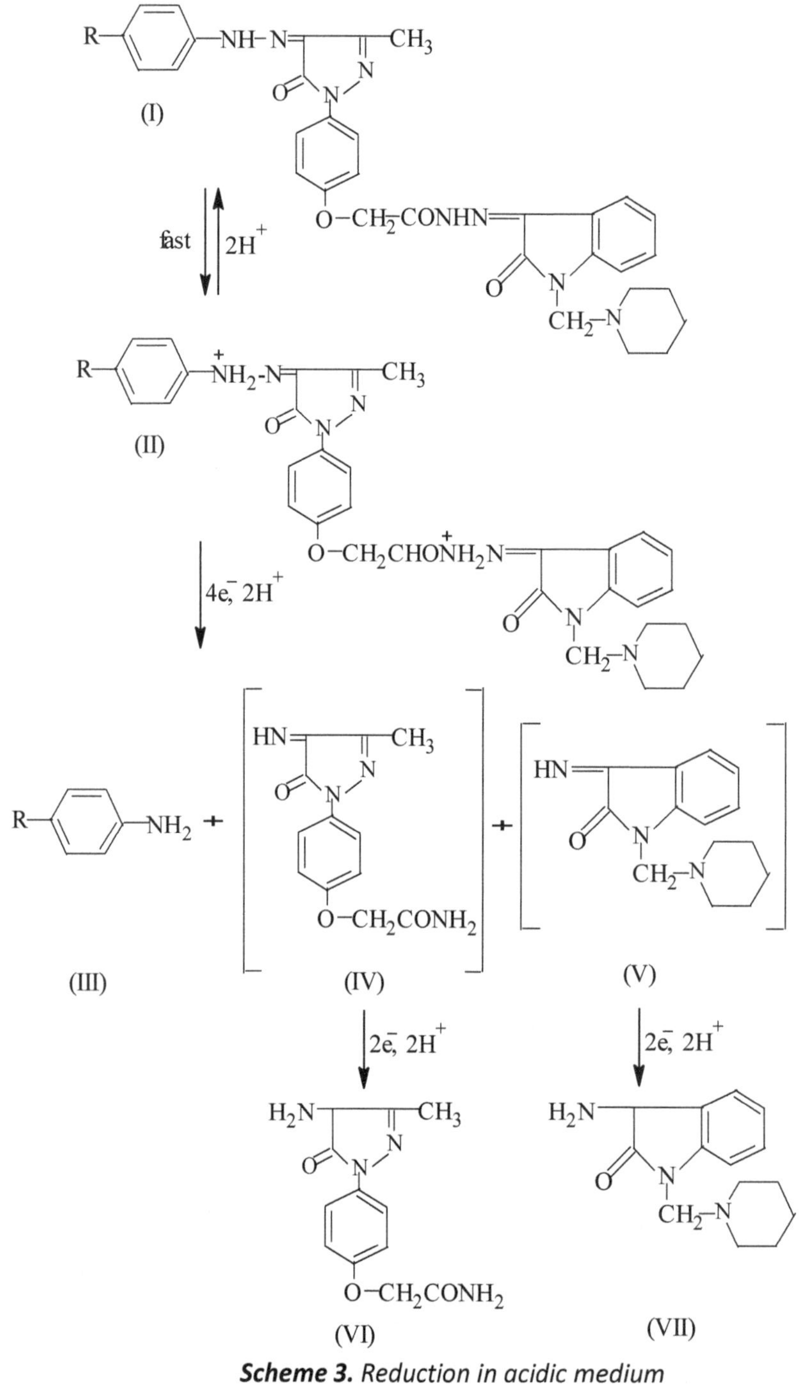

Scheme 3. *Reduction in acidic medium*

Reduction pattern in alkaline media

The position of peaks on potential axis in alkaline medium indicates that the first and second peaks were due to two 4-electron reduction processes of azomethine anionic form and third peak was due to reduction of carbonyl group to carbinol. The protons required for the ketone reduction were obtained from aqueous solution (Scheme 4).

Inverted peaks

Except in solutions of pH 8.1-10.1 and in 0.1 M NaOH, the compounds of V series exhibit inverted peaks (Figure 11). The inverted peak is a common phenomenon in electrochemical reduction of organic compounds [44-46].

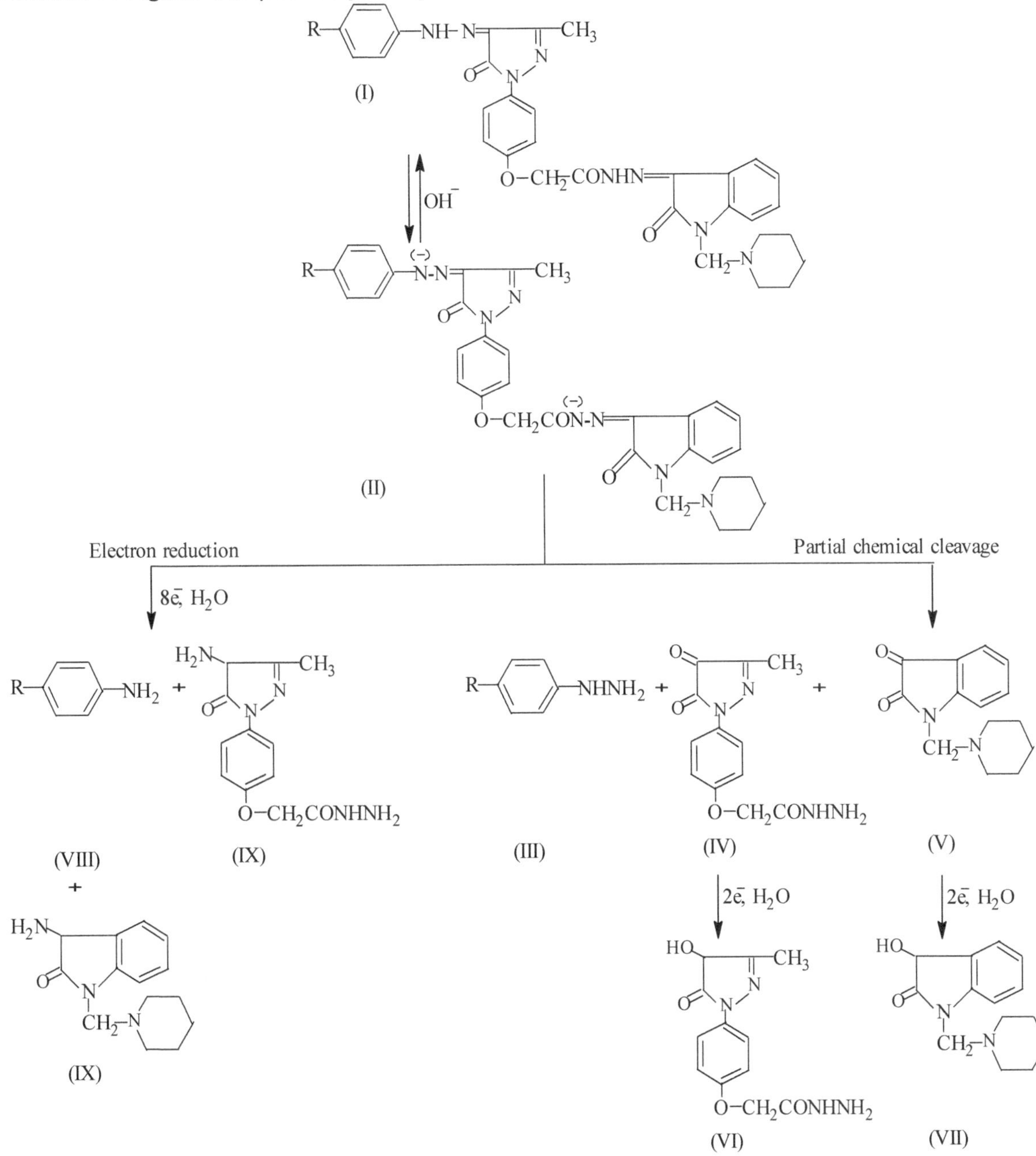

***Scheme 4.** Reduction in basic medium*

In present studies, the peak potentials were unaltered at lower sweep rates (10-50 V s^{-1}). At higher sweep rates (100-500 V s^{-1}), peak current increases with increases in sweep rate. However a cathodic peak during anodic cycle (Inverted peak) was noticed in buffer solutions of pH 2.1-6.10.

Inverted peaks are generally attributed to

- The movement of mercury surface due to uneven drop polarisation. This bears similarity to the explanation of polarographic maximum of the first kind [47-49] or
- the inhibition of electrode reaction as a consequence of coverage of the electrode by a product of the electrode reaction [50-52]

If the peak is attributed to the second possibility, its intensity is expected to increase with decrease in scan rate. Hence the results in the present studies were not attributed to the second possibility. Moreover, a significant polarographic maximum was observed for the compounds under the experimental conditions of study. It is clear therefore that the inverted peak was attributed to the movement of the mercury surface due to uneven drop polarization.

Effect of substituents on the cyclic voltammetric behaviour of {4-[3- methyl-5-oxo-4(4^{l}-substituted phenyl hydrazono)-4,5-dihydro-pyrazol-1-yl]phenoxy}-acetic acid (2-oxo-1-piperidene-1-ylmethyl- 1,2-dihydro indol-3-ylidene)-hydrazide.

To investigate the influence of the substituent on cathodic peaks corresponding to the azo group reduction, Hammett's linear free energy relations were applied. The plots drawn between the cathodic peak potential and the Hammett's substituent constant were presented in Figure 12. The slope (ρ) of the linear plots so obtained was positive and confirms the nucleophilic reduction. The values are presented in Table 12.

Table 12. *Effect of pH on the reaction constant for the reduction of {4-[3-Methyl-5-oxo-4-(4^{l}-substituted phenyl hydrazono)-4,5-dihydro-pyrazol-1-yl] phenoxy}-acetic acid (2-oxo-1-piperidine-1-yl methyl)-1.2 di-hydro-indol-3-ylidene)- hydrazide.*

pH	ρ	
	I wave	II wave
2.1	0.33	0.33
4.1	0.25	0.25
6.1	0.25	0.25
8.1	0.33	0.33
10.1	0.33	0.33

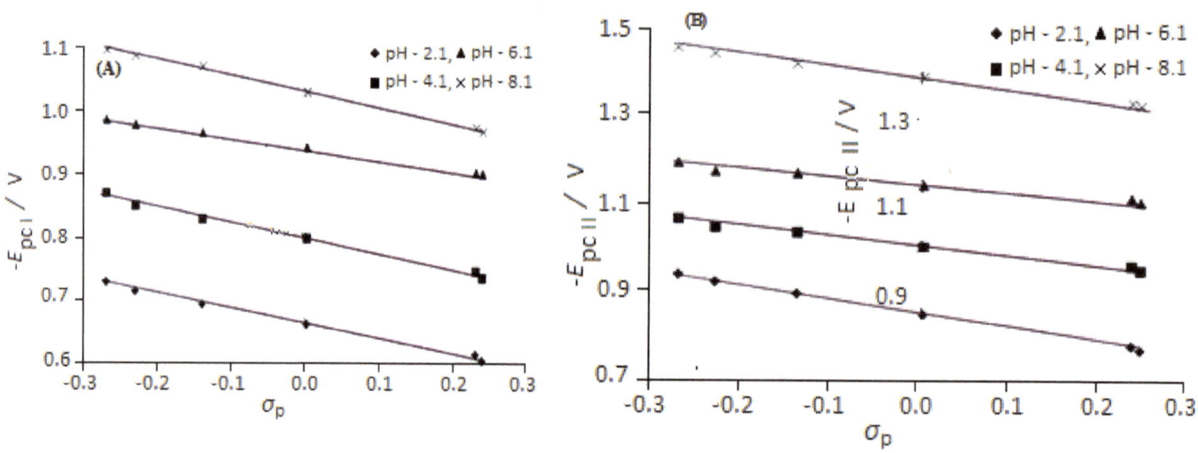

Figure 12. *$-E_{1/2}$–σ_p plots of Mannich bases (A: First wave, B: Second wave).* c = *1 mM; Medium = Dimethylformamide (40% v/v).*

Cyclic voltammetric studies of {4-[3-methyl-5-oxo-4-(4'-substituted phenyl hydrazono)-4,5-dihydro-pyrazol-1-yl]-phenoxy}-acetic acid (2-oxo-1-piperidine-1-ylmethyl-1,2-dihydro–indol-3-ylidene)-hydrazide (Mannich bases V a-f) with a crown–ether modified carbon paste electrode.

Cyclic voltammetric studies of {4-[3-methyl-5-oxo-4-(4'-substituted phenyl hydrazono)-4,5-dihydro-pyrazol-1-yl]-phenoxy}-acetic acid (2-oxo-1-piperidine-1-ylmethyl-1,2-dihydro–indol-3-ylidene)-hydrazide (V a-f) were carried out at modified carbon paste electrode in buffer solutions of pH value 2.1, 4.1, 6.1, 8.1 and 10.1 at different scan rates *i.e.* 10 mV s^{-1}, 20 mV s^{-1}, and 50 mV s^{-1}, 100 mV s^{-1}, 200 mV s^{-1}, 300 mV s^{-1}, 500 mV s^{-1}. At all scan rates, the compounds exhibit two well defined cathodic peaks in solutions of pH 2.1-6.1 and three cathodic peaks in alkaline solutions of pH 8.1 - 10.1. An extra anodic peak was noticed in the acidic solutions of pH 2.1 – 6.1.

Effect of scan rate on peak potential and peak current

Peak potentials and peak currents change with the change of scan rates. The cathodic peak potentials become more negative and cathodic peak currents increase with increase in scan rate (Tables 2-7 in Supplementary material).

Effect of pH on peak potentials and peak current

The cathodic peak potentials were shifted to more negative values with the increase in pH. The peak currents decrease with increase in pH of the medium. The data is given in the Tables 2-7 in Supplementary material.

Nature of electrode process

The irreversible nature of electrode process was established by the fact that a plot of $i_{pc}/\upsilon^{1/2}$ versus sweep rate was a straight line parallel to sweep rate axis. The irreversible nature of electrode process was confirmed by

- dependence of peak potential on sweep rate,
- the separation of anodic peak and cathodic peak potential (E_{pa}–E_{pc} = ΔE_p) was not equal to 60 / *n* mV at 25°C[53], where *n* represents number of electronics involved in the corresponding electrode process.
- the shape of $i_{pc}/\upsilon^{1/2}$ versus υ plot was in accordance with the Nicholson and Shain criteria [43].

The linear plots of i_{pc} versus $\upsilon^{1/2}$ suggest the diffusion controlled nature of the electrode process. This was further supported by the negative shift of peak potential with increase in sweep rate. The plots of i_{pc} *vs* concentration and i_{pc} *vs* $\upsilon^{½}$ fulfil the criteria for the diffusion controlled nature of the electrode process. The plots of Epc *vs* pH were similar to $E_{½}$ *vs.* pH plots and this supports the findings and conclusion drawn from DC polorography.

Comparison between polarographic studies and cyclic voltammetric studies

Equal number of polarographic waves and cyclic voltammetric peaks noticed in DC polarographic and cyclic voltammetric studies respectively suggest similar behavior in both the studies. However, an additional anodic peak was noticed in solutions of pH 2.1 – 6.1 may be attributed to oxidation of amine formed in the previous reduction step.

Conclusions

A series of six novel Mannich bases were synthesized and characterized by IR, ^{1}H NMR, mass spectral data and elemental analysis. The compounds under study gave two well defined polarographic waves in the media pH 1.1-7.1 and three waves in the media pH 8.1-10.1. The results were compared with those obtained in cyclic voltammetry at hanging mercury drop electrode and modified carbon paste electrode. The mechanism for the reduction process in acidic as well as basic medium is proposed.

References

[1] H. S. El-Kashef, M. A. Abd-Alla, B. E. Bayoumi, A. A. M. El-Timawy, *J. Chem. Tech. Biotech.* **33** (1983) 294-298.

[2] K. Guckian, M.B. Carter, E. Y. Lin, M. Choi, L. Sun, P. A. Boriack-Sjodin, C. Chuaqui, B. Lane, K. Cheung, L. Ling, W. C. Lee, *Bioorg. Med. Chem. Lett.* **20** (2010) 326-329.

[3] P. Manojkumar, T. K. Ravi, S. Gopalakrishnan, *Eur. J.Med. Chem.* **44** (2009) 4690-4694.

[4] J. P. Raval, A. B. Shah, N. H. Patel, H. V. Patel, P. S. Patel, K. K. Bhatt, K. R. Desai, *Eur. J. Chem* **2** (2011) 238-242.

[5] X. Fan, X. Zhang, L. Zhou, K.A. Keith, E. R. Kernb, P. F. Torrencea, *Bioorg. Med. Chem. Lett.* **16** (2006) 3224-3228.

[6] R. Robinson, *J. Chem. Soc.* **111** (1917) 762-768.

[7] E. J. Corey, R. D. Balanson, *J. Am. Chem. Soc.* **96** (1974) 6516-6517.

[8] S. D. Knight, L .E. Overman, G. Pairaudeau, *J. Am. Chem. Soc.* **115** (1993) 9293-9294.

[9] S. Joshi, A. D. Manikpuri, P. Tiwari, *Bioorg. Med. Chem. Lett.* **17** (2007) 645-648.

[10] S. Joshi, N. Khosla, P. Tiwari, *Bioorg. Med. Chem. Lett.* **12** (2004) 571-576.

[11] B. Shivarama Holla, BS. Sreenivasa, B. Kalluraya, *Boll. Chim. Farm.* **133** (1994) 527-531.

[12] T. B. Shah, A. Gupta, M. R. Patel, V. S. Chaudhari, H. Patel, V. C. Patel, *Ind. J. Chem. B* **48B** (2009) 88-96.

[13] A. Tambo-ong, S. Chopra, B. T. Glasar, K. Matsuyama, T. Tran, P. B. Madrid, *Bioorg. Med. Chem. Lett.* **21** (2011) 5697-5700.

[14] G. Q. Hu, L. L. Hou, S. Q. Xie, W. L. Huang, H. B. Zhang, *Yao Xue Xue Bao* **43** (2008) 926-929.

[15] M. A. La-Scalea, C. M. de Souza Menezes, M. S. da Silva Juliao, M. C. Chung, S. H. Pires Serrano, E. I. Ferreira, *J. Braz. Chem. Soc.* **16** (2005) 774-782.

[16] D. I. Edwards, Comprehensive Medicinal Chemistry. The Rational Design, Mechanistic Study & Therapeutic Application of Chemical Compounds, Pergamon Press, Oxford, 1990, 546-554.

[17] C. Viode, C. De Albuquerque, G. Chauviere, C. Houée-Levin, J. Perie, *New J. Chem.* **21** (1997) 1331-1338.

[18] A. Guissani, Y. Henry, N. Lougmani, B. Hickel, *Free Rad. Biol. Med.* **8** (1990) 173-189.

[19] J. Rozenski, C.J. De Ranter, H. Verplanken, *Quant. Struct-Act. Rel.* **14** (1995) 134-141.

[20] H. Cerecetto, R. Di Maio, M. González, M. Risso, G.Sagrera, G. Seoane, A. Denicola, G. Peluffo, C. Quijano, A. O. M. Stoppani, M. Paulino, C. Olea-Azar, M. A. Basombrio, *Eur. J. Med. Chem.* **35** (2000) 343-350.

[21] K. R. Kumar, A. R. G. Prasad, V. Srilalitha, G. N. Swamy, L.K. Ravindranath, *Sci. Iran. Trans. C* **19** (2012) 605-618.

[22] K. R. Kumar, A. R. G. Prasad, V. Srilalitha, G. N. Swamy, L.K. Ravindranath, *Ovidius University Annals of Chemistry* **23** (2012) 5-20.

[23] L. K. Ravindranath, K. Srikanth, M. Dastagiri Reddy, S. D. Ishrath Begum, *Orient. J. Chem.* 25 (2009) 993-998.

[24] C. S. Marvel, G. S. Heirs, *Organic Synthesis, Vol. I*, John Wiley and Sons, New York, 1941, p. 342.

[25] J.W. Ross, R.D.Demars, I. Shain. *Anal. Chem.* **28** (1956) 1768-1772.

[26] J. Tomes, *Coll. Czech. Chem. Commun.* **9** (1937) 12-18.

[27] W. V. Malik, R. N. Goyal, R. Jain, *J. Electroanal. Chem.* **87** (1978) 129-135.

[28] K. R. Kumar, A. R.G. Prasad, V. Srilalitha, G. N.Swamy, L.K. Ravindranath, *Rev. Colomb. Quim*. **40** (2011) 165-184.
[29] J.J. Lingane, *J. Am. Chem. Soc.* **67** (1945) 1916-1922.
[30] F. Feigl, *Spot Test, Vol. II*, Elsevier, Amsterdam, 1954, 231.
[31] R. N. Adams, Electrochemistry *at Solid Electrodes*, Marce Dekkar Inc., New York, 1969, 332.
[32] R. N. Adams, *Anal. Chem.* **30** (1958) 1576-1576.
[33] Saul Patai, *The chemistry of hydrazo, azo and azoxy* groups (Part – I), John Wiley and Sons, Interscience Publications, London, 1975, 389.
[34] P. Zuman, *Topics in Organic Polarography,* Plenum Press, New York, 1970, 231.
[35] J. Heyrovsky, A polarographic study of the electrokinetic phenomena of adsorption, electro-reduction and overpotential displayed at the dropping mercury cathode, Hermann, Paris, 1934, 448. (Actualités Scientifiques et Industrielles, No. 90: Réunion Internationale de Chimie-Physique, 1933, No. 10)
[36] J. Heyrovsky, *Polarography*, Springer Verlag, Wien, 1941, 239.
[37] T. De Vries, J. L. Kroon*, J. Am. Chem. Soc.* **75** (1953) 2484-2486.
[38] A. R.G. Prasad, V. Seshagiri, L.K. Ravindranath, *Jordan J. Chem.* **6** (2011) 51-64.
[39] L. B. Hammett, *Physical Organic Chemistry*, McGraw-Hill, New York, 1940, 787.
[40] H. H. Jaffe, *Chem. Rev.* **53** (1953) 191-261.
[41] P. Zuman, *Substituent effects in Polarography*, Plenum Press: New York, 1976, 213.
[42] L. K. Ravindranath, S. R. Ramadas, S. Brahmaji Rao, *Electrochim. Acta* **28** (1983) 601-603.
[43] R. S. Nicholson, I. Shain, *Anal. Chem.* **36** (1964) 706-723.
[44] B. J. H. Wang, K .L. Lee. *J. App. Electrochem.* **26** (1996) 153-159.
[45] P. Krtil, L. Kavan, I. Moskovcova, K. Kratochvicova, *J. Appl. Electrochem.* **26** (1996) 523-527.
[46] M. Neol, C. Ravichandran, P. N. Anantharaman, *J. Appl. Electrochem.* **25** (1995) 690-698.
[47] P. Delahay, *New Instrumental Methods in Electrochemistry*, Interscience Publishers Inc., New York, 1966, 389.
[48] M. Pourobaix, *Atlass of Electrochemical Equilibrie in Aqueous Solutions*, Pergamon Press, New York, 1966, 571.
[49] R. S. Nicholson, I. Shain, *Anal. Chem.* **36** (1964) 706-723.
[50] D. A. Tyssee, M.M. Baizer. *J. Electrochem. Soc.* **118** (1971) 1420-1425.
[51] P. Tissot, P. Margaretha, *Electrochim. Acta* **23** (1978) 1049-1052.
[52] K. S. V. Santhanam, A. J. Bard, *J. Am. Chem. Soc.* **88** (1966) 2669-2675.
[53] P. Vanysek, *Electrochim. Acta* **40** (1995) 2841-2847.

Table 1. *Polarographic characteristics and kinetic parameters of{4-[3-methyl-5-oxo-4-(4^{l}-substituted phenyl hydrazono)-4,5-dihydro-pyrazol-1-yl]-phenoxy}-acetic acid (2-oxo-1-piperidine-1-ylmethyl-1,2-dihydro–indol-3-ylidene)- hydrazide (1 mM) at pH 4.1 in various treatments, Medium : Aqueous dimethyl formamide (40% V/V) (I W = First wave II W = Second wave III W = Third wave)*

#	pH	($\Delta E_{1/2}/\Delta$pH)/ mV		α_{na}			No. of protons p		$D \times 10^{-6}$ / cm sec^{-1}			$i \times 10^{3}$/ μA			K^{0}_{fh}/ cm sec^{-1}			ΔG^* / kJ mol^{-1}		
		I W	II W	I W	II W	III W	I W	II W	I W	II W	III W	I W	II W	III W	I W	II W	III W	I W	II W	III W
Va	2.10	0.06	0.09	0.47	0.43	-	0.476	0.654	7.50	6.10	-	3.33	3.00	-	3.085	1.333	-	11.682	12.083	-
	4.10	0.06	0.09	0.47	0.43	-	0.476	0.654	6.10	4.79	-	3.00	2.66	-	2.579	1.134	-	11.765	12.159	-
	6.10	0.06	0.09	0.45	0.39	-	0.456	0.593	2.68	1.87	-	2.00	1.66	-	2.264	1.060	-	11.828	12.188	-
	8.10	0.06	0.09	0.41	0.36	0.42	0.415	0.547	1.51	0.92	5.85	1.50	1.16	2.94	0.704	1.050	0.032	12.388	12.196	11.401
	10.10	0.06	0.09	0.41	0.36	0.42	0.415	0.547	1.51	0.92	5.85	1.50	1.16	2.94	0.704	1.050	-0.032	12.388	12.196	11.401
Vb	2.10	0.09	0.08	0.52	0.48	-	0.791	0.649	8.82	7.02	-	3.61	3.22	-	3.001	3.758	-	11.694	11.585	-
	4.10	0.09	0.08	0.52	0.48	-	0.791	0.649	7.23	5.61	-	3.27	2.88	-	2.787	3.144	-	11.728	11.673	-
	6.10	0.09	0.08	0.50	0.44	-	0.760	0.595	3.84	2.13	-	2.38	1.77	-	2.660	1.416	-	11.753	12.054	-
	8.10	0.09	0.08	0.46	0.41	.0.43	0.699	0.554	1.98	1.10	6.76	1.72	1.27	3.16	1.055	1.395	3.145	12.192	12.062	11.012
	10.10	0.09	0.08	0.46	0.41	0.43	0.699	0.554	1.98	1.10	6.76	1.72	1.27	3.16	1.055	1.395	3.145	12.192	12.062	11.012
Vc	2.10	0.11	0.09	0.58	0.52	-	1.078	0.791	8.29	7.23	-	3.50	3.27	-	1.898	1.597	-	11.912	11.996	-
	4.10	0.11	0.09	0.58	0.52	-	1.078	0.791	6.76	5.85	-	3.16	2.94	-	1.562	1.366	-	12.008	12.071	-
	6.10	0.11	0.09	0.56	0.48	-	1.041	0.730	3.16	2.28	-	2.16	1.83	-	1.523	1.008	-	12.016	12.217	-
	8.10	0.11	0.09	0.52	0.45	0.47	0.967	0.684	1.87	1.20	6.56	1.66	1.33	3.11	1.281	0.991	1.025	12.100	12.226	11.377
	10.10	0.11	0.09	0.52	0.45	0.47	0.967	0.684	1.87	1.20	6.56	1.66	1.33	3.11	1.281	0.991	1.025	12.100	12.226	11.377
Vd	2.10	0.11	0.09	0.55	0.49	-	1.022	0.745	8.05	6.80	-	3.44	3.16	-	1.549	6.679	-	12.012	11.314	-
	4.10	0.11	0.09	0.55	0.49	-	1.022	0.745	6.56	5.66	-	3.11	2.88	-	1.480	1.115	-	12.033	12.167	-
	6.10	0.11	0.09	0.53	0.45	-	0.985	0.684	3.02	2.14	-	2.11	1.77	-	0.808	0.744	-	12.322	12.343	-
	8.10	0.11	0.09	0.51	0.42	0.49	0.948	0.639	1.76	1.10	7.04	1.61	1.27	3.22	0.706	0.698	0.145	12.385	12.393	12.121
	10.10	0.11	0.09	0.51	0.42	0.49	0.948	0.639	1.76	1.10	7.04	1.61	1.27	3.22	0.706	0.698	0.145	12.385	12.393	12.121
Ve	2.10	0.06	0.08	0.56	0.58	-	0.568	0.784	8.84	7.28	-	3.61	3.27	-	12.462	5.753	-	11.012	11.385	-
	4.10	0.06	0.08	0.56	0.58	-	0.568	0.784	7.53	5.88	-	3.33	2.94	-	3.025	4.511	-	11.690	11.498	-
	6.10	0.06	0.08	0.53	0.52	-	0.537	0.703	3.52	2.14	-	2.27	1.77	-	1.117	3.239	-	12.167	11.657	-
	8.10	0.06	0.08	0.48	0.44	0.41	0.486	0.595	2.14	1.10	6.56	1.77	1.27	3.11	0.982	2.350	0.374	12.229	11.811	11.016
	10.10	0.06	0.08	0.48	0.44	0.41	0.486	0.595	2.14	1.10	6.56	1.77	1.27	3.11	0.982	2.350	0.374	12.229	11.811	11.016
Vf	2.10	0.08	0.09	0.58	0.60	-	0.784	0.912	8.57	7.04	-	3.55	3.22	--	1.687	1.579	-	11.970	11.999	-
	4.10	0.08	0.09	0.58	0.60	-	0.784	0.912	7.04	5.66	-	3.22	2.88	-	1.276	1.325	-	12.104	12.083	-
	6.10	0.08	0.09	0.55	0.54	--	0.743	0.821	3.35	2.14	-	2.22	1.77	-	1.238	1.116	-	12.117	12.167	-
	8.10	0.08	0.09	0.50	0.49	0.43	0.676	0.745	2.01	1.10	6.80	1.72	1.27	3.16	1.146	1.113	0.215	12.555	12.167	11.447
	10.10	0.08	0.09	0.50	0.49	0.43	0.676	0.745	2.01	1.10	6.80	1.72	1.27	3.16	1.146	1.113	0215	12.555	12.167	11.447

Table 2. *Cyclic voltammetric results of {4-[3-methyl-5-oxo-4-(phenyl hydrazono)-4,5-dihydro-pyrazol-1-yl]-phenoxy}-acetic acid (2-oxo-1-piperidine-1-ylmethyl-1,2-dihydro–indol-3-ylidene)- hydrazide (1 mM), Medium: Aqueous dimethylformamide (40% V/V)*

pH	Scan rate, V s^{-1}	Hanging mercury drop electrode								Crown–ether modified carbon paste electrode							
		$-E_{PCI}$ / V	$-E_{PCII}$ / V	$-E_{PCIII}$ / V	$-E_{PC}$ inv, V	i_{PCI} / μA	i_{PCII}/ μA	i_{PCIII}/ μA	i_{PC} inv/ μA	E_{pcI} / V	E_{pcII} / V	E_{pcIII} / V	E_{pa} / V	i_{pcI} / μA	i_{pcII} / μA	E_{pcIII}, / V	i_{pa},/ μA
2.1	0.010	0.54	0.74		0.64	1.3	1.1		1.0	0.44	0.64		0.86	1.9	1.7		1.2
	0.020	0.58	0.78		0.68	1.8	1.6		1.4	0.48	0.68		0.90	2.6	2.3		1.6
	0.050	0.62	0.82		0.72	2.9	2.4		2.2	0.52	0.72		0.94	4.8	3.7		2.6
	0.100	0.66	0.86		0.76	4.1	3.5		3.1	0.56	0.76		0.98	6.0	5.3		3.7
	0.200	0.70	0.90		0.80	5.8	4.9		4.7	0.60	0.80		10.2	8.4	7.5		5.3
	0.300	0.76	0.96		0.86	7.2	6.0		5.4	0.66	0.86		1.08	10.3	9.2		6.5
	0.500	0.82	1.02		0.92	9.1	7.8		7.1	0.72	0.92		1.14	13.4	12.0		8.4
4.1	0.010	0.67	0.88		0.77	0.9	0.8		0.7	0.57	0.78		1.00	1.6	1.5		1.1
	0.020	0.71	0.92		0.81	1.2	1.2		1.0	0.61	0.82		1.04	2.2	2.1		1.5
	0.050	0.75	0.96		0.85	1.9	1.7		1.5	0.65	0.86		1.08	3.5	3.3		2.4
	0.100	0.79	1.00		0.79	2.8	2.5		2.2	0.69	0.90		1.12	5.0	4.7		3.4
	0.200	0.83	1.04		0.83	4.0	3.5		3.3	0.73	0.94		1.16	7.1	6.7		4.9
	0.300	0.89	1.10		0.89	5.0	4.3		3.8	0.79	1.00		1.22	8.7	8.2		6.0
	0.500	0.95	1.16		0.95	6.3	5.6		4.9	0.85	1.06		1.28	11.3	10.6		7.8
6.1	0.010	0.80	1.06		0.90	0.6	0.5		0.4	0.70	0.96		1.18	1.3	1.2		1.0
	0.020	0.84	1.10		0.94	0.8	0.7		0.6	0.74	1.00		1.22	1.8	1.6		1.4
	0.050	0.88	1.14		0.98	1.3	1.1		0.9	0.78	1.04		1.26	2.9	2.6		2.2
	0.100	0.92	1.18		1.02	1.8	1.5		1.3	0.82	1.08		1.30	4.1	3.7		3.1
	0.200	0.96	1.22		1.06	2.6	2.3		1.8	0.86	1.12		1.34	5.8	5.3		4.4
	0.300	1.02	1.26		1.12	3.2	2.7		2.3	0.92	1.18		1.40	7.1	6.5		5.4
	0.500	1.08	1.32		1.16	4.2	3.5		2.8	0.98	1.24		1.46	9.1	8.4		7.1
8.1	0.010	0.90	1.26	1.61		0.5	0.3	0.8		0.80	1.18	1.51		1.0	0.8	1.1	
	0.020	0.94	1.30	1.65		0.7	0.4	1.2		0.84	1.22	1.55		1.4	1.2	1.5	
	0.050	0.98	1.34	1.69		1.1	0.7	1.7		0.88	1.26	1.59		2.2	1.7	2.4	
	0.100	1.02	1.38	1.73		1.5	0.9	2.5		0.92	1.30	1.63		3.1	2.5	3.4	
	0.200	1.06	1.42	1.77		2.3	1.5	3.5		0.96	1.34	1.67		4.4	3.6	4.9	
	0.300	1.12	1.48	1.83		2.7	1.6	4.3		1.02	1.40	1.73		5.4	4.3	6.0	
	0.500	1.18	1.54	1.89		3.5	2.2	5.6		1.08	1.46	1.79		7.1	5.6	7.8	
10.1	0.010	0.90	1.26	1.61		0.5	0.3	0.8		0.80	1.18	1.51		1.0	0.8	1.1	
	0.020	0.94	1.30	1.65		0.7	0.4	1.2		0.84	1.22	1.55		1.4	1.2	1.5	
	0.050	0.98	1.34	1.69		1.1	0.7	1.7		0.88	1.26	1.59		2.2	1.7	2.4	
	0.100	1.02	1.38	1.73		1.5	0.9	2.5		0.92	1.30	1.63		3.1	2.5	3.4	
	0.200	1.06	1.42	1.77		2.3	1.5	3.5		0.96	1.34	1.67		4.4	3.6	4.9	
	0.300	1.12	1.48	1.83		2.7	1.6	4.3		1.02	1.40	1.73		5.4	4.3	6.0	
	0.500	1.18	1.54	1.89		3.5	2.2	5.6		1.08	1.46	1.79		7.1	5.6	7.8	

Table 3. *Cyclic voltammetric results of {4-[3-methyl-5-oxo-4-(4ˡ-methyl phenyl hydrazono)-4,5-dihydro-pyrazol-1-yl]-phenoxy}-acetic acid (2-oxo-1-piperidine-1-ylmethyl-1,2-dihydro–indol-3-ylidene)- hydrazide (1 mM), Medium: Aqueous dimethylformamide (40% V/V)*

pH	Scan rate, V s^{-1}	Hanging mercury drop electrode								Crown–ether modified carbon paste electrode							
		$-E_{PCI}$ / V	$-E_{PCII}$ / V	$-E_{PCIII}$ / V	$-E_{PC}$ inv, V	i_{PCI} / µA	i_{PCII}/ µA	i_{PCIII}/ µA	i_{PC} inv/ µA	E_{pcI} / V	E_{pcII} / V	E_{pcIII} / V	E_{pa} / V	i_{pcI} / µA	i_{pcII} / µA	E_{pcIII}, / V	i_{pa}/ µA
	0.010	0.64	0.77		0.67	1.1	1.0		0.9	0.54	0.67		0.88	1.6	1.2		1.1
	0.020	0.68	0.83		0.73	1.6	1.4		1.2	0.58	0.71		0.92	2.2	1.6		1.5
	0.050	0.72	0.87		0.77	2.4	2.2		1.9	0.62	0.75		0.96	3.5	2.6		2.4
2.1	0.100	0.76	0.91		0.81	3.5	3.1		2.8	0.66	0.79		1.00	5.0	3.7		3.4
	0.200	0.80	0.95		0.85	4.9	4.7		4.0	0.70	0.83		1.04	7.1	5.3		4.9
	0.300	0.86	1.01		0.91	6.0	5.4		5.0	0.76	0.89		1.10	8.7	6.5		6.0
	0.500	0.92	1.07		0.97	7.8	7.1		6.3	0.82	0.95		1.16	11.3	8.4		7.8
	0.010	0.54	1.02		0.92	1.0	0.9		0.7	0.88	0.92		1.13	1.3	1.0		1.0
	0.020	0.58	1.06		0.96	1.4	1.2		1.0	0.92	0.96		1.17	1.8	1.4		1.4
	0.050	0.62	1.10		1.00	2.2	1.9		1.5	0.96	1.00		1.21	2.9	2.2		2.2
4.1	0.100	0.66	1.14		1.04	3.1	2.8		2.2	1.0	1.04		1.25	4.1	3.1		3.1
	0.200	0.70	1.18		1.08	4.7	4.0		3.3	1.04	1.08		1.29	5.8	4.4		4.4
	0.300	0.76	1.24		1.14	5.4	5.0		3.8	1.10	1.14		1.35	7.1	5.4		5.4
	0.500	0.82	1.30		1.20	7.1	6.3		4.9	1.16	1.20		1.41	9.1	7.1		7.1
	0.010	1.11	1.10		1.00	0.8	0.6		0.5	1.01	1.00		1.21	1.1	0.8		0.6
	0.020	1.15	1.14		1.04	1.2	0.8		0.7	1.05	1.04		1.25	1.5	1.2		0.9
	0.050	1.19	1.18		1.08	1.7	1.3		1.1	1.09	1.08		1.29	2.4	1.7		1.3
6.1	0.100	1.23	1.22		1.12	2.5	1.8		1.5	1.13	1.12		1.33	3.4	2.5		1.9
	0.200	1.27	1.26		1.16	3.5	2.6		2.3	1.17	1.16		1.37	4.9	3.6		2.7
	0.300	1.33	1.32		1.22	4.3	3.2		2.3	1.23	1.22		1.43	6.0	4.3		3.2
	0.500	1.39	1.38		1.28	5.6	4.2		3.5	1.29	1.28		1.49	7.8	5.6		4.3
	0.010	1.21	1.28	1.58		0.6	0.4	1.0		1.11	1.18	1.48		0.9	0.6	1.0	
	0.020	1.25	1.32	1.62		0.8	0.6	1.4		1.15	1.22	1.52		1.2	0.9	1.4	
	0.050	1.29	1.36	1.66		1.3	0.9	2.2		1.19	1.26	1.56		1.9	1.3	2.2	
8.1	0.100	1.33	1.40	1.70		1.8	1.3	3.1		1.23	1.30	1.60		2.8	1.9	3.1	
	0.200	1.37	1.44	1.74		2.6	1.8	4.7		1.27	1.34	1.64		4.1	2.7	4.4	
	0.300	1.43	1.50	1.80		3.2	2.3	5.4		1.33	1.40	1.70		5.0	3.2	5.4	
	0.500	1.49	1.56	1.86		4.2	2.8	7.1		1.39	1.46	1.76		6.3	4.3	7.1	
	0.010	1.21	1.28	1.58		0.6	0.4	1.0		1.11	1.18	1.48		0.9	0.6	1.0	
	0.020	1.25	1.32	1.62		0.8	0.6	1.4		1.15	1.22	1.52		1.2	0.9	1.4	
	0.050	1.29	1.36	1.66		1.3	0.9	2.2		1.19	1.26	1.56		1.9	1.3	2.2	
10.1	0.100	1.33	1.40	1.70		1.8	1.3	3.1		1.23	1.30	1.60		2.8	1.9	3.1	
	0.200	1.37	1.44	1.74		2.6	1.8	4.7		1.27	1.34	1.64		4.1	2.7	4.4	
	0.300	1.43	1.50	1.80		3.2	2.3	5.4		1.33	1.40	1.70		5.0	3.2	5.4	
	0.500	1.49	1.56	1.86		4.2	2.8	7.1		1.39	1.46	1.76		6.3	4.3	7.1	

Table 4. *Cyclic voltammetric results of {4-[3-methyl-5-oxo-4-(4'-methoxy phenyl hydrazono)-4,5-dihydro-pyrazol-1-yl]-phenoxy}-acetic acid (2-oxo-1-piperidine-1-ylmethyl-1,2-dihydro–indol-3-ylidene)- hydrazide (1 mM), Medium: Aqueous dimethylformamide (40% V/V)*

pH	Scan rate, V s^{-1}	Hanging mercury drop electrode								Crown–ether modified carbon paste electrode							
		$-E_{PCI}$ / V	$-E_{PCII}$ / V	$-E_{PCIII}$ / V	$-E_{PC}$ inv, V	i_{PCI} / μA	i_{PCII}/ μA	i_{PCIII}/ μA	i_{PC} inv/ μA	E_{pcI} / V	E_{pcII} / V	E_{pcIII} / V	E_{pa} / V	i_{pcI} / μA	i_{pcII} / μA	E_{pcIII}, / V	i_{pa},/ μA
2.1	0.010	0.73	0.78		0.68	1.1	1.0		0.8	0.63	0.68		0.89	2.1	2.3		1.9
	0.020	0.77	0.82		0.72	1.6	1.4		1.2	0.67	0.72		0.93	2.9	3.2		2.6
	0.050	0.81	0.86		0.76	2.4	2.2		1.7	0.71	0.76		0.97	4.6	5.0		4.8
	0.100	0.85	0.90		0.80	3.5	3.1		2.5	0.75	0.80		1.01	6.6	7.2		6.0
	0.200	0.89	0.94		0.84	4.9	4.7		3.6	0.79	0.84		1.05	9.3	10.2		8.4
	0.300	0.95	1.00		0.90	6.0	5.4		4.3	0.85	0.90		1.11	11.4	12.5		10.3
	0.500	1.01	1.06		0.96	7.8	7.1		5.6	0.91	0.96		1.17	14.8	16.2		13.4
4.1	0.010	0.98	1.05		0.90	1.0	0.9		0.7	0.90	0.94		1.15	2.0	2.1		1.7
	0.020	1.02	1.09		0.94	1.4	1.3		0.9	0.94	0.98		1.19	2.8	2.9		2.3
	0.050	1.06	1.13		0.98	2.2	1.9		1.5	0.98	1.02		1.23	4.4	4.6		3.7
	0.100	1.10	1.17		1.02	3.1	2.8		2.3	1.02	1.06		1.27	6.3	6.6		5.3
	0.200	1.14	1.21		1.06	4.7	4.1		3.2	1.06	1.10		1.31	8.9	9.3		7.5
	0.300	1.20	1.27		1.12	5.4	4.9		3.9	1.12	1.16		1.37	10.9	11.4		9.2
	0.500	1.26	1.33		1.18	7.1	6.4		4.9	1.18	1.22		1.43	14.1	14.8		12.0
6.1	0.010	1.26	1.18		1.08	0.7	0.6		0.6	1.08	1.16		1.37	1.7	1.8		1.5
	0.020	1.30	1.22		1.12	0.9	0.9		0.9	1.12	1.20		1.41	2.1	2.5		2.1
	0.050	1.34	1.26		1.16	1.5	1.3		1.3	1.16	1.24		1.45	3.3	3.9		3.3
	0.100	1.38	1.30		1.20	2.3	1.9		1.9	1.20	1.28		1.49	4.7	5.6		4.7
	0.200	1.42	1.34		1.24	3.2	2.7		2.7	1.24	1.32		1.53	6.7	8.0		6.7
	0.300	1.48	1.40		1.30	3.9	3.2		3.2	1.30	1.38		1.59	8.2	9.8		8.2
	0.500	1.54	1.46		1.36	4.9	4.3		4.3	1.36	1.44		1.65	10.6	12.7		10.6
8.1	0.010	1.43	1.33	1.64		0.5	0.4	0.9		1.23	1.33	1.54		1.2	1.4	1.6	
	0.020	1.47	1.37	1.68		0.7	0.6	1.3		1.27	1.37	1.58		1.6	1.9	2.2	
	0.050	1.51	1.41	1.72		1.1	0.9	1.9		1.31	1.41	1.62		2.6	3.0	3.5	
	0.100	1.55	1.45	1.76		1.5	1.3	2.8		1.35	1.45	1.66		3.7	4.4	5.0	
	0.200	1.39	1.49	1.80		2.3	1.8	4.1		1.39	1.49	1.70		5.3	6.2	7.1	
	0.300	1.65	1.55	1.86		2.8	2.3	4.9		1.45	1.55	1.76		6.6	7.6	8.7	
	0.500	1.71	1.61	1.92		3.5	2.8	6.4		1.51	1.61	1.82		8.4	9.8	11.3	
10.1	0.010	1.43	1.33	1.64		0.5	0.4	0.9		1.23	1.33	1.54		1.2	1.4	1.6	
	0.020	1.47	1.37	1.68		0.7	0.6	1.3		1.27	1.37	1.58		1.6	1.9	2.2	
	0.050	1.51	1.41	1.72		1.1	0.9	1.9		1.31	1.41	1.62		2.6	3.0	3.5	
	0.100	1.55	1.45	1.76		1.5	1.3	2.8		1.35	1.45	1.66		3.7	4.4	5.0	
	0.200	1.59	1.49	1.80		2.3	1.8	4.1		1.39	1.49	1.70		5.3	6.2	7.1	
	0.300	1.65	1.55	1.86		2.8	2.3	4.9		1.45	1.55	1.76		6.6	7.6	8.7	
	0.500	1.71	1.61	1.92		3.5	2.8	6.4		1.51	1.61	1.82		8.4	9.8	11.3	

Table 5. *Cyclic voltammetric results of {4-[3-methyl-5-oxo-4-(4'-ethoxy phenyl hydrazono)-4,5-dihydro-pyrazol-1-yl]-phenoxy}-acetic acid (2-oxo-1-piperidine-1-ylmethyl-1,2-dihydro–indol-3-ylidene)- hydrazide (1 mM), Medium: Aqueous dimethylformamide (40 % V/V)*

pH	Scan rate, V s^{-1}	Hanging mercury drop electrode								Crown–ether modified carbon paste electrode							
		$-E_{PCI}$ / V	$-E_{PCII}$ / V	$-E_{PCIII}$ / V	$-E_{PC}$ inv, V	i_{PCI} / μA	i_{PCII}/ μA	i_{PCIII}/ μA	i_{PC} inv/ μA	E_{pcI} / V	E_{pcII} / V	E_{pcIII} / V	E_{pa} / V	i_{pcI} / μA	i_{pcII} / μA	$E_{pcIII,}$ / V	$i_{pa,}$/ μA
2.1	0.010	0.76	0.73		0.63	1.2	1.0		0.9	0.75	0.78		0.99	1.9	1.7		1.5
	0.020	0.80	0.77		0.67	1.7	1.4		1.3	0.79	0.82		1.03	2.6	2.3		2.1
	0.050	0.84	0.81		0.71	2.6	2.2		1.9	0.83	0.86		1.07	4.8	3.7		3.3
	0.100	0.88	0.85		0.75	3.7	3.1		2.8	0.87	0.90		1.11	6.0	5.3		4.7
	0.200	0.92	0.89		0.79	5.3	4.4		4.1	0.91	0.94		1.15	8.4	7.5		6.7
	0.300	1.06	0.95		0.85	6.6	5.4		5.0	0.97	1.00		1.21	10.3	9.2		8.2
	0.500	1.12	1.01		0.91	8.5	7.1		6.4	1.03	1.06		1.27	13.4	12.0		10.6
4.1	0.010	1.08	0.94		0.84	1.0	0.9		0.7	0.96	1.1		1.31	1.6	1.4		1.3
	0.020	1.12	0.98		0.88	1.4	1.3		1.0	1.00	1.14		1.35	2.2	1.9		1.8
	0.050	1.16	1.02		0.92	2.2	1.9		1.5	1.04	1.18		1.39	3.5	3.0		2.9
	0.100	1.20	1.06		0.96	3.1	2.8		2.2	1.08	1.22		1.43	5.0	4.4		4.1
	0.200	1.24	1.10		1.00	4.4	4.1		3.3	1.12	1.26		1.47	7.1	6.2		5.8
	0.300	1.30	1.16		1.06	5.4	5.0		3.8	1.18	1.32		1.53	8.7	7.6		7.1
	0.500	1.36	1.22		1.12	7.1	6.4		4.9	1.24	1.38		1.59	11.3	9.8		9.1
6.1	0.010	1.23	1.12		1.02	0.8	0.7		0.6	1.14	1.25		1.46	1.2	1.1		1.0
	0.020	1.27	1.16		1.06	1.2	1.0		0.9	1.18	1.29		1.50	1.6	1.6		1.4
	0.050	1.31	1.20		1.10	1.8	1.5		1.3	1.22	1.33		1.54	2.6	2.4		2.2
	0.100	1.35	1.24		1.14	2.5	2.2		1.9	1.26	1.37		1.58	3.7	3.4		3.1
	0.200	1.39	1.28		1.18	3.6	3.3		2.7	1.30	1.41		1.62	5.3	4.9		4.4
	0.300	1.45	1.34		1.24	4.3	3.8		3.2	1.36	1.47		1.68	6.6	6.0		5.4
	0.500	1.51	1.40		1.30	5.6	4.9		4.3	1.42	1.53		1.74	8.4	7.8		7.1
8.1	0.010	1.49	1.30	1.59		0.6	0.5	1.0		1.32	1.51	1.61		0.9	0.7	1.2	
	0.020	1.53	1.34	1.63		0.9	0.7	1.4		1.36	1.55	1.65		1.2	1.0	1.6	
	0.050	1.57	1.38	1.67		1.3	1.1	2.2		1.40	1.59	1.69		1.9	1.5	2.6	
	0.100	1.61	1.42	1.71		1.9	1.5	3.1		1.44	1.63	1.73		2.8	2.2	3.7	
	0.200	1.65	1.46	1.75		2.7	2.3	4.4		1.48	1.67	1.77		4.0	3.3	5.3	
	0.300	1.71	1.52	1.81		3.2	2.8	5.4		1.54	1.73	1.83		4.9	3.8	6.6	
	0.500	1.77	1.58	1.87		4.3	3.5	7.1		1.60	1.79	1.89		6.3	4.9	8.4	
10.1	0.010	1.49	1.30	1.59		0.6	0.5	1.0		1.32	1.51	1.61		0.9	0.7	1.2	
	0.020	1.53	1.34	1.63		0.9	0.7	1.4		1.36	1.55	1.65		1.2	1.0	1.6	
	0.050	1.57	1.38	1.67		1.3	1.1	2.2		1.40	1.59	1.69		1.9	1.5	2.6	
	0.100	1.61	1.42	1.71		1.9	1.5	3.1		1.44	1.63	1.73		2.8	2.2	3.7	
	0.200	1.65	1.46	1.75		2.7	2.3	4.4		1.48	1.67	1.77		4.0	3.3	5.3	
	0.300	1.71	1.52	1.81		3.2	2.8	5.4		1.54	1.73	1.83		4.9	3.8	6.6	
	0.500	1.77	1.58	1.87		4.3	3.5	7.1		1.60	1.79	1.89		6.3	4.9	8.4	

Table 6. *Cyclic voltammetric results of {4-[3-methyl-5-oxo-4-(4^{l}-chloro phenyl hydrazono)-4,5-dihydro-pyrazol-1-yl]-phenoxy}-acetic acid (2-oxo-1-piperidine-1-ylmethyl-1,2-dihydro–indol-3-ylidene)- hydrazide (1 mM), Medium : Aqueous dimethylformamide (40% V/V)*

pH	Scan rate, V s^{-1}	Hanging mercury drop electrode								Crown–ether modified carbon paste electrode							
		$-E_{PCI}$ / V	$-E_{PCII}$ / V	$-E_{PCIII}$ / V	$-E_{PC}$ inv, V	i_{PCI} / µA	i_{PCII}/ µA	i_{PCIII}/ µA	i_{PC} inv/ µA	E_{pcI} / V	E_{pcII} / V	E_{pcIII} / V	E_{pa} / V	i_{pcI} / µA	i_{pcII} / µA	$E_{pcIII,}$ / V	$i_{pa,}$/ µA
2.1	0.010	0.34	0.78		0.68	1.3	1.1		1.0	0.24	0.70		0.91	1.7	1.4		1.2
	0.020	0.38	0.82		0.72	1.8	1.6		1.4	0.28	0.74		0.95	2.3	1.9		1.6
	0.050	0.42	0.86		0.76	2.9	2.4		2.2	0.32	0.78		0.99	3.7	3.0		2.6
	0.100	0.46	0.90		0.80	4.1	3.4		3.1	0.36	0.82		1.03	5.3	4.4		3.7
	0.200	0.50	0.94		0.84	5.8	4.9		4.4	0.40	0.86		1.07	7.5	6.2		5.3
	0.300	0.56	1.00		0.90	7.2	6.0		5.4	0.46	0.92		1.13	9.2	7.6		6.6
	0.500	0.62	1.06		0.96	9.1	7.8		7.1	0.52	0.98		1.19	12.0	9.8		8.4
4.1	0.010	0.51	0.99		0.89	1.1	0.9		0.8	0.40	0.89		1.1	1.5	1.2		1.0
	0.020	0.55	1.03		0.93	1.6	1.2		1.2	0.44	0.93		1.14	2.1	1.6		1.4
	0.050	0.59	1.07		0.97	2.4	1.9		1.7	0.48	0.97		1.18	3.3	2.6		2.2
	0.100	0.63	1.11		1.01	3.4	2.8		2.5	0.52	1.01		1.22	4.7	3.7		3.1
	0.200	0.67	1.15		1.05	4.9	4.1		3.6	0.56	1.05		1.26	6.7	5.3		4.4
	0.300	0.73	1.21		1.11	6.0	5.0		4.3	0.62	1.11		1.32	8.2	6.6		5.4
	0.500	0.79	1.27		1.17	7.8	6.3		5.6	0.68	1.17		1.38	10.6	8.4		7.1
6.1	0.010	0.66	1.17		1.07	0.9	0.7		0.6	0.56	1.07		1.28	1.2	1.0		0.7
	0.020	0.70	1.21		1.11	1.2	1.0		0.9	0.60	1.11		1.32	1.6	1.4		1.0
	0.050	0.74	1.25		1.15	1.9	1.5		1.3	0.64	1.15		1.36	2.6	2.2		1.5
	0.100	0.78	1.29		1.19	2.8	2.2		1.9	0.68	1.19		1.40	3.7	3.1		2.2
	0.200	0.82	1.33		1.23	4.1	3.3		2.7	0.72	1.23		1.44	5.3	4.4		3.3
	0.300	0.88	1.39		1.29	5.0	3.8		3.2	0.78	1.29		1.50	6.6	5.4		3.8
	0.500	0.94	1.45		1.35	6.3	4.9		4.3	0.84	1.35		1.56	8.4	7.1		4.9
8.1	0.010	0.81	1.32	1.63		0.6	0.5	1.0		0.71	1.23	1.53		0.8	0.6	1.0	
	0.020	0.85	1.36	1.67		0.9	0.7	1.4		0.75	1.27	1.57		1.2	0.9	1.4	
	0.050	0.89	1.40	1.71		1.3	1.1	2.2		0.79	1.31	1.61		1.7	1.3	2.2	
	0.100	0.93	1.44	1.75		1.9	1.5	3.1		0.83	1.35	1.65		2.7	1.9	3.1	
	0.200	0.97	1.48	1.79		2.7	2.3	4.4		0.87	1.39	1.69		3.6	2.7	4.4	
	0.300	1.03	1.54	1.85		3.2	2.8	5.4		0.93	1.45	1.75		4.3	3.2	5.4	
	0.500	1.09	1.60	1.91		4.3	3.5	7.1		0.99	1.51	1.81		5.6	4.3	7.1	
10.1	0.010	0.81	1.32	1.63		0.6	0.5	1.0		0.71	1.23	1.53		0.8	0.6	1.0	
	0.020	0.85	1.36	1.67		0.9	0.7	1.4		0.75	1.27	1.57		1.2	0.9	1.4	
	0.050	0.89	1.40	1.71		1.3	1.1	2.2		0.79	1.31	1.61		1.7	1.3	2.2	
	0.100	0.93	1.44	1.75		1.9	1.5	3.1		0.83	1.35	1.65		2.7	1.9	3.1	
	0.200	0.97	1.48	1.79		2.7	2.3	4.4		0.87	1.39	1.69		3.6	2.7	4.4	
	0.300	1.03	1.54	1.85		3.2	2.8	5.4		0.93	1.45	1.75		4.3	3.2	5.4	
	0.500	1.09	1.60	1.91		4.3	3.5	7.1		0.99	1.51	1.81		5.6	4.3	7.1	

Table 7. *Cyclic voltammetric results of {4-[3-methyl-5-oxo-4-(4ˡ-bromo phenyl hydrazono)-4,5-dihydro-pyrazol-1-yl]-phenoxy}-acetic acid (2-oxo-1-piperidine-1-ylmethyl-1,2-dihydro–indol-3-ylidene)- hydrazide (1 mM), Medium: Aqueous dimethylformamide (40% V/V)*

pH	Scan rate, V s^{-1}	Hanging mercury drop electrode								Crown–ether modified carbon paste electrode							
		$-E_{PCI}$ / V	$-E_{PCII}$ / V	$-E_{PCIII}$ / V	$-E_{PC}$ inv, V	i_{PCI} / µA	i_{PCII}/ µA	i_{PCIII}/ µA	i_{PC} inv/ µA	E_{pcI} / V	E_{pcII} / V	E_{pcIII} / V	E_{pa} / V	i_{pcI} / µA	i_{pcII} / µA	$E_{pcIII,}$ / V	$i_{pa,}$/ µA
	0.010	0.43	0.82		0.72	1.6	1.4		1.1	0.33	0.72		0.93	2.0	1.8		1.4
	0.020	0.47	0.86		0.76	2.2	2.0		1.6	0.37	0.76		0.97	2.8	2.5		1.9
	0.050	0.51	0.90		0.80	3.5	3.0		2.4	0.41	0.80		1.01	4.4	3.9		3.0
2.1	0.100	0.55	0.94		0.84	5.1	4.4		3.4	0.45	0.84		1.05	6.3	5.6		4.4
	0.200	0.59	0.98		0.88	7.2	6.2		4.9	0.49	0.88		1.09	8.9	8.0		6.2
	0.300	0.65	1.04		0.94	8.7	7.6		6.0	0.55	0.94		1.15	10.9	9.8		7.6
	0.500	0.71	1.10		1.00	11.3	9.9		7.8	0.61	1.00		1.21	14.1	12.7		9.8
	0.010	0.54	1.02		0.92	1.3	1.1		0.9	0.44	0.92		1.13	1.7	1.5		1.1
	0.020	0.58	1.06		0.96	1.8	1.6		1.2	0.48	0.96		1.17	2.3	2.1		1.6
	0.050	0.62	1.10		1.00	2.9	2.4		1.9	0.52	1.00		1.21	3.7	3.3		2.4
4.1	0.100	0.66	1.14		1.04	4.1	3.4		2.8	0.56	1.04		1.25	5.3	4.7		3.4
	0.200	0.70	1.18		1.08	5.8	4.9		4.1	0.60	1.08		1.29	7.5	6.7		4.9
	0.300	0.76	1.24		1.14	7.2	6.0		5.0	0.66	1.14		1.35	9.2	8.2		6.0
	0.500	0.82	1.30		1.20	9.1	7.8		6.3	0.72	1.20		1.41	12.0	10.6		7.8
	0.010	0.75	1.19		1.09	1.1	0.9		0.7	0.65	1.09		1.30	1.2	1.1		0.9
	0.020	0.79	1.23		1.13	1.6	1.2		1.0	0.69	1.13		1.34	1.6	1.5		1.2
	0.050	0.83	1.27		1.17	2.4	1.9		1.5	0.73	1.17		1.38	2.6	2.4		1.9
6.1	0.100	0.87	1.31		1.21	3.4	2.8		2.2	0.77	1.21		1.42	3.7	3.4		2.8
	0.200	0.91	1.35		1.25	4.9	4.1		3.3	0.81	1.25		1.46	5.3	4.9		4.0
	0.300	0.97	1.41		1.31	6.0	5.0		3.8	0.87	1.31		1.52	6.6	6.0		4.9
	0.500	1.03	1.47		1.37	7.8	6.3		4.9	0.93	1.37		1.58	8.4	7.8		6.3
	0.010	0.90	1.38	1.64		0.8	0.6	1.0		0.80	1.28	1.54		1.0	0.6	0.9	
	0.020	0.94	1.42	1.68		1.2	0.9	1.4		0.84	1.32	1.58		1.4	0.9	1.2	
	0.050	0.98	1.46	1.72		1.7	1.3	2.2		0.88	1.36	1.62		2.2	1.3	1.9	
8.1	0.100	1.02	1.50	1.76		2.5	1.9	3.1		0.92	1.40	1.66		3.1	1.9	2.8	
	0.200	1.06	1.54	1.80		3.6	2.7	4.4		0.96	1.44	1.70		4.4	2.7	4.0	
	0.300	1.12	1.60	1.86		4.3	3.2	5.4		1.02	1.50	1.76		5.4	3.2	4.9	
	0.500	1.18	1.66	1.92		5.6	4.3	7.1		1.08	1.56	1.82		7.1	4.3	6.3	
	0.010	0.90	1.38	1.64		0.8	0.6	1.0		0.80	1.28	1.54		1.0	0.6	0.9	
	0.020	0.94	1.42	1.68		1.2	0.9	1.4		0.84	1.32	1.58		1.4	0.9	1.2	
	0.050	0.98	1.46	1.72		1.7	1.3	2.2		0.88	1.36	1.62		2.2	1.3	1.9	
10.1	0.100	1.02	1.50	1.76		2.5	1.9	3.1		0.92	1.40	1.66		3.1	1.9	2.8	
	0.200	1.06	1.54	1.80		3.6	2.7	4.4		0.96	1.44	1.70		4.4	2.7	4.0	
	0.300	1.12	1.60	1.86		4.3	3.2	5.4		1.02	1.50	1.76		5.4	3.2	4.9	
	0.500	1.18	1.66	1.92		5.6	4.3	7.1		1.08	1.56	1.82		7.1	4.3	6.3	

10

Binder-less activated carbon electrode from gelam wood for use in supercapacitors

NIRWAN SYARIF*,**✉, IVANDINI A. TRIBIDASARI* and WIDAYANTI WIBOWO*

*Department of Chemistry, Faculty of Mathematics and Natural Sciences, University of Indonesia, Indonesia

**Department of Chemistry, Faculty of Mathematics and Natural Sciences, Sriwijaya University, Indonesia

✉ Corresponding Author

Abstract

This work focused on the relation between the porous structure of activated carbon and its capacitive properties. Three types of activated carbon monoliths were used as the electrodes in a half cell electrochemical system. One monolith was produced from activated carbon and considered to be a binder-less electrode. Two others were produced from acid and high pressure steam oxidized activated carbon. The micrographs clearly indicate that three electrodes have different porous structures. Both porosity and surface area of carbons increased due to the formation of grains during oxidation. This fact specified that an acid oxidized carbon monolith will have relatively higher capacitance compared to non-oxidized and steam oxidized monoliths. Maximum capacitance values for acid, steam oxidized and non-oxidized electrodes were 27.68, 2.23 and 1.20 F g^{-1}, respectively.

Keywords

pyrolysis; voltammogram; surface oxidation; capacitance; micropore.

Introduction

Supercapacitors have become the focus attention of many researchers for decades due to their potential application as energy storage devices in relation to their high energy density, great power density and long cycle life. Much effort has been done to develop advanced carbons to meet the demand for high performance supercapacitors [1]. Fossil- and organic-based carbonaceous materials are amongst the most widely used starting or precursor materials for supercapacitor electrodes because they are relatively inexpensive and easy to fabricate. These electrode materials also have large specific surface areas, good electrical conductivity and excellent chemical stability. The carbonaceous materials extracted from biomass such as fruit peels, beans, plant shells, bamboo, woods, natural fibers and bagasse have been studied for supercapacitor carbon electrodes.

The fabrication of carbon electrodes for supercapacitors, normally requires an addition of a polymer matrix in order to bind together the carbon particles into monolithic (pellet) form. This addition blocks a part of the carbon pores [2] and additionally causes an increase in the electrical resistivity [3]. Furthermore, the use of polymer binder makes the product more expensive and environmentally less friendly. Therefore, the carbon electrode without the use of any binding agent is highly recommended in order to be potentially used in supercapacitor applications, because they might have lower resistance in the absence of a polymeric binder and they are also of lower cost [4].

In most cases, the physicochemical properties of the carbon electrode, such as its structure, electrical conductivity and porosity, was improved by an activation process. The physical activation is normally carried out by treatment of the carbonized sample using gas atmosphere (such as CO_2, N_2 and water steam) under a moderately higher temperature of 800 – 1000 °C to improve the internal porosity [5].

In this paper, we report the preparation and characterization of free binder activated carbon monolith electrodes prepared from Gelam wood, which were activated by physical process under N_2/CO_2 gas and followed by surface oxidation using nitric acid and high pressure steam. It should be emphasized that the use of this activated carbon for supercapacitor was not yet reported.

Experimental

Activated carbon and electrode preparation

Activated carbons were prepared from the gelam wood piles obtained from local timber industry. Non-bark wood piles sizes 10 cm length and 5 cm diameter was used in the production of activated carbon rod. Pyrolysis was conducted after preheating process and followed by second pyrolysis for activation process. Wood was preheated in the oven at 230 – 260 °C for a half hour. Wood were carbonized by heating from 40 to 700 °C at rate of 3 °C per minute under atmosphere of nitrogen flow and controlled for ½ - 1 hour at peak temperature. Nitrogen was flowed to the reactor at rate of 1 mL min^{-1}. The heating procedure was repeated but using CO_2 as the atmosphere to produce activated carbon. This activated carbon was labeled as CCP.

CCP was cut into pellets of 5 mm in thickness by using a mini-circular electric knife. CCP was subjected for surface oxidation using concentrated nitric acid and high pressure steam, both labeled as $CCPHNO_3$ and CCPWOT respectively. Activated carbons were oxidized in reflux system for 3 hours using concentrated nitric acid at 60 °C. The high pressure steam oxidation was conducted in the hydrothermal reactor for 16 hours at 200 °C. The products were washed with abundant de-mineralized water and dried. These three types of activated carbon pallets were used as electrodes.

Surface characterization

The characterization of activated carbons was carried out through physical adsorption of nitrogen gas at 77 K, using a conventional volumetric system (Micromeritics ASAP 2400 instrument). The surface area based on nitrogen isotherms was calculated using the BET equation, and using the "t" method. Scanning Electron Microscopy (SEM) and Energy Dispersive X-ray analysis (EDX) were used to observe the structural morphology and composition of activated carbon, respectively. Hitachi S-4000 field emission SEM was used for this analysis, operated at 30 keV. Uncoated samples were subjected for SEM observation.

The Boehm titration method was applied to determine the acid and basic sites in the activated carbons [6]. The total acid sites were neutralized with 0.1 M NaOH solutions, and the basic sites were neutralized with a 0.1 M HCl solution. The reaction between reagents and acidic oxygenated-functional groups on the surface is based on the difference strength of the acid/base. The strength order of acidic and basic groups is as follow: carboxyl > lactone > phenol. The carboxylic and lactonic sites were titrated with a 0.1 MNa_2CO_3 solution, the carboxylic sites were titrated with a 0.1 M $NaHCO_3$ solution and the phenolic sites were determined by the difference. For each determination, 50 mL of the solution was added to one gram of activated carbon in a glass container. The container was inserted into vortex shaker for over one day. A 10 mL sample was titrated with 0.1 M of HCl or NaOH solutions.

Electrochemical measurement

Electrochemical tests of the electrodes were conducted using the cyclic voltammetry technique using a three electrode configuration on a potentiostat instrument (eDAQ Pty Ltd, model EA161). Ag/AgCl electrode was used as reference electrode and a platinum rod was used as the counter electrode. Cyclic voltammetry tests were conducted at scan rates of 2, 5, 10, 20, 50, and 100 mV s^{-1} with potential windows ranging from -1 to 1 V in 1, 2, and 4 mol L^{-1} KOH or H_2SO_4 aqueous solution as electrolytes. Considering that the anodic voltammetric charges and cathodic voltammetric charges are not the same in the shape of voltammogram curve, the average specific capacitance of the electrode was calculated using integral area approximation

$$C_{avg} = \frac{\Delta Q}{w \Delta V} = \frac{\int I dV}{s \Delta V w} \tag{1}$$

Where ΔQ is the total amount of the charge accumulated over a potential window, ΔV, w is the mass of active material in one electrode, I is the current, and s is the potential scan rate. Hence $\int I\, dV$ is integral area of voltammogram curve.

Results and Discussion

Pore structures and chemical characteristics

Three carbon electrodes, namely CCP, CCPHNO3 and CCPWOT, were prepared from Gelam wood activated carbon. The pores and grains were formed during preparation affecting the properties of electrodes. Pores act as the entrance of small molecules penetrated deeper into the micro pores [7]. Penetrating water steam and gases made some cracks and led to changed porous structures [8].

The porous structure of carbon can be characterized by the adsorption isotherms and pore size distribution plot [9]. The adsorption isotherms plot for the three electrodes (Figure 1A) are type I represented by a flat curve, indicating micropores. The broadening of the knee in the very low relative pressures range in CCPHNO3 indicates that the microporosity of the carbon became wider upon acid oxidation treatment (Figure 1A (b)). Acid oxidation burst opening pore in carbon surface and increasing pore volume two times larger. In the other hand, some micro pores vanished (Figure 1A (c)) and pore size distribution became narrow (Figure 1B (c)) upon hydrothermal treatment due to fracture of the bulky network and the pore walls [10]. This fracturing blocked some pores and affected the pore structures and surface functional groups of carbon.

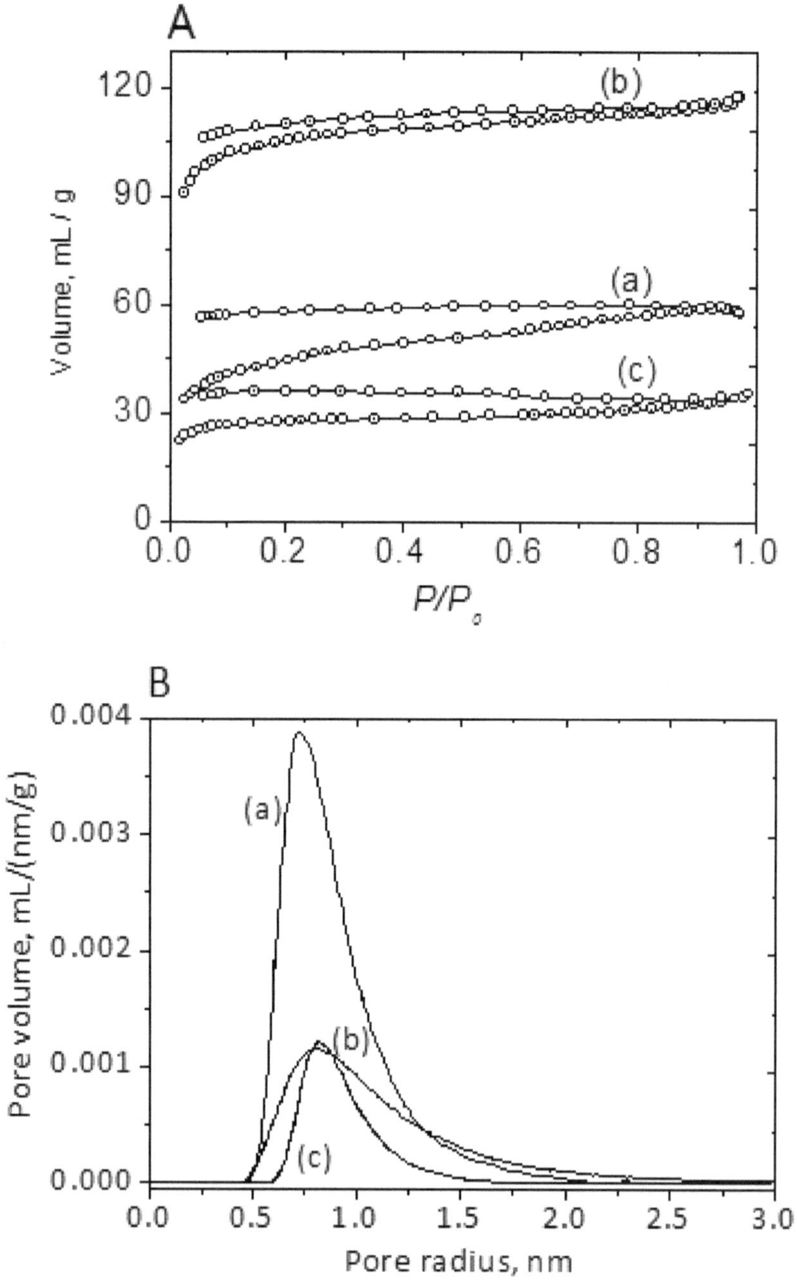

Figure 1. *(A) Nitrogen adsorption-desorption on the surface of electrodes (B) Pore size distribution for **(a)** CCP, **(b)** CCPHNO3 and **(c)** CCPWOT.*

Table 1 shows the changes in surface area, pore volume, and surface functional groups as a result of oxidation treatments. CCPHNO3 had the highest BET surface (337.1 $m^2 g^{-1}$), and a total pore volume measured at P/P_0 = 0.95 of 0.185 $cm^3 g^{-1}$, most of the pores being micropores (81.50 %), with an average pore width of 1.1 nm. On the other hand, CCPWOT exhibited the lowest BET surface (88.800 $m^2 g^{-1}$), with a much lower total pores volume (0.014 $cm^3 g^{-1}$) and a narrower pore width (1.408 nm). A pore size distribution in the range of micropore in which larger than the size of two solvated ions, was then identified as a way of improving the energy density and the power capability [11], whereas mesopores improved the ionic mass transport inside the bulk of carbon. Therefore, it was assumed that a good balance between micro- and mesoporosity was needed to maximize capacitance [12]. Some capacitance values from the literature are provided to clarify the relationship between pore structure and capacitance. It can also be seen from the Table 1 that the some capacitance values obtained in this study are smaller than those obtained from the literatures.

Table 1. *Surface and electrochemical properties of the CCP, CCPHNO3 and CCPWOT electrodes.*

Electrode	S_{BET} / $m^2 g^{-1}$	V_{total} / $cm^3 g^{-1}$	micro pore content, %	Functional groups concentration, mmol g^{-1}				$C_{sp,max}$
				fenolic	lactonic	carboxylic	base	F g^{-1}
CCP	149.848	0.093	56.249	0.200	3.85	0.00	0.24	1.22
CCPHNO3	337.116	0.185	81.499	0.100	3.08	0.20	0.19	27.68
CCPWOT	88.800	0.014	28.306	0.200	3.40	0.10	0.22	2.23
Carbon from sucrose [13]	2749.0	2.01						233
Carbon from sago waste [14]	1408.2	0.897						64.1
Carbon from coal [14]	179.2	0.073						16.2

S_{BET} – specific surface area; V_{total} - total pore volume, measured at P/P_0=0.995.

Furthermore, the specific surface area of the CCPHNO3 electrode as listed in Table 1 is 337.116 $m^2 g^{-1}$, which is higher than that of the CCP and CCPWOT electrode, *i.e.* 149.85 and 88.80 $m^2 g^{-1}$, respectively. The Boehm analysis reveals that the three electrodes have similar amounts of functional groups, except for the carboxylic and lactone groups. A computational study was undertaken on the effect of the two functional groups on the structure of graphite basal. The study found that the COOH functional groups that fill the vertical plane will bend graphite layer led to reduced effectiveness of electron transfer between layers. These functional groups improved capacitive processes of the carbon [15], and facilitated the penetration of the electrolyte ions into the interior pores. Although the CCPWOT electrode has many more functional groups, however, many more pores were vanished and the surface area was decreased, leading to a decrease in the capacitance of electrode. Furthermore, changes in the carbon pore structure can be clearly seen as surface morphology by using electron microscope.

The micrographs clearly indicate that the three electrodes have different surface morphologies. Figure 2a and 2d show the analysis of the CCP microstructure by SEM. They revealed continuous and smooth surface of the CCP with less grain boundaries wood (Figure 2a and 2d) due to the limited process of opening pore in line with the nature of the wood. The other SEM pictures show the opening pores in the surface of the carbon which occurred when CCP was treated with nitric acid (CCPHNO3), especially the oxidation of low molecular weight carbon. These SEM images of CCPHNO3 show irregular sizes and flaky grains in which the structures tend to form voids between the grains (Figure 2b and e). The number of grains and micropores of CCPHNO3 are higher than those of CCPWOT and CCP, due to the reaction between low molecular weight carbon fragments with nitrogen. The oxidation reaction generated volatile molecules that leave the pores behind after volatilization. Furthermore, SEM images of CCPWOT show more regular size and relatively uniform (Figure 2c and f). This porous structure, *i.e.* pore volume, surface area and pore size distribution along with functional groups on the surface of carbon are regarded as the important factors which determined the properties of capacitive processes, *i.e.* ionic accessibility, electrical double layer and redox reactions [16-18].

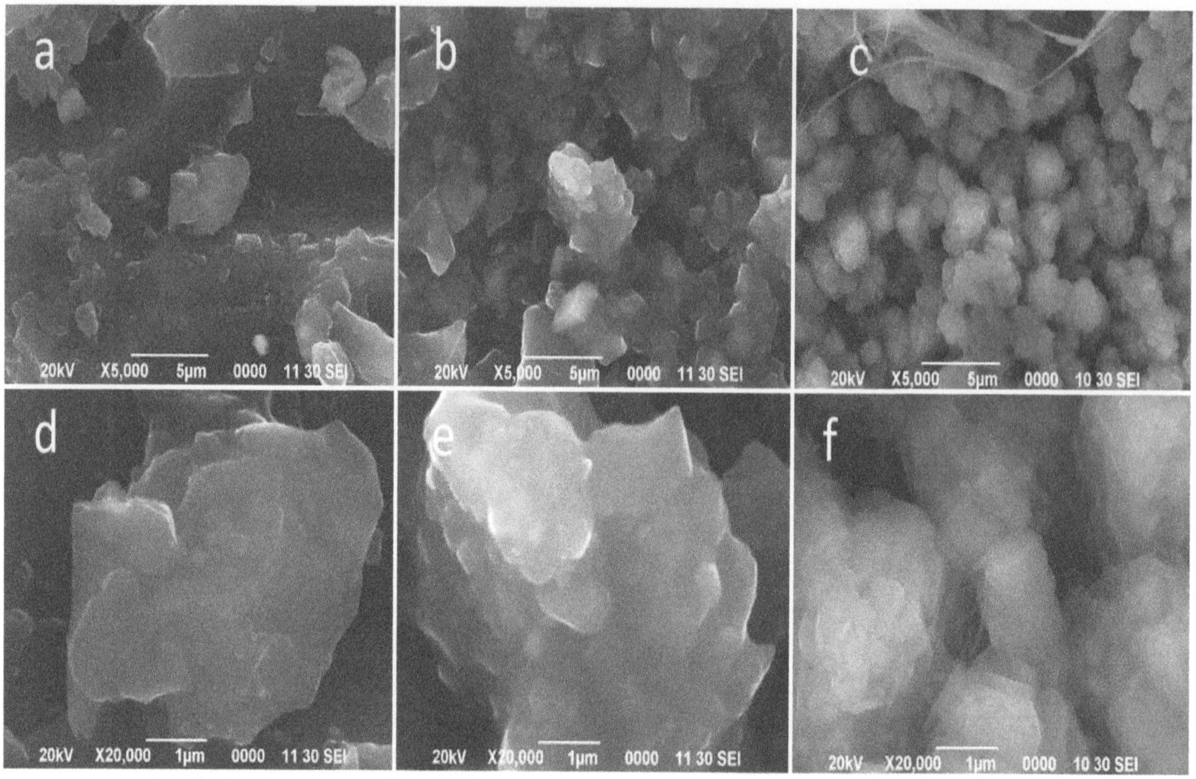

Figure 2. *SEM images for CCP **(a and d)**, CCPHNO3 **(b and e)**, and CCPWOT **(c and f)***

Electrochemical performance

Cyclic voltammetry tests were conducted at scan rates of 2, 5, 10, 20, 50, and 100 mV s^{-1} with potential windows ranging from -1 to 1 V versus Ag/AgCl in 1, 2, and 4 mol L^{-1} KOH or H_2SO_4 aqueous solution to quantify the electrochemical performance of CCP, $CCPHNO_3$ and CCPWOT electrodes. As shown in Figure 3 and 4, no pronounced reversible reduction oxidation peak can be observed from voltammograms for the three electrodes. The shape of the voltammograms was different from one to another.

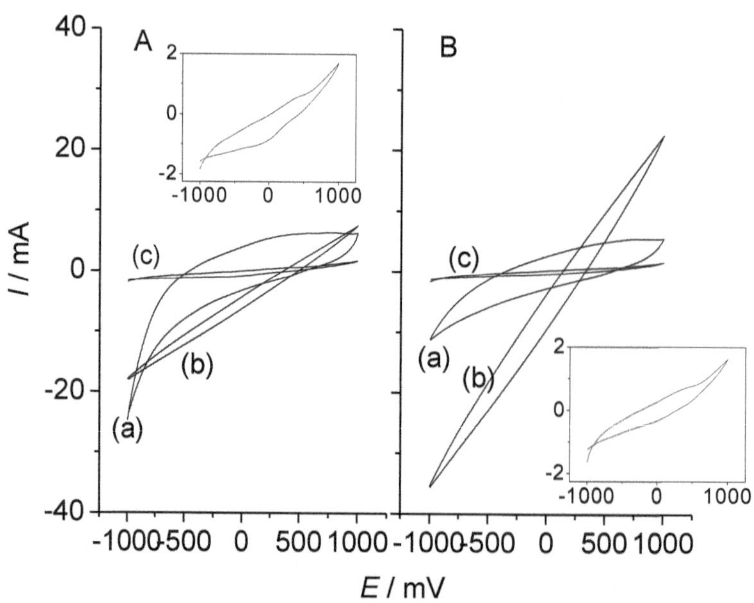

Figure 3. *CV Curve of carbon electrodes **(a)** CCP, **(b)** CCPHNO3 and **(c)** CCPWOT in the H_2SO_4 1M **(A)** and 4M **(B)**. Scan rate was 20 mV s^{-1}. **Inset:** magnification of voltammogram **(c)**.*

The voltammogram of the CCPWOT electrode was relativelylower compared to the other two curves. Therefore the oxidation treatment under water steam reduced the capacitance. On the other hand, the highest current of CCPHNO3 electrode points to the largest capacitance. The changes in capacitance occurred on the electrode after the oxidation process was associated with changes in pore volume of carbon, as previously described. Oxidation using strong acid clarified the existence of the Faraday process (redox reaction) that occurred at the carbon as described by other authors [19]. The voltammograms shape of CCPHNO3 in both acid and base electrolytes are much the same, which means that the tendency to establish a redox reaction was similar. The voltammogram area in Figure 3 shows the decrease in capacitance at high concentrations of H_2SO_4. By contrast, the capacitance of the electrodes increased at a low concentration KOH (Figure 4).

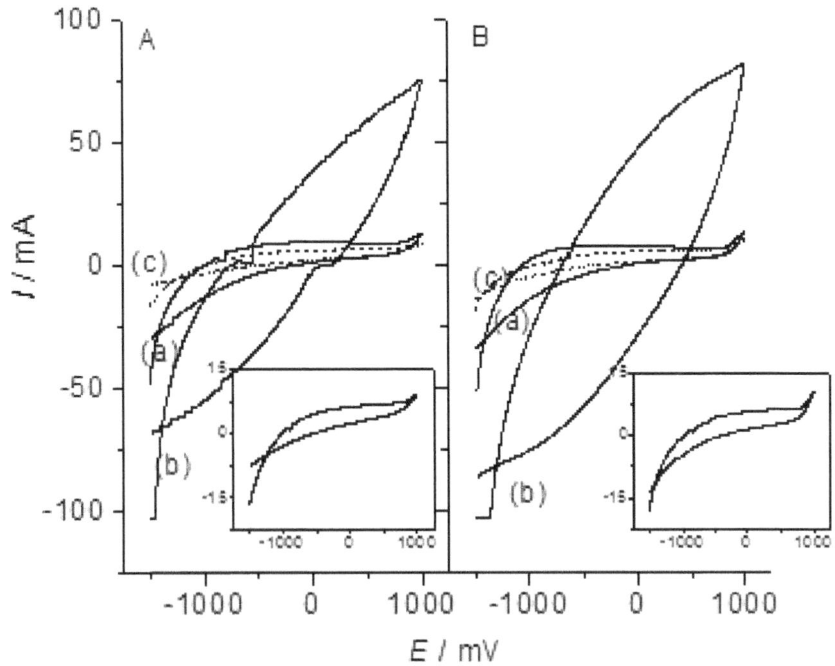

Figure 4. *Voltammogram of carbon electrodes* ***(a)*** *CCP,* ***(b)*** *CCPHNO3 and* ***(c)*** *CCPWOT in the KOH 2M* ***(A)*** *and 4M* ***(B)****. Scan rate was 5mV s-^{1}.* ***Inset****: magnification of voltammogram* ***(c)****.*

Calculations of specific capacitances for all electrodes at various scan rates in the H_2SO_4 solution are presented in Figure 5. At the 5 mV s^{-1} scans rate in H_2SO_4, the CCP electrode had a specific capacitance of 1.2 F g^{-1}, while at a scan rate of 100 mV s^{-1} this was 0.22 F g^{-1}. Capacitance values of the CCPWOT electrode (Figure 5 B) at 5 – 10 mV s^{-1} were lower than that of CCP (Figure 5 A), but were higher at 20 – 100 mV s^{-1}. The capacitance values of CCP and CCPWOT electrodes were much smaller than those of the CCPHNO3 electrode (Figure 5 C). Maximum capacitance for non-oxidized, steam and acid oxidized electrodes were 1.22, 0.63 and 2.57 F g^{-1}, respectively. These values occurred at 5 mV s^{-1}.

Plots of the electrodes specific capacitances *vs.* scan rate in KOH are presented in Figure 6. Capacitances of CCP in KOH were much the same as those in H_2SO_4 both at low and high scan rates. Contrary to what occurred in the H_2SO_4, the capacitance values of the CCPWOT electrode (Figure 6B) were higher than those of CCP (Figure 6A), both at low and high scan rates. Capacitance values of CCP and CCPWOT electrodes were also much smaller than those of the CCPHNO3 electrode. Plot of specific capacitance of CCPHNO3 in KOH was discrete (Figure 6C) due to instability of the system, especially at a scan rate higher than 10 mV s^{-1}. A stable voltammogram was only obtained at a scan rates higher than 40 mV s^{-1} at 1M KOH. Capacitance value for acid and steam oxidized electrodes were 2.23 and 27.68 F g^{-1}, respectively.

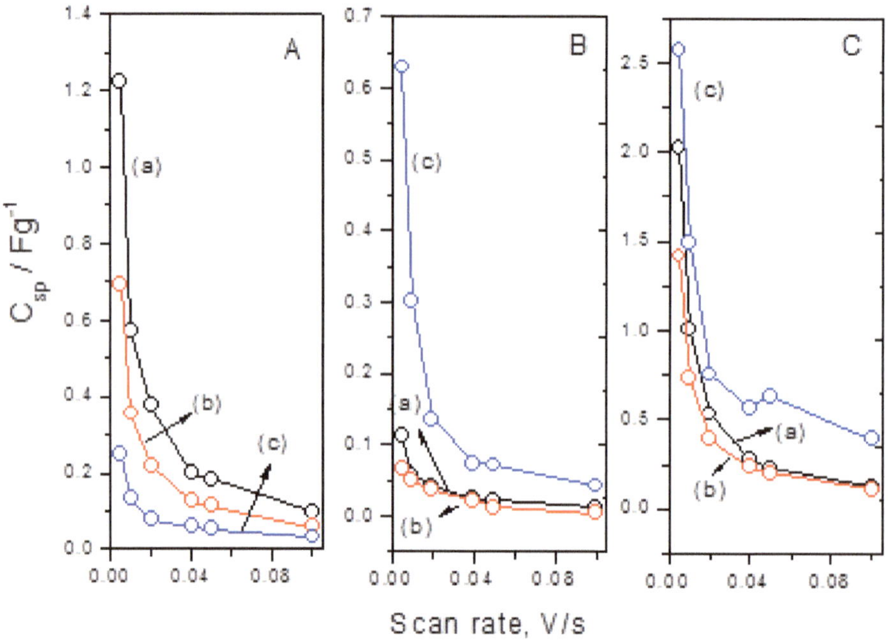

Figure 5. *Plot of specific capacitance of carbon electrode* ***(A)*** *CCP,* ***(B)*** *CPPWOT and* ***(C)*** *CCPHNO3 versus scan rate in* H_2SO_4 ***(a)*** *4M,* ***(b)*** *2M and* ***(c)*** *1M.*

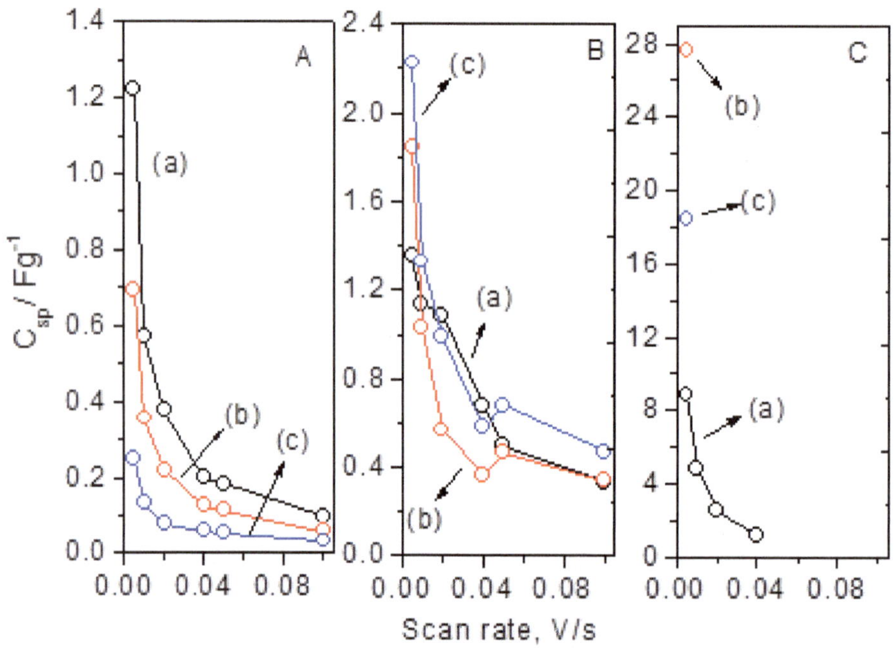

Figure 6. *Plot of specific capacitance (C_{sp})of carbon electrode* ***(A)*** *CCP,* ***(B)*** *CPPWOT and* ***(C)*** *CCPHNO3 versus scan rate in KOH* ***(a)*** *4M,* ***(b)*** *2M and* ***(c)*** *1M.*

The presence of the micropores in the activated carbon are unique for Gelam wood. Combining their uniqueness, simplicity of the preparation method and the abundance availability, makes these electrode materials relatively inexpensive and environmentally friendly. Thus Gelam wood activated carbon has the potential to be developed as electrodes in electrochemical capacitors.

Conclusions

The three types of electrodes, namely CCP, CCPHNO3 and CCPWOT, prepared from Gelam wood activated carbon, have different surface morphologies which basically contained many micropores. CCP has continuous - smooth surface and contains very few grain boundaries. On the other hand CCPWOT and CCPHNO3 have rough surfaces and contain many cracks. The different oxidation methods using nitric acid and water steam gave different effects on the surface area of the carbons. Acid nitric oxidation led to a larger pore volume and surface area, while water steam oxidation led the opposite effect. In addition, CCPHNO3 had relatively large amounts of micropores. The presence of surface oxides was important factor in improving the capacitance of electrode and increased the effective utilization of the pore structure. CCPHNO3 had the highest capacitance value of 27.68 F g^{-1} while the specific capacitance of the electrode CCPWOT was similar to CCP, at 2.23 and 1.22 F g^{-1}, respectively.

References

[1] M. Jayalakshmi, K. Balasubramanian, *Int. J. Electrochem. Sci.* **3** (2008) 1196–1217

[2] W. Li, H. Probstle, J. Fricke, *J. Non-Cryst. Solids* **325** (2003) 1–5

[3] A.G. Pandolfo, A.F. Hollenkamp, *J. Power Sources* **157** (2006) 11–27

[4] E. Taer, M. Deraman, I.A. Talib, A. Awitdrus, S.A. Hashmi, A.A. Umar, *Int. J. Electrochem. Sci.* **6** (2011) 3301 – 3315

[5] C. Merino, P. Soto, E. Vilaplana-Ortego, J.M.G.d. Salazar, F. Pico, J.M. Rojo, *Carbon* **43** (2005) 551–557

[6] H.P. Boehm, Carbon 32 (1994) 759-769

[7] C. Nieto-Delgado, M. Terrones, J.R. Rangel-Mendez, *Biomass Bioenerg.* **35** (2011) 103–112

[8] S. Villar-Rodil, A. Martinez-Alonso, J.A. Pajares, J.M.D. Tascon, M. Jasienko-Hałat, E. Broniek, J. Kaczmarczyk, A. Jankowska, A. Albiniak, T. Siemieniewska, *Micropor. Mesopor. Mat.* **64** (2003) 11-19

[9] K.P. Gadkaree, M. Jaroniec, *Carbon* **38** (2000) 983–993

[10] K. Ramakrishnan, C. Namasivayam, *Sustain. Environ. Res.* **19** (2009) 173 - 178

[11] C. Du, N. Pan, *Nanotechnology* **17** (2006) 5314–5318

[12] R. Nickolov, D. Kovacheva, M. Mladenov, N.Velichkova, R. Raicheff, P. Tzvetkova, *J. Univ. Chem. Tech. Metall.* **46** (2011) 275–282

[13] K. Xia, Q. Gao, J. Jiang, J. Hu, *Carbon* **46** (2008) 1718-1726

[14] H. Aripin, L. Lestari, D. Ismail, S. Sabchevski, *Open Mat. Sci. J.* **4** (2010) 117–124

[15] M.J. Bleda-Martınez, J.A. Macia-Agullo, D. Lozano-Castello, E. Morallon, D. Cazorla-Amoros, A. Linares-Solano, *Carbon* **43** (2005) 2677–2684

[16] L.L. Zhang, X. Zhao, H. Ji, M.D. Stoller, L. Lai, S. Murali, S. Mcdonnell, B. Cleveger, R.M. Wallace, R.S. Ruoff, *Energy Environ. Sci.* **5** (2012) 9618-9625

[17] B.G. Choi, Y.S. Huh, W.H. Hong, H.J. Kim, H.S. Park, *Nanoscale* **4** (2012) 5394-5400

[18] E. Frackowiak, F. Beguin, *Carbon* **39** (2001) 937-950

[19] M. Seredych, D. Hulicova-Jurcakov, G.Q. Lu, T.J. Bandosz, *Carbon* **46** (2008) 1475-1388

11

On the suppression of cathodic hypochlorite reduction by electrolyte additions of molybdate and chromate ions

JOHN GUSTAVSSON, GONGZHUO LI, CHRISTINE HUMMELGÅRD*, JOAKIM BÄCKSTRÖM* and ANN CORNELL✉

Applied Electrochemistry, School of Chemical Science and Engineering, KTH Royal Institute of Technology, SE 100 44 Stockholm, Sweden

**Department of Natural Sciences, Engineering and Mathematics, Mid Sweden University, SE 851 70 Sundsvall, Sweden*

✉Corresponding Author: E-mail: amco@kth.se

Abstract

The goal of this study was to gain a better understanding of the feasibility of replacing Cr(VI) in the chlorate process by Mo(VI), focusing on the cathode reaction selectivity for hydrogen evolution on steel and titanium in a hypochlorite containing electrolyte. To evaluate the ability of Cr(VI) and Mo(VI) additions to hinder hypochlorite reduction, potential sweep experiments on rotating disc electrodes and cathodic current efficiency (CE) measurements on stationary electrodes were performed. Formed electrode films were investigated with scanning electron microscopy and energy-dispersive X-ray spectroscopy. Cathodic hypochlorite reduction is hindered by the Mo-containing films formed on the cathode surface after Mo(VI) addition to the electrolyte, but much less efficient compared to Cr(VI) addition. Very low levels of Cr(VI), in the μM range, can efficiently suppress hypochlorite reduction on polished titanium and steel. Phosphate does not negatively influence the CE in the presence of Cr(VI) or Mo(VI) but the Mo-containing cathode films become thinner if the electrolyte during the film build-up also contains phosphate. For a RuO_2-TiO_2 anode polarized in electrolyte with 40 mM Mo(VI), the anode potential increased and increased molybdenum levels were detected on the electrode surface.

Keywords

Current efficiency; hydrogen evolution; in-situ additives; cathode; electrolysis; EDS; SEM; potential sweeps; galvanostatic polarization

Introduction

Sodium chlorate is industrially produced from sodium chloride in an electrolytic process according to reaction 1 below.

$$NaCl + 3H_2O \rightarrow NaClO_3 + 3H_2 \quad (1)$$

Chlorine is produced on the anode and forms hypochlorite (here defined as ClO^- + ClOH) (reactions 2-4) [1], and chlorate is formed in a chemical reaction (reaction 5) that shows an optimal reaction rate at pH 6-7. To prevent that the intermediate hypochlorite and the product chlorate are reduced on the cathode (reactions 6, 7 and 8) hexavalent chromium, Cr(VI), is added to the electrolyte. The hexavalent chromium is reduced on the cathode surface to a film of chromium (III)hydroxide [2]. The $Cr(OH)_3$-film formed has been shown to be very thin and efficient at inhibiting the reduction of hypochlorite, chlorate [3], oxygen and ferricyanide [2].

$$2Cl^- \leftrightarrow Cl_2 + 2e^- \quad (2)$$

$$Cl_2 + H_2O \leftrightarrow HClO + Cl^- + H^+ \quad (3)$$

$$HClO \leftrightarrow ClO^- + H^+ \quad (4)$$

$$2HClO + ClO^- \rightarrow ClO_3^- + 2H^+ + 2Cl^- \quad (5)$$

$$ClO^- + H_2O + 2e^- \rightarrow Cl^- + 2OH^- \quad (6)$$

$$HClO + H^+ + 2e^- \rightarrow Cl^- + H_2O \quad (7)$$

$$ClO_3^- + 3H_2O + 6e^- \rightarrow Cl^- + 6OH^- \quad (8)$$

Unfortunately Cr(VI) is toxic and carcinogenic, which motivates investigations of possible alternatives.

Molybdenum is in the same group as chromium in the periodic table and the two elements have similarities. Molybdate (MoO_4^{2-}) is opposed to chromate not toxic and is used as an environmentally friendly alternative to Cr(VI) in corrosion applications [4]. Mo-containing cathode films are formed during electrolysis of an electrolyte with Mo(VI), and these films can hinder the cathodic reduction of oxygen on copper [5]. This inhibition was shown to be efficient in electrolytes of neutral pH, where films of MoO_2 are stable. Mo(VI) also buffers in a pH region close to that of Cr(VI) [6] and can activate the desired cathode reaction of hydrogen evolution [6,[7]. Not surprisingly Mo(VI) has been considered as a possible alternative to Cr(VI) in the chlorate process [6],[8]. A potential problem observed with Mo(VI) electrolyte additions to chlorate electrolyte is an increased oxygen level in the hydrogen cell gas [6], as this would affect the overall current efficiency (CE) for chlorate production as well as being a safety risk if the oxygen/hydrogen mixture reaches an explosive composition. Other results showed though that at low Mo(VI) concentrations (typically < 1 mM Mo(VI) compared to 39 mM Mo(VI) in [6]) no increased oxygen levels were reached and still the hydrogen evolution reaction on MAXTHAL® 312 (Ti_3SiC_2) was substantially activated [8], by the Mo(VI) addition. Whether the increased oxygen levels related to homogeneous reaction in the electrolyte or to anode reactions is not clear, and no measurements of the anode potential were made in [6] or [8].

At low levels of Mo(VI) or Cr(VI) an additional buffer may be needed to stabilize the electrolyte pH in the chlorate process at 6-7. Phosphate is a possible buffer alternative, which may also form cathode films that can affect the CE [9] and thus the CE in electrolytes containing combinations of Mo(VI), Cr(VI) and phosphate are of interest.

The goal of this study is to gain a better understanding of the feasibility of replacing Cr(VI) in the chlorate process by Mo(VI), focusing on the cathode reaction selectivity in the presence of a Mo-containing film. Reaction 6 or 7, the reduction of hypochlorite, was chosen as an indicator of the ability of different films to increase the CE for hydrogen evolution. Experiments with Cr(VI) additions to the electrolyte were made to compare the effects of Mo(VI) and Cr(VI). Cathodes of low carbon steel and titanium were used, as these are the most common cathode materials in chlorate electrolysis today. Both rotating disc electrodes (RDE) and stationary electrodes were used. The possibility of effects on the anode reactions of electrolyte additives were studied on a dimensionally stable anode (DSA)-type RDE.

Experimental

Water purified in a Millipore Direct-Q system was used in the preparation of all electrolytes, and to rinse the electrodes prior to experiments. Sodium hydroxide, sodium chloride, hydrochloric acid, sodium molybdate dihydrate, sodium dihydrogen phosphate all from Merck and of Pro Analysi grade; sodium hypochlorite solution 0.5 M in 0.1 M NaOH from VWR BDH Prolabo. In the text the term "hypochlorite" refers to the sum of [ClO^-] and [HClO]. If not stated otherwise, the given hypochlorite concentration was the initial concentration. The terms "phosphate concentration" or "total phosphate concentration" here refers to the sum of phosphate species $[PO_4^{3-}]_{tot}=[H_3PO_4]+ +[H_2PO_4^-]+ [HPO_4^{2-}]+ [PO_4^{3-}]$.

For potential sweeps and polarization experiments RDEs were used and the measurements were performed using a PAR 273A potentiostat and an electrode rotator Model 616 from Pine Instrument Company. The Fe-RDE (5 mm diameter) and Ti-RDE (diameter 4 mm) were shielded by Teflon sheaths, polished by grit 4000 SiC grinding paper and rinsed in water before all experiments. The Fe-RDE was prior to the measurements corroded by immersion at OCP for 3 min with no rotation, in an electrolyte containing 5.2 M $NaClO_3$, 1.9 M NaCl and 10 mM NaClO, pH 6.5 at 70°C. The potential sweep and polarization experiments were performed in a jacketed glass cell with 200 ml electrolyte. The temperature was controlled by a water bath and the electrolyte was purged with nitrogen for at least 15 min prior to and during the experiments. The reference electrodes connected to the electrolyte by an ion bridge filled with 5 M NaCl and a Luggin capillary. As the counter electrode a platinum grid was used.

For investigation of effects on the anode reactions polarization experiments were performed on a DSA-type anode. The anode was produced by spin-coating to produce a well-defined and relatively smooth coating of $Ti_{0.7}Ru_{0.3}O_2$. A solution of 5.351 g $RuCl_3{\cdot}nH_2O$ (Ru content 35.51 % by mass) and 14.929 g Ti(IV)n-butoxide in n-propanol to a final volume of 50 ml was prepared. The solution was thoroughly stirred for several days. A disc of commercially pure titanium with smooth surface and a diameter of 59 mm was pickled in an aqueous solution of 27 ml concentrated HF per liter and 71 ml concentrated HNO_3 per liter for two min at room temperature. The disc was placed in a spin coater (Electronic Micro Systems), an amount of solution was applied to the surface of the disc, and the solution was evenly distributed by spinning at 550 min^{-1}. The disc was then dried at 80 °C for 10 min and calcined at 470 °C for 10 min. Three layers of coating were applied the same way. The final layer was dried for 40 min and calcined for one hour. The mass difference between the bare disc and the final sample was 0.0883 g corresponding to a Ru loading of 10.2 g m^{-2}. To make DSA-RDEs, small discs (diameter 11.3 mm) were punched from the larger disc and put into titanium holders. Exposed titanium was shielded by silicon tubing and epoxy. For the anodic polarization experiments the electrodes were pre-polarized for 30 min at 3 kA m^{-2} in a separate

2 M NaCl electrolyte at 70 °C. The recorded potentials were IR-corrected using a current interrupt technique [10].

Measurements of the CE for hydrogen evolution were performed in a divided cell [11] separated by a Gore Select ion-exchange membrane. The cell was immersed in a water bath for temperature control and a magnetic stirrer, set at the same level for all experiments, was used for the cathode compartment. The length of time needed to produce a given volumes of hydrogen gas (between 5 and 20 ml) were recorded. The CE was determined as by calculation from a calibration curve where hydrogen was evolved on a Pt-wire in Na_2SO_4 solution under the assumption of 100 % CE for hydrogen evolution. For the CE measurements stationary electrodes of titanium and low carbon steel were used. The electrodes were polished with grit 4000 SiC grinding paper and rinsed in water before all experiments. For CE measurements in neutral and alkaline electrolyte the hypochlorite level was measured using arsenite titration. The arsenite titration was performed with 5 ml electrolyte diluted by 200 ml water and pH adjusted to neutral pH. Mo(VI) was found to interfere with the titration, giving unreasonable high "hypochlorite levels". Therefore hypochlorite levels in the electrolyte were only measured in experiments where no Mo(VI) had been added to the electrolyte. The hypochlorite concentration typically decreased from 80 to 40 mM after 20 min of electrolysis at -3 kA m^{-2}, and the corresponding pH increased from 6.5 to 9-11.

X-Ray Fluorescence (XRF) measurements were performed using a Niton XLT898 with an accelerator voltage of 35 kV and a silver anode. Two different instruments for Scanning Electron Microscopy (SEM) were used; a Zeiss EvO 50 SEM with EDS for the elemental analyses and a Hitachi S-4800 field emission SEM for the micrographs presented.

Results and discussion

Current efficiency for hydrogen evolution in the presence of molybdate

Potential sweep experiments on a Ti-RDE with and without additions of Mo(VI) and hypochlorite to the electrolyte are shown in Figure 1. The sweeps were made in anodic direction after 5 min of pre-polarization at -1.5 V *vs.* Ag/AgCl, which enabled the formation of Mo-containing films on the cathode whenever the electrolyte contained Mo(VI). In the presence of 15 mM hypochlorite, the dotted line, a cathodic limiting current for hypochlorite reduction of about 350 A m^{-2} can be seen. This limiting current was suppressed by the addition of Mo(VI), and the higher the Mo(VI) concentration the more efficient the hindering effect – see curves with addition of 1, 10 and 100 mM Mo(VI), respectively. From the Levich equation and data for viscosity and diffusivity from Byrne et al. [12] a limiting current density for hypochlorite reduction of 340 A m^{-2} was calculated, which agrees well with the experimental result.

Looking at the cathodic end of the graph an activation of hydrogen evolution about 300 mV is seen for the cases of 1 and 10 mM Mo(VI). As high Mo(VI) level as 100 mM did not give a similar decrease in overpotential, likely caused by resistive properties of the Mo(VI) film formed [7].

In Figure 2 potential sweeps on slightly corroded iron are shown, and the electrode had been pre-polarized as above. Again a limiting current for hypochlorite reduction is visible at -300 to -400 A m^{-2}, and addition of 100 mM Mo(VI) only partly suppressed this current. A comparison with the result in Figure 1 indicates that it is easier to form a hindering film on a titanium substrate than on corroded iron. The high currents appearing at potentials > -0.6 V *vs.* Ag/AgCl in the Mo(VI) free case likely relate to oxidation of the iron electrode, which is hindered by the presence of a molybdenum containing film. Similar as in the case of titanium cathodes (Figure 1) the addition of as high concentration as 100 mM Mo(VI) was less effective at activating the HER than lower Mo(VI) concentration.

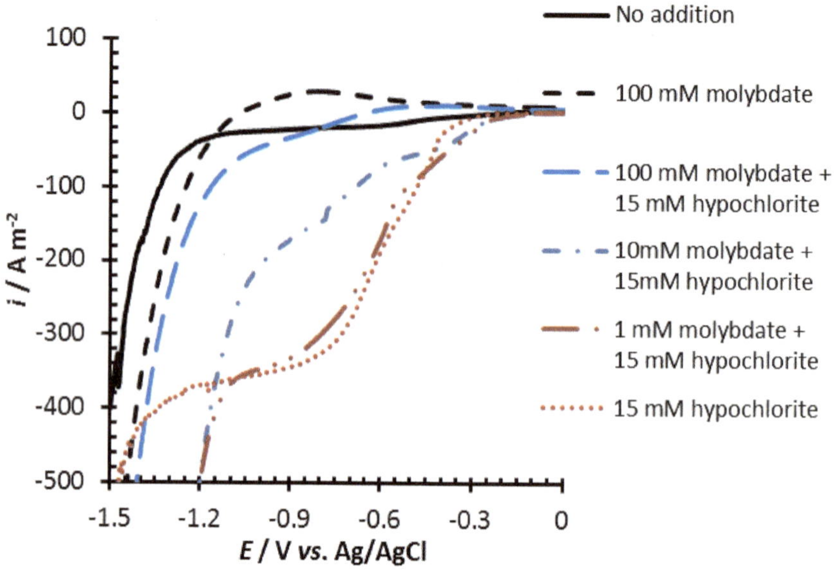

Figure 1. *Potential sweeps from -1.5 V to 0 V* vs. *Ag/AgCl at a sweep rate of 50 mV/s in 5 M NaCl, 70 °C, pH 6.5 with a Ti-RDE at a rotation rate of 3000 rpm. Pre-polarization at -1.5 V* vs. *Ag/AgCl for 5 min.*

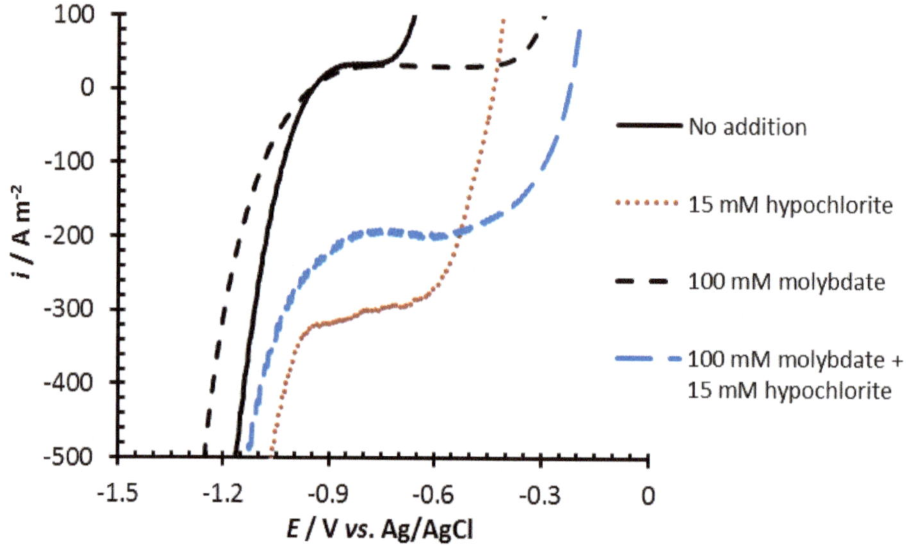

Figure 2. *Potential sweeps from -1.5 V to 0 V* vs. *Ag/AgCl at a sweep rate of 50 mV s^{-1} in 5 M NaCl, pH 6.5, 70 °C with a corroded Fe-RDE at a rotation rate of 3000 rpm. Pre-polarization at -1.5 V* vs. *Ag/AgCl for 5 min.*

Measurements of the CE for hydrogen evolution were made to evaluate how hypochlorite reduction as a side reaction could be suppressed at an industrially relevant current density of -3 kA m^{-2}, which corresponds to a potential region where hydrogen evolution prevails. In pH neutral electrolyte the CE on titanium was around 80 % in the presence of 80 mM hypochlorite (Figure 3). When adding 80 mM Mo(VI) the CE increased to 91-94 %. Thus, similar as shown in the potential sweeps (Figure 1), addition of Mo(VI) to a pH neutral electrolyte can hinder hypochlorite reduction on a titanium cathode. In alkaline pH the CE was lower (Figure 3), and some possible explanations for this are (i) a higher and more stable hypochlorite concentration of 80 mM,

(ii) a better mass transport of hypochlorite to the cathode surface due to a lower viscosity of the sodium hydroxide electrolyte compared to 5 M NaCl and (iii) at alkaline pH hypochlorite is present as the ion ClO^-, which is possibly more easily reduced compared to HClO [12]. The CE was still low after addition of Mo(VI), 55-60%, although there seems to be some minor hindrance of hypochlorite reduction.

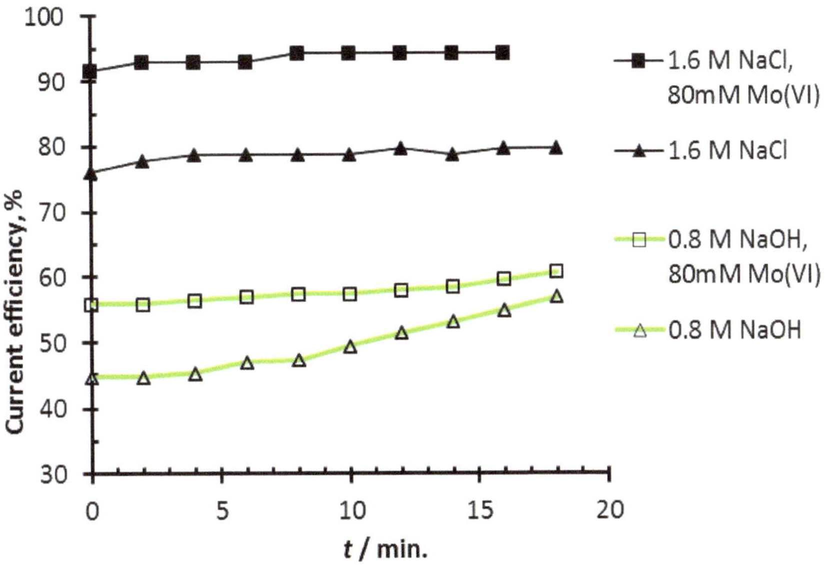

Figure 3. *Current efficiency measurements for hydrogen evolution on stationary titanium electrode polarized at -3 kA m^{-2}, 70 °C. Initially: [hypochlorite] ≈80 mM, pH 6.5 respectively pH ≈13*

The relatively high concentration of 80 mM Mo(VI) was chosen to see an immediate response of the addition, though at a realistic process a lower concentration is probably needed to avoid high cathode overpotentials and high oxygen levels in the off gas [8]. An experiment with 4 mM Mo(VI) was made, showing that such low concentrations are required over 30 min to give any effect at all on the CE, see Figure 4. The effect would probably be increased if the experiment was continued, giving the Mo-containing cathode film a longer time to build up.

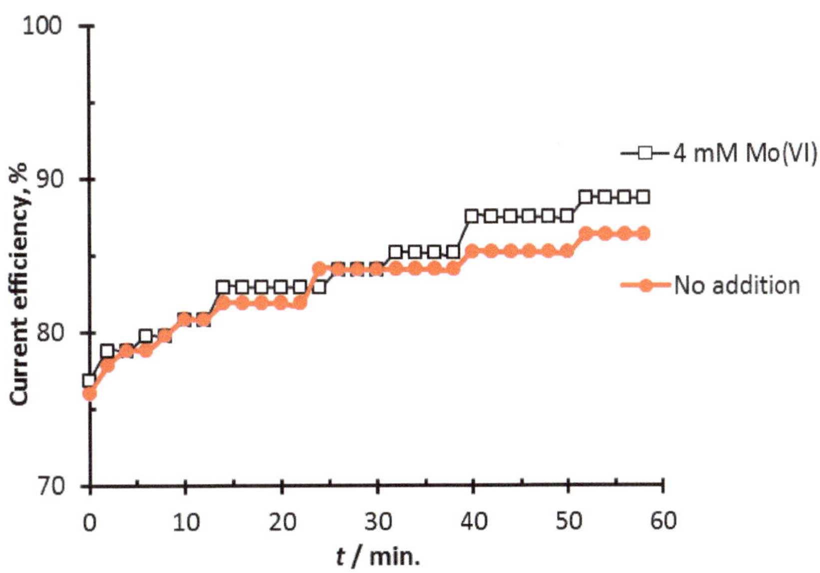

Figure 4. *Current efficiency measurements for hydrogen evolution on stationary titanium electrode polarized at -3 kA m^{-2} 1.6 M NaCl, 70 °C. Initially: [hypochlorite] ≈80 mM, pH 6.5*

As mentioned in the introduction the addition of Cr(VI) to industrial chlorate electrolyte has several functions, one of which is to buffer the electrolyte pH. If replacing Cr(VI) by low levels of Mo(VI), or if lowering considerably the levels of Cr(VI), an additional buffer such as phosphate would be needed. Measurements of the CE on titanium in the presence of phosphate were made to see if possible phosphate films would interfere with the Mo-containing cathode films. As seen in Figure 5 the addition of phosphate had no negative effect on the CE. In contrast, it showed a small positive effect. It has earlier been found that molybdate and phosphate have a synergistic effect on preventing corrosion of an Mg alloy [13].

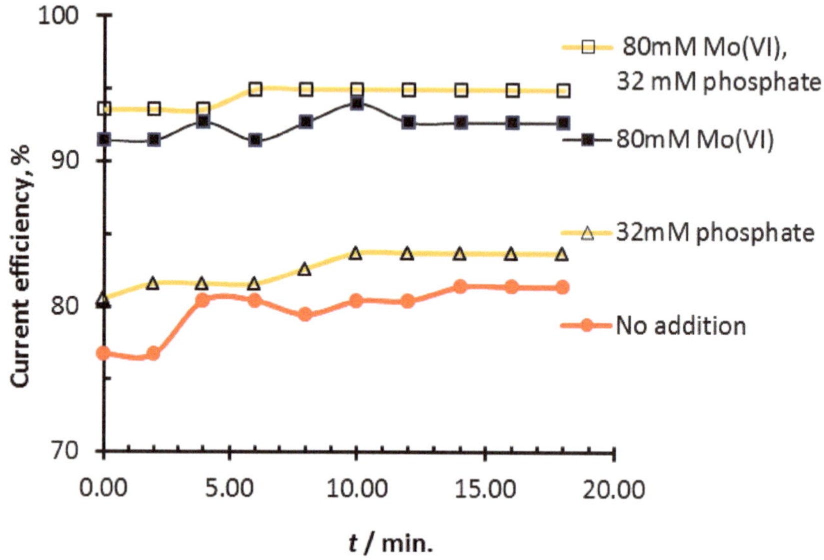

Figure 5. *Current efficiency measurements for hydrogen evolution on a stationary titanium electrode at -3 kA m^{-2} in 1.6 M NaCl, 70 °C. Initially: [hypochlorite]≈80 mM, pH 6.5*

Results from measurements of CE on a steel electrode can be seen in Figure 6. Lower CE-values were obtained on steel compared to on titanium, which is consistent with our results from the potential sweeps in Figures 1 and 2. Furthermore, also on steel the phosphate additions had no negative effects on the CE, but could possibly increase it a little. The CE values in Figure 6 are relatively low in the beginning of the experiments, and then gradually increase – a pattern not seen as clearly for the titanium electrodes.

Current efficiency for hydrogen evolution at low chromate concentrations

In the chlorate process the product chlorate salt is separated from the electrolyte by crystallization, and as the electrolyte contains Cr(VI), small amounts of chromate will co-precipitate or adsorb onto the chlorate crystals and end up in the chlorate product. A CrO_4^- concentration of as low as 9 mg dm^{-3} (80 μM Cr) in the electrolyte has been shown to be enough to form a chromium hydroxide film that hindered the reduction of nitrate and nitrite [14].

Potential sweep experiments were made on titanium in electrolytes with varying Cr(VI) concentrations (Figure 7) to study the effect of low Cr(VI) concentrations in chlorate electrolyte. In the 5 M NaCl electrolyte with 15 mM hypochlorite there was a limiting current for hypochlorite reduction. This electrolyte was Cr(VI) free and there was no inhibition of hypochlorite reduction. A similar sweep was made in chlorate electrolyte made from commercial grade $NaClO_3$ and p.a. grade NaCl, and showed that hypochlorite reduction was inhibited in the presence of 550 g dm^{-3} $NaClO_3$.

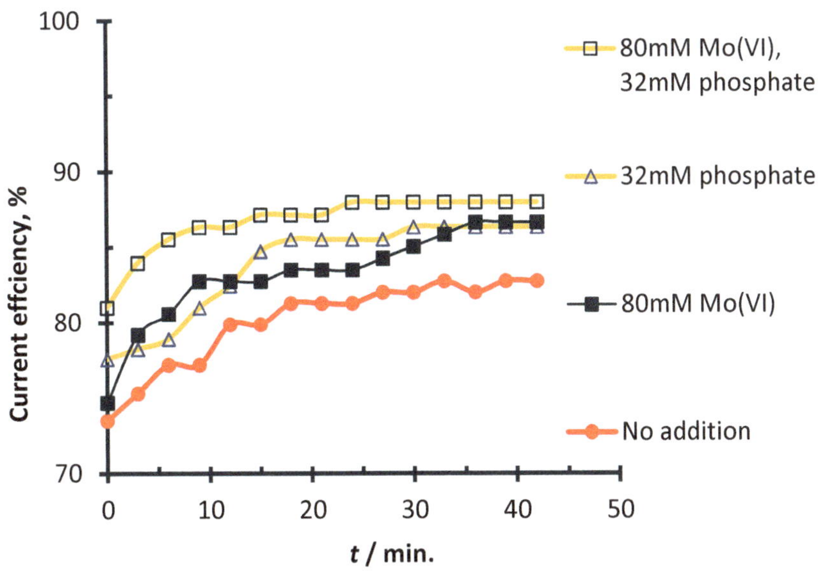

Figure 6. *Current efficiency measurements for hydrogen evolution on a stationary steel electrode at -3 kA m^{-2} in 1.6 M NaCl, 70 °C. Initially: [hypochlorite] ≈80 mM, pH 6.5*

From ICP analyses of the chlorate crystals it was calculated that this electrolyte contained also about 20 μM Cr(VI) and this inhibition was reproduced by adding 20 μM Cr(VI) to 5 M NaCl, see Figure 7. Thus, the inhibition of hypochlorite reduction on titanium in chlorate electrolyte without added Cr(VI) was caused by the Cr(VI) traces in non-recrystallized chlorate salt. Experiments were also made with an electrolyte made from $NaClO_3$ that had been recrystallized by cooling crystallization in the lab. For electrolytes with recrystallized chlorate the limiting current for hypochlorite reduction was higher, but the chromate effect was still there. This makes it difficult to evaluate Cr(VI) free alternatives using chlorate salt produced from a conventional industrial process with Cr(VI) containing electrolyte and the present study thus only considers hypochlorite reduction (reactions 6 and 7) and not chlorate reduction (reaction 8).

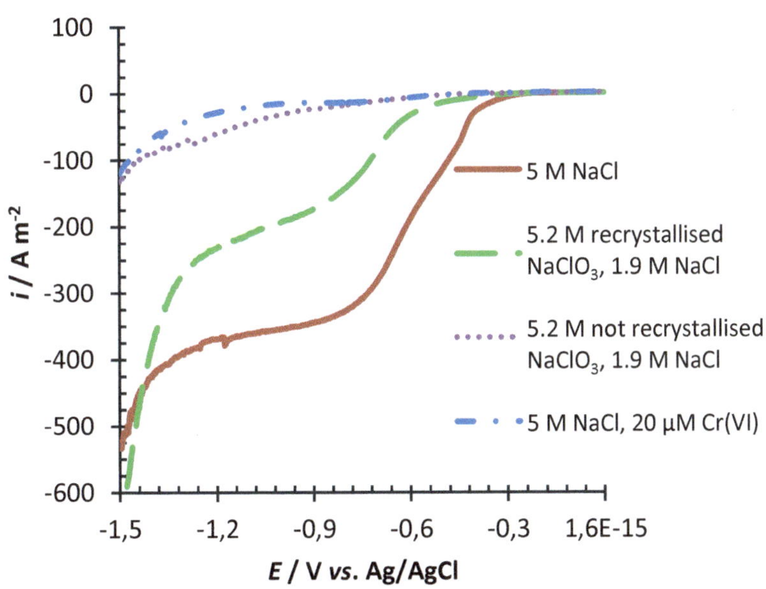

Figure 7. *Potential sweeps (anodic direction) using a Ti-RDE at 3000 rpm and 70 °C in electrolytes containing 15 mM hypochlorite performed with a sweep rate of 50 mV s^{-1}. Pre-polarization at -1.5 V vs. Ag/AgCl for 5 min.*

The effect of low levels of Cr(VI) was also studied in CE measurements on a steel electrode, see Figure 8 where a curve representing Mo(VI) additions has been added for comparison. The addition of 2.7 μM Cr(VI) increased the CE, as measured after 20 min of electrolysis, from about 80 % to over 94 %. This can be compared to 80 mM Mo(VI), which only increased the CE to about 83 % despite being present in a concentration about 30000 times higher. Increasing the Cr(VI) level to 27 μM increased the CE to 97 % after 20 min of electrolysis. With the lower Cr(VI) concentration it took over 10 min to reach steady state CE values, whereas for 27 μM Cr(VI) a high CE was reached instantaneously (in less than 1 min). Adding 80 mM Mo(VI) to the Cr(VI) containing electrolyte did not have a negative effect on the CE, but instead a small positive effect. Note that the high CE values obtained in these experiments relate to a steel cathode that had been polished prior to experiments and submerged into the electrolyte under cathodic polarization to avoid steel corrosion. In an earlier study it was found that the selectivity for hydrogen evolution on steel in chlorate electrolytes was highly dependent on whether the steel surface had been allowed to corrode prior to the measurements [11]. Hydrogen evolution was suppressed on corroded steel, in favor of electrochemical chlorate reduction. Compared to non-corroded steel, much higher concentrations of Cr(VI) was needed in the electrolyte to obtain high current efficiencies for hydrogen evolution. Thus, in the present study, if the electrode had been allowed to corrode on open circuit in hypochlorite containing electrolyte, the CE values would have been lower and probably the very low Cr(VI) levels in the present study would have had a negligible effect on the CE. Not only corroded steel but also RuO_2 prepared by thermal decomposition have shown a low selectivity for hydrogen evolution in chlorate electrolytes [15]. However, if a smooth and relative stable substrate material, free from oxides catalyzing chlorate reduction, is used the Cr(VI) level in the chlorate process may be lowered substantially while maintaining a high CE. This seems true also for Mo(VI) addition and therefore it will be difficult to replace Cr(VI) in the existing process without replacing the steel cathodes with a more dimensionally stable material. We have earlier shown that Mo(VI) addition can activate a substrate so it will have a similar activity for HER as iron [7], thus less active materials can be used instead of iron/low alloyed carbon steel.

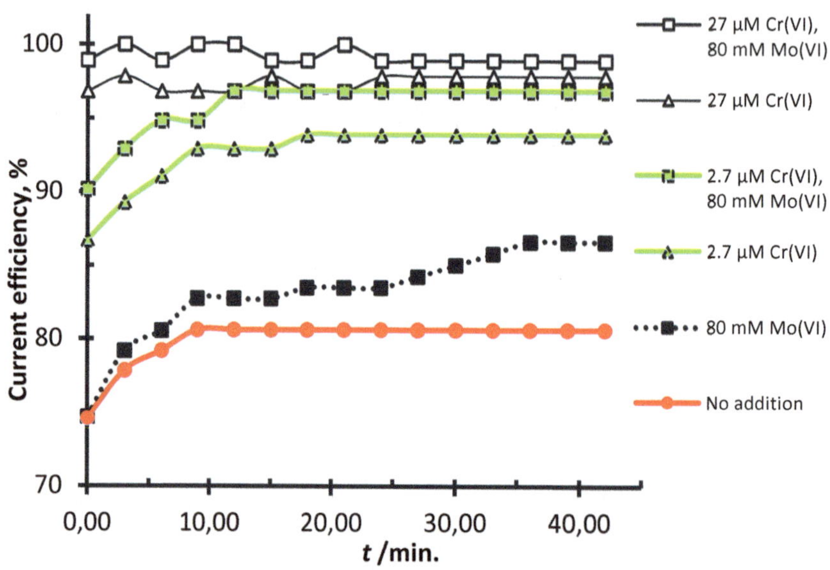

Figure 8. *Current efficiency measurements for hydrogen evolution on a stationary steel electrode at -3 kA m^{-2} in 1.6 M NaCl at 70 °C. Initially: [hypochlorite]≈80 mM, pH 6.5*

Surface analysis by SEM and EDS

In Figure 9 SEM micrographs of films obtained on a titanium cathode are shown. Figure 9 a, shows the bare titanium surface prior to any contact with the electrolyte and is given for comparison. The other micrographs show the cathode surface after 20 min of electrolysis at -3 kA m^{-2} in 5 M NaCl with different additives. With 80 mM Mo(VI) in the electrolyte a dense Mo-containing film, covering the cathode surface, was formed (Figure 9 b). The cracks seen are most probably due to drying of the film. With 32 mM phosphate in the electrolyte a totally different surface appears with needle-like crystals, see Figures 9 c and 9 d (larger magnification). In the presence of both Mo(VI) and phosphate in the electrolyte there is again a dense film (figures 9 e and 9 f), now with a spotted appearance.

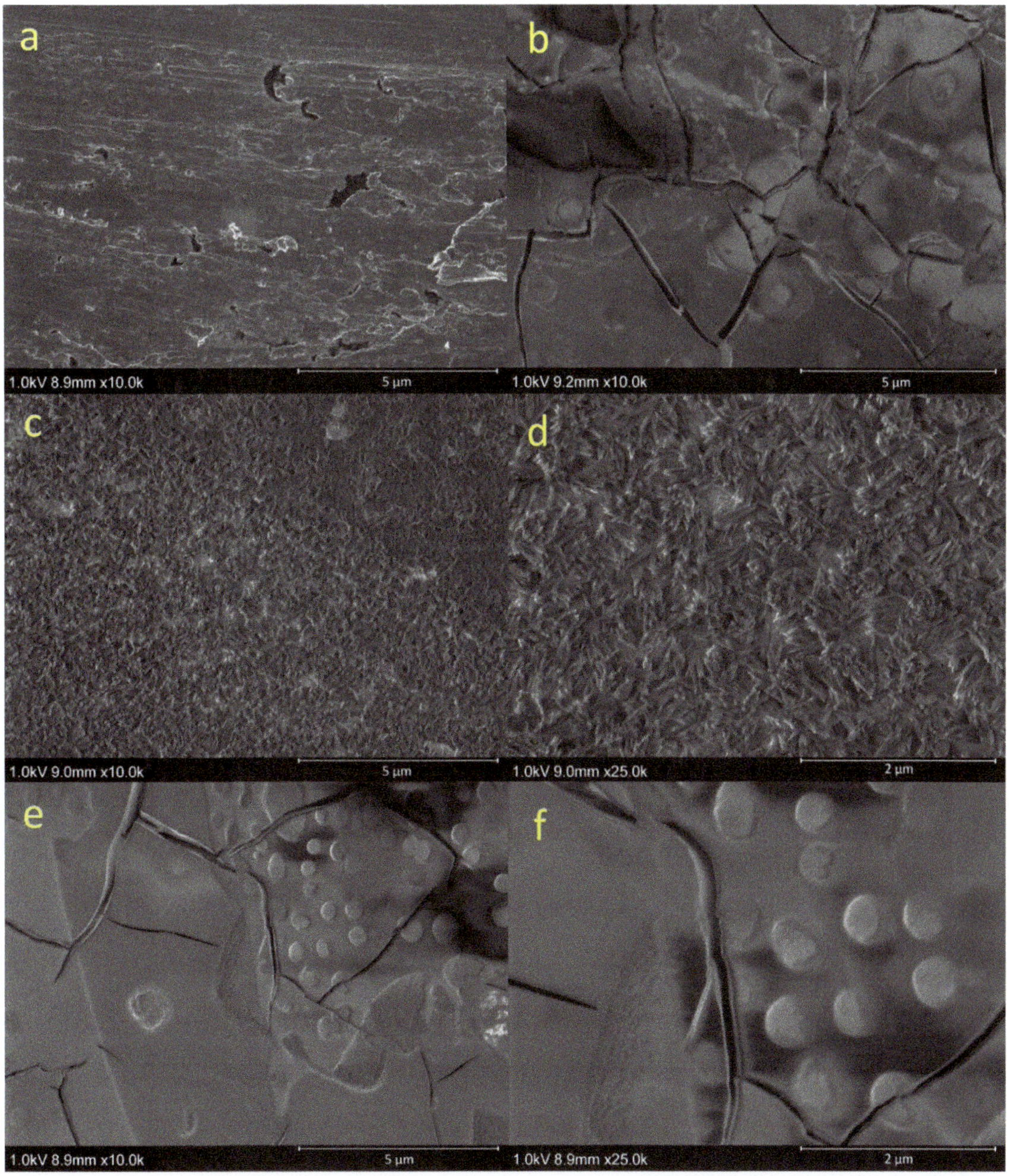

Figure 9: SEM images of titanium electrodes polarized at -3 kA m^{-2} for 20 min at 70 °C in 1.6 M NaCl with 100 mM hypochlorite and the following electrolyte additions: 80 mM molybdate (9 b); 32 mM phosphate (9 c-d); 80 mM molybdate + 32 mM phosphate (9 e-f). The sample in 9 a shows, for comparison, a polished titanium surface not exposed to electrolyte.

Elemental analyses with EDS showed that the films formed in the presence of Mo(VI) all contained Mo to a large extent (Table 1). The film formed in the presence of Mo(VI) and phosphate also contained P and was thinner than the film formed when only Mo(VI) was present. Looking at the sample with only phosphate addition, Figures 9 c,d and Table 1 the compounds phosphorous and oxygen were present in a molar ratio of 1:4 and thus phosphorous was likely present as phosphate in the cathode film. For comparison a sample was just dipped, with no cathodic polarization, into the electrolyte with 32 mM phosphate. No needle-like crystals were observed when examining the sample in the SEM and no phosphorous could be detected by EDS (not shown in Table 1).

Table 1. *Elemental analysis with EDS on titanium electrodes polarized at -3 kA m^{-2} for 20 min at 70 °C in 1.6 M NaCl with 100 mM hypochlorite and electrolyte additions according to table. The values are given in atomic %.*

Element	*80 mM Mo(VI)	*80 mM Mo(VI)+32 mM PO_4^{3-}	**32 mM PO_4^{3-}
O	56.6	61.7	64.9
Ti	14.0	18.4	-
Na	12.7	10.7	5.5
Cl	8.8	2.3	-
Mo	7.5	5.1	-
P	-	1.8	15.9

* Acceleration voltage: 15 kV **Acceleration voltage: 5 kV

In an earlier publication we found that if phosphate was added to a Mo(VI)-containing electrolyte, no molybdenum containing films could be detected with EDS on the cathode surface [7]. In that case the phosphate concentration was 2.5-10 times higher than the molybdate concentration. In the CE measurements in the present work, 20 times higher molybdate concentration (80 mM Mo(VI)) was used than in the previous study (4 mM Mo(VI), Ti-RDE 3000 rpm, and no added hypochlorite [7]) and therefore molybdate can be favored in a competitive adsorption between molybdate and phosphate.

In the presence of both molybdate and phosphate in the electrolyte there were also some larger circular spots on the electrode surface (Figure 10). The spots were around 10 µm in diameter and darker than the surrounding film (image taken in back-scatter mode). EDS analysis showed that the molybdenum level was higher in the spots that in other areas of the film. The spot can be an elevated structure, dramatically increasing film thickness.

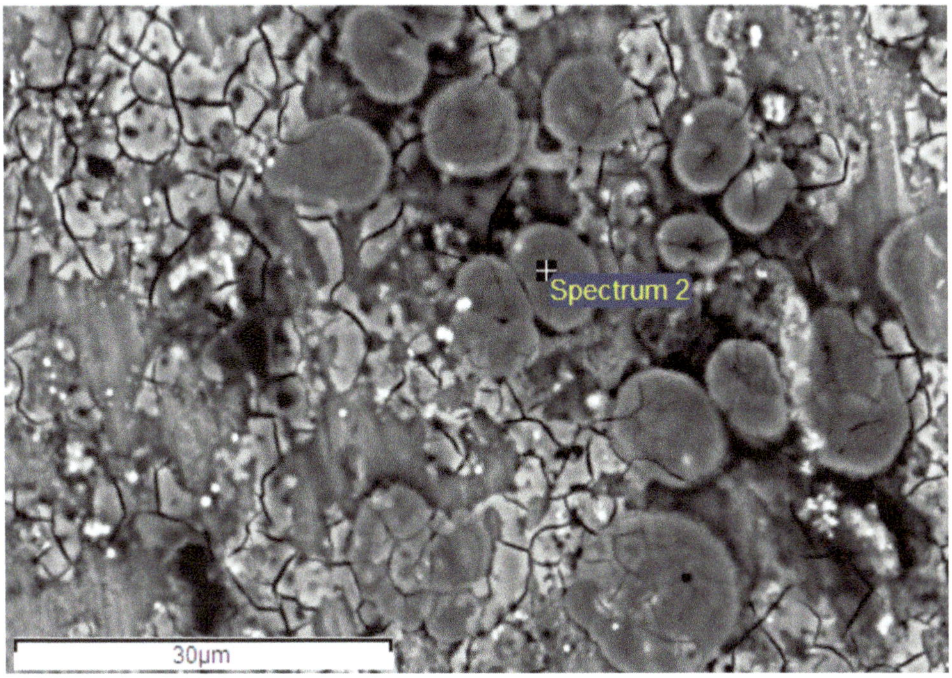

Figure 10. *SEM image in back-scatter mode of a titanium electrode polarized at -3 kA m^{-2} for 20 min at 70 °C in 1.6 M NaCl with 100 mM hypochlorite, 80 mM Mo(VI) and 32 mM phosphate.*

Effects of molybdate on the anode potential

Since chlorate cells are undivided, the anode will encounter the same electrolyte as the cathode. Thus any effects on the anode reactions that the electrolyte additives may have are important to study, and Mo(VI) at high concentrations have shown to increase the oxygen levels in the off gas [6],[8]. Oxygen may form from several different reactions on the anode or in homogeneous reactions in the electrolyte [1] and experiments were made to evaluate if the high oxygen values are accompanied by an increase in anode potential. The anions $Cr_2O_7^{2-}$, PO_4^{3-} and NO_3^- have been observed to increase the oxygen levels in the off gas in chlorate electrolysis, probably caused by anodic oxygen evolution and the likely mechanism suggested was an adsorption of the anions on the anode surface [16]. In another study an increased anode potential of about 15 mV was observed when increasing the Cr(VI) concentration from 3 to 9 g dm^{-3} $Na_2Cr_2O_7$ for a DSA in an electrolyte of pH 2 [10].

To study anode effects of Mo(VI), RuO_2/TiO_2 electrodes produced by spin coating and shaped into RDEs were polarized at 3 kA m^{-2} for 30 min with and without 40 mM Mo(VI) in the electrolyte. The addition of Mo(VI) resulted in an increase in anode potential around 40 mV, see Figure 11. Using SEM, no clear morphology change due to film formation or deposits could be seen on the RuO_2/TiO_2 surfaces, although Mo/Na ratios 2-10 times larger than expected from electrolyte composition were measured with EDS. Also XRF measurements indicated increased molybdenum content on the surface. Furthermore predominance diagrams show $MoO_3{\cdot}2H_2O(s)$ as the dominant phase in 10 mM Mo(VI) solution if the conditions are acidic and oxidizing [17], such as close to the anode. It is thus likely that the increased oxygen is caused by precipitation or adsorption of Mo-containing species on the anode surface.

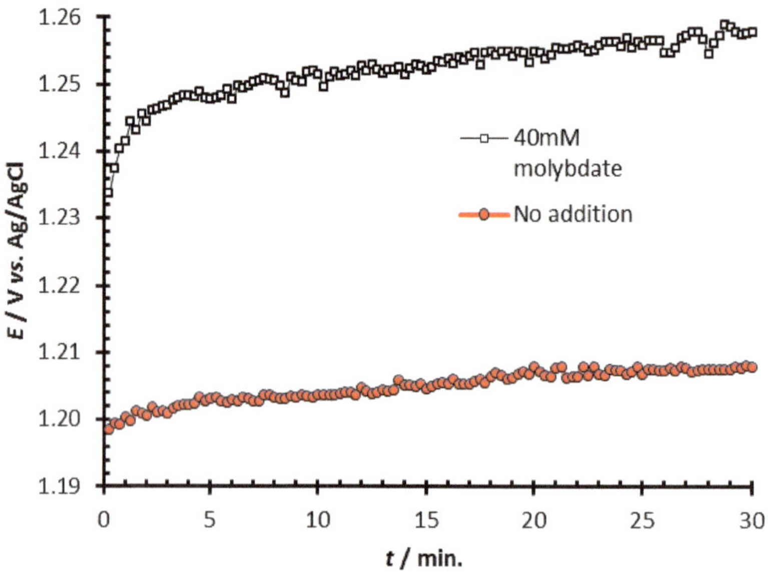

Figure 11. *Galvanostatic polarizations of DSA-RDEs (spincoated 3-layers) at 3 kA m^{-2} in 2 M NaCl, pH 7, 70°C at a rotation rate of 3000 rpm.*

Conclusions

The presence of Mo(VI) ions in an electrolyte of near neutral pH can increase the selectivity for hydrogen evolution as the Mo-containing film formed on the cathode surface hinders the side reaction of hypochlorite reduction.

Potential sweeps and current efficiency experiments in neutral electrolytes show that on a smooth substrate such as polished titanium the suppression of hypochlorite reduction by Mo(VI) addition is more efficient than on steel or on corroded iron.

Even very low levels of Cr(VI), in the μM range, can efficiently suppress hypochlorite reduction on polished titanium and steel. Thus if using smooth and stable cathode materials the chromate concentration on the chlorate process may be substantially reduced compared to the levels used today (1-6 g dm^{-3} $Na_2Cr_2O_7$ corresponding to about 10-50 mM Cr(VI)). Note though that the reaction of electrochemical chlorate reduction has not been addressed in the present study. The inhibiting action of very low Cr(VI) concentrations also means that when studying alternatives to Cr(VI) in the chlorate process possible Cr(VI) present in the chlorate salt used to make the electrolyte in the experiments should be considered. Mo(VI) on the other hand needed higher concentrations and longer polarization times to show an effect on the current efficiency.

Phosphate, a possible buffer needed in the chlorate process if removing, or substantially lowering the concentration of, Cr(VI) does not negatively influence the current efficiency in the presence of Cr(VI) or Mo(VI). Instead a slight increase in current efficiency was noticed when adding 32 mM NaH_2PO_4 to the electrolyte. The Mo-containing cathode films formed during cathodic polarization become thinner if the electrolyte during the film build-up also contained phosphate, but the films still appeared as effective at hindering hypochlorite reduction.

The presence of Mo(VI) in the electrolyte can increase the potential of a RuO_2-TiO_2 anode. The anode potential increased 40 mV after adding Mo(VI) corresponding to 40 mM, a concentration that in a previous study [6] resulted in increased oxygen levels in the off gas. Thus the increased oxygen production is likely produced in an anodic side reaction promoted by the adsorption of Mo(VI) species on the anode.

Considering increased oxygen levels and increased cathode and anode potentials, if Mo(VI) is to be used in the chlorate process it should be used at a low concentration to minimize the negative side effects.

Acknowledgements: *The Swedish Energy Agency, the Swedish Research Council, Permascand, Eka Chemicals and the European Regional Development Fund through the project "EnergyWise" are acknowledged for financing this study.*

References

[1] J.E. Colman, *American Institute of Chemical Engineers Symposium Series* **77** (1981) 244-263
[2] G. Lindbergh and D. Simonsson, *Electrochim. Acta* **36** (1991) 1985-1994
[3] A. Cornell, G. Lindbergh and D. Simonsson, *Electrochim. Acta* **37** (1992) 1873-1881
[4] M.S. Vukasovich, *Polyhedron* **5** (1986) 551-559
[5] F.J. Presuel-Moreno, M.A. Jakab and J.R. Scully, *J. Electrochem. Soc.* **152** (2005) B376-B387
[6] M. Li, Z. Twardowski, F. Mok, N. Tam, *J. Appl. Electrochem* **37** (2007) 499–504
[7] J. Gustavsson, C. Hummelgård, J. Bäckström, I. Odnevall Wallinder, S. Mohammed Habibur Rahman, G. Lindbergh, S. Eriksson and A. Cornell, *J. Electrochem. Sci. Eng.* **2(3)** (2012) 105-120
[8] M. Rosvall, K. Hedenstedt, A. Sellin, J. Gustavsson and A. Cornell, (Akzo Nobel Chemicals International B.V.), WO 2010/130546A1 (2010)
[9] N. Krstajic, V. Jovic and G.N. Martelli, (Industrie De Nora S.p.A.), WO 2007/063081A2 (2007)
[10] A. Cornell, B. Håkansson and G. Lindbergh, *J. Electrochem. Soc.* **150** (2003) D6-12.
[11] J. Wulff and A. Cornell, *J. Appl. Electrochem*. **37** (2007)181.
[12] P. Byrne, E. Fontes, O. Parhammar and G. Lindbergh, *J. Electrochem. Soc.* **148** (2001) D125-D132
[13] Z. Yong, J. Zhu, C. Qiu and Y. Liu, *Appl. Surf. Sci.* **255** (2008) 1672–1680
[14] H.A. Duarte, K. Jha and J.W. Weidner, *J. Appl. Electrochem.* **28** (1998) 811–817
[15] A. Cornell and D. Simonsson, *J. Electrochem. Soc.* **140** (1993) 3123-3129
[16] M.M. Jakšić, A. Despić, B. Nikolić and S. Maksić, *Croat. Chem. Acta* **44** (1972) 61-66
[17] P. Wang, L. L. Wilson, D. J. Wesolowski, J. Rosenqvist and A. Anderko, *Corr. Sci.* **52** (2010) 1625-1634

12

Morphology and cyclic voltammetry analysis of in situ polymerized polyaniline/graphene composites

DEVESH KUMAR MAHLA, SUBHENDU BHANDARI, MOSTAFIZUR RAHAMAN*✉, DIPAK KHASTGIR✉

Rubber Technology Centre, Indian Institute of Technology Kharagpur, Kharagpur, India
**Presently at Chemical Engineering Department, College of Engineering Sciences, King Fahd University of Petroleum and Minerals (KFUPM), Dhahran, Saudi Arabia*

✉Corresponding Authors

Abstract

Graphene (G) was synthesized from normal graphite powder via graphene oxide (GO). Graphene was also produced from finer particles of graphite subjected to ball milling to check the effect of particle size of graphene on the properties of composite. PANI/graphene composites of different compositions were prepared by in situ polymerization of aniline to polyaniline in the presence of graphene powder. Graphene, graphene oxide and PANI/graphene composites were characterized by UV, IR, TEM, and cyclic voltammetry. PANI/graphene composites exhibit higher current in cyclic voltammetry study compared to either neat PANI or neat graphene. The value of capacitance achieved for PANI/graphene composites is found to depend on the size of graphene particles, finer the particle higher is the capacitance for the composites. However, the effect of composition on CV characteristics of composite is relatively less pronounced compared to the size of graphene sheets coated with PANI.

Keywords

Polyaniline; Graphene; Conductive composites; Cyclic voltammetry.

Introduction

Conductive polymer composites filled with different carbon fillers have been used in different electrical and electronic applications [1–7]. Among the carboneous fillers, graphene based polymer composites have attracted much attention in the recent years [8–10]. Graphene, one-atom-thick planar sheet of carbon atoms, is densely packed in a honeycomb crystal lattice of graphite. Graphene, at large scale, is produced by reducing graphene oxide by some reducing agent. Graphene is having large aspect ratio (ratio of surface area to thickness) [11], high mechanical strength [12], high electrical conductivity [13], ability to disperse in polymers well [14], high thermal conductivity and stability [15]. It can be effectively exploited for many applications [16,17]. Graphene oxide derived from graphite through oxidation has many polar groups on its surface like –C=O, –COOH, –OH, which give its properties like hydrophilicity, dispersibility, and compatibility with many polymers.

Among conductive polymers, polyaniline (PANI) has attracted considerable attention because of its ease of preparation, low cost of monomer, good environmental stability, good conductivity control through doping as well as a good control of oxidation level. All these properties give PANI the potential for wide applications [18–22]. It can effectively be used as opto-electronic sensors [23–26], conductive paints and adhesive [27–29].

PANI has also good power and energy density, so it has been analyzed for the development of new supercapacitor materials [30–32]. Generally, electrochemical super capacitors are of two types; type-1 is electrical double layer capacitor (EDLC) due to charge separation at electrode/electrolyte interface, and type-2 is pseudocapacitor, where capacitance is due to both reversible faradic reaction and charge separation [33]. The latter category of capacitor materials can be further divided in to two different types of material based on their origin, for example i) metal oxide like RuO_2 and ii) conducting polymer like PANI.

The energy and power density of power sources like batteries and capacitors can be shown by Ragone plot. Batteries have good energy density and EDLC type capacitors have good power density whereas pseudocapacitors work on the principle of both batteries and capacitors, simultaneously. PANI can act as a pseudocapacitor and it has the power to store energy by reversible chemical reactions, as observed in batteries, and also like EDLC. In polyaniline/graphene composite, graphene not only increases the EDLC capacity of PANI but also helps in facilitating charge transfer and consequently enhances the utilization of PANI as a capacitor material by increasing available surface area.

The scrutiny of available literatures reveals that there exist some reports on PANI/graphene composites. However, all these reports mainly deal with only one or at best two compositions of PANI/graphene system. The effect of size of graphene particles on the properties of PANI/graphene composites has not been extensively investigated. So the present study deals with synthesis and characterization of PANI/graphene composites with different compositions. The preparation of PANI/graphene composites with different size of graphene particles has also been reported.

Experimental Details

Polyaniline was synthesized in aqueous medium as reported in the literature [34]. The graphite oxide was prepared by modified Hummers method [35]. During the preparation of graphite oxide, 50 mg of graphite was mixed with 25 mg of $NaNO_3$ and 2ml H_2SO_4. After stirring of the mixture for 10 minute in ice bath, 150 mg of $KMnO_4$ was added to it with constant stirring for 15 minutes. Then the resultant mixture was heated up to 98 °C and deionised water was added slowly with constant stirring. H_2O_2 was added and then the suspension was filtered followed by washing of filtrate with deionised water for 8 times to remove any water soluble residues. The filtrate so prepared is graphite oxide which was used in further synthesis. During the preparation of polyaniline/graphene composites the following steps have been adopted. In the first step, the graphite oxide was added (as per the ratio desired, explained below) to deionised water. The mixture was ultrasonicated for one hour followed by addition of aniline to it, and the mixture was again ultrasonicated for half an hour. While ultrasonicated, layers in graphite oxide were delaminated to produce graphene oxide. This mixture of graphene oxide and aniline was then kept as such for a day at room temperature. In this mixed mass the requisite amount of 0.2 M solution of HCl was added at room temperature followed by requisite amount of 0.25 M solution of ammonium persulfate was added. Continuous stirring was maintained during the reaction. After addition of ammonium persulfate the stirring was continued for 10 minutes and the conical flask was covered with lid and kept as such for 24 hours. The whole reaction was carried out between 5 - 10 °C with the help of ice bath. The polyaniline/graphene composite so prepared was filtered. To remove the unreacted reactant from the mixture and to dope PANI adequately the filtrate was repeatedly washed with 2 M HCl solution. The so prepared composite was dried in vacuum oven at 50 °C temperature until the final weight became constant.

The graphite oxide in polyaniline/graphene composites was varied in different ratio with its weight percent being 4, 8, 12 and 16 of combined weight of initially taken aniline and graphite oxide. The nomenclature of the so varied weight has been given as PG_4, PG_8, PG_{12} and PG_{16}.

PANI/graphene composites were also prepared in a little different way. In this preparation the graphite powder as received was ball milled for 60 hours at 300 rpm to form finer particles of graphite and used for preparation of graphene oxide followed by preparation of PANI/graphene composites. This was done to check the effect of size of graphene sheet on composite properties. The so prepared nano graphite powder was converted into graphite oxide by modified Hummers method [19]. The procedure adopted for the preparation of polyaniline/nano-graphene composite is the same as described earlier. The weight per cent of nano-graphite oxide has also been varied in the same way as described above in the proportion of 4, 8, 12 and 16, and identified as PnG_4, PnG_8, PnG_{12} and PnG_{16}.

Different polyaniline/graphene composites were characterized by different techniques like cyclic voltammetry (CV) and transmission electron microscopy (TEM). The CV study was performed with different scan rates of 10 mV/s, 50 mV/s, 80 mV/s and 100 mV/s with electrolyte being 2 M H_2SO_4 solution. The reference electrode used was Ag/AgCl and counter electrode being platinum wire. The working electrode used in the experiment was modified glassy carbon electrode. The polyaniline/graphene composite was mixed in the Nafion solution of chloroform. The mixture was then put on the electrode with the help of glass rod. The CV was performed within the potential range of -0.2 to 1.0 V at room temperature.

For TEM analysis, the samples were used after ultrasonication for one hour in deionised water. IR and UV techniques were used for characterization, and sample of graphene and graphene oxide were ultrasonicated in acetone for half an hour before testing. This solution was spread over KBr pellet and IR was performed. For UV analysis the G and GO samples were ultrasonicated in deionised water for 1 hour before test.

Results and Discussion

UV and FT-IR

Synthesized graphene oxide and graphene were characterized by UV and FT-IR spectroscopy. From UV spectra (Figure 1), one can detect the $\pi - \pi^*$ transition at 260 nm (38461 cm^{-1}) for the graphene while graphene oxide shows $n - \pi^*$ transition at 305 nm (32786 cm^{-1}) and $\pi - \pi^*$ at 240 nm (41667 cm^{-1}). From IR spectra (Figure 2) for graphene a very few peaks can be detected as neat graphene contains very few organic groups. FT-IR peak in between 3500 to 3100 cm^{-1} is due to the absorbed moisture and the peak around 2200 cm^{-1} is due to atmospheric carbon dioxide often seen if the experiment is not done in nitrogen environment. There is a little hump around 1500 cm^{-1} because of C=C stretching in benzene ring of graphene. In the case of GO, due to oxidation there are several absorption bands corresponding to large number of chemical groups attached to the graphene sheets.

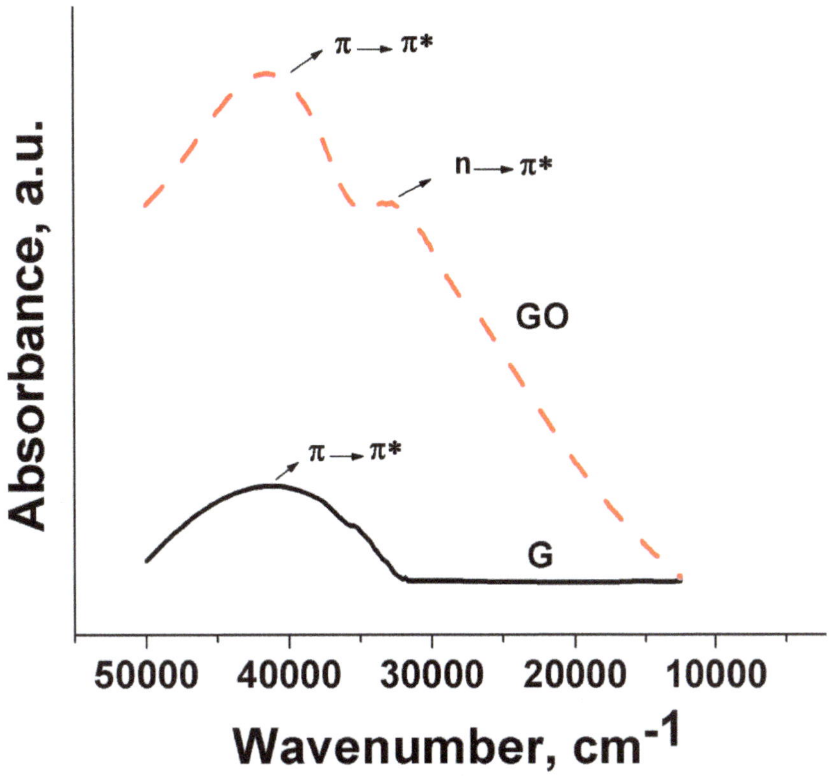

***Figure 1**. UV spectra of graphene and graphene oxide.*

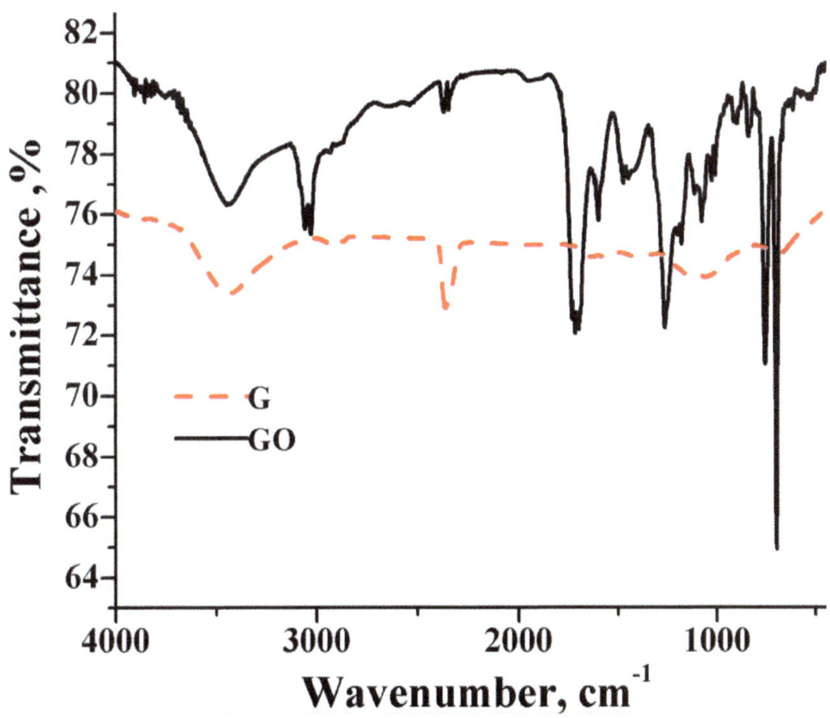

***Figure 2**. FT-IR spectra of graphene and graphene oxide.*

TEM Analysis

In TEM images the presence of single layer graphene (Figure 3) and multi layer graphene sheets stacked together (Figure 4) can be seen. From other TEM images (Figures 5 and 6) the formation of polyaniline particles (Figure 5) and polyaniline coating on graphene sheets (Figure 6) can also be detected. Graphite to graphene was routed through formation of graphene oxide followed by ultrasonication when different degree of delamination of original graphite occured. If the extent of delamination is very less, the original layered structure of graphite is almost retained. All these studies reveal that successful preparation of graphene from graphite can be achieved by the method described in experimental section.

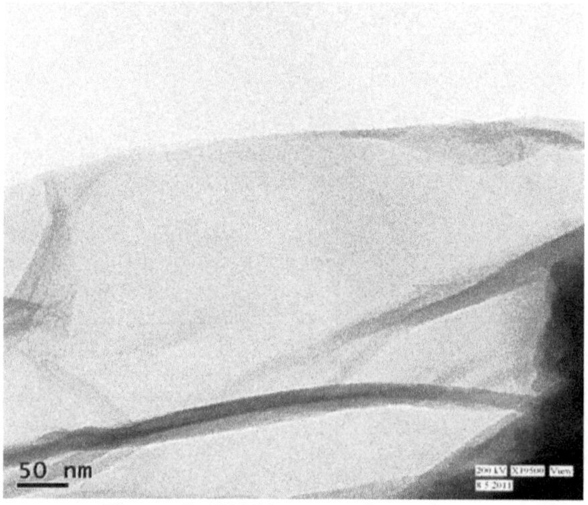

***Figure 3.** TEM image of graphene sheet*

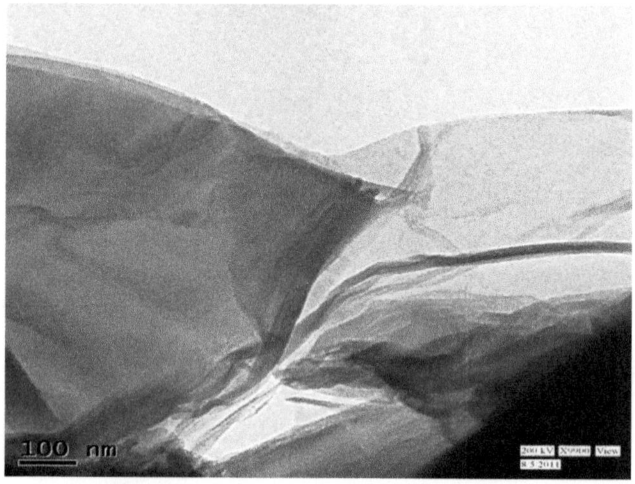

***Figure 4.** TEM image of few graphene sheets stacked together*

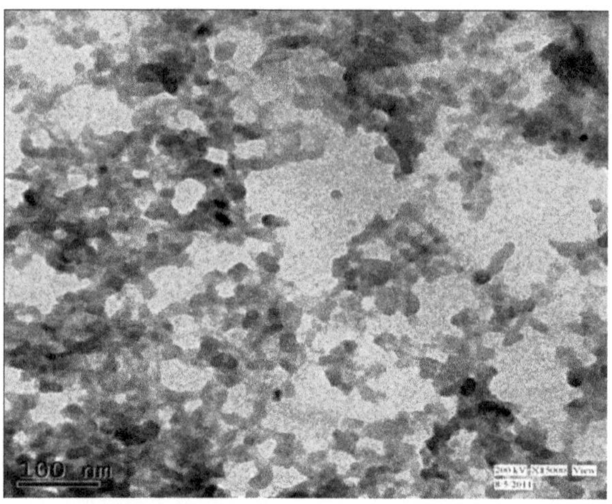

***Figure 5.** TEM image of polyaniline particles*

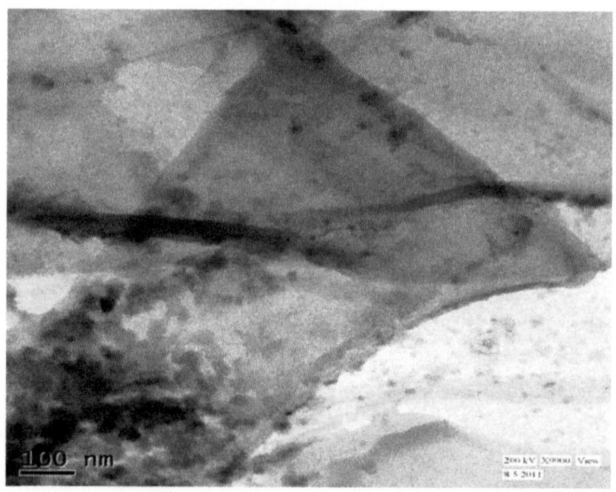

***Figure 6.** TEM image of polyaniline/graphene composite.*

Cyclic voltammetry analysis

The study of CV (current-voltage) for the polyaniline/graphene composite was carried out to understand the electrochemical behavior of these composites. Figure 7 shows the CV of PANI, G, GO and PG_4. The shapes of CV curves show the reversible charge-discharge behavior of composite electrode [36,37]. For PANI the first anodic peak appears around 0.29 V, which is associated with the oxidation of leucoemeraldine to emeraldine [38–41]. The second anodic peak around 0.84 V is attributed to the emeraldine to pernigraniline transition (Scheme 1). Figure 7 also shows that with the addition of graphene the current flow increases and the peak also broadens as compared to neat PANI and neat graphene as shown in the inset. The differences in qualitative and quantitative shape of CV curve for PANI and polyaniline/graphene composites have been attributed to the dispersion of PANI particles on graphene sheet which reduces the migration length of electrolyte ions during the charge-discharge process where graphene sheets provide an excellent path for the charge transfer [42]. Hence, the graphene increases the efficacy of PANI by providing a larger surface area and reducing the resistance to current flow.

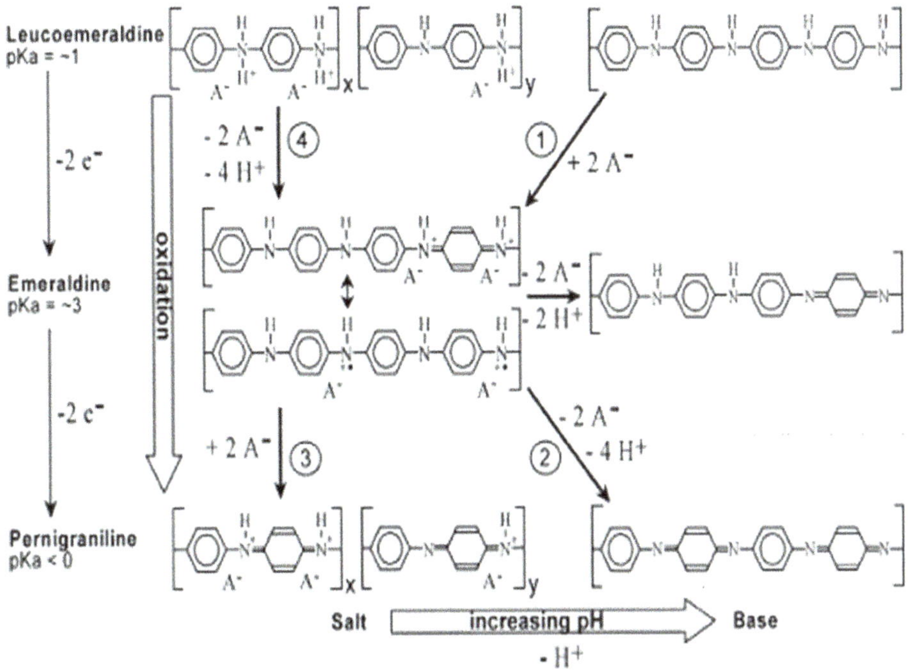

***Scheme 1** Oxidation/reduction chemistry of polyaniline.*

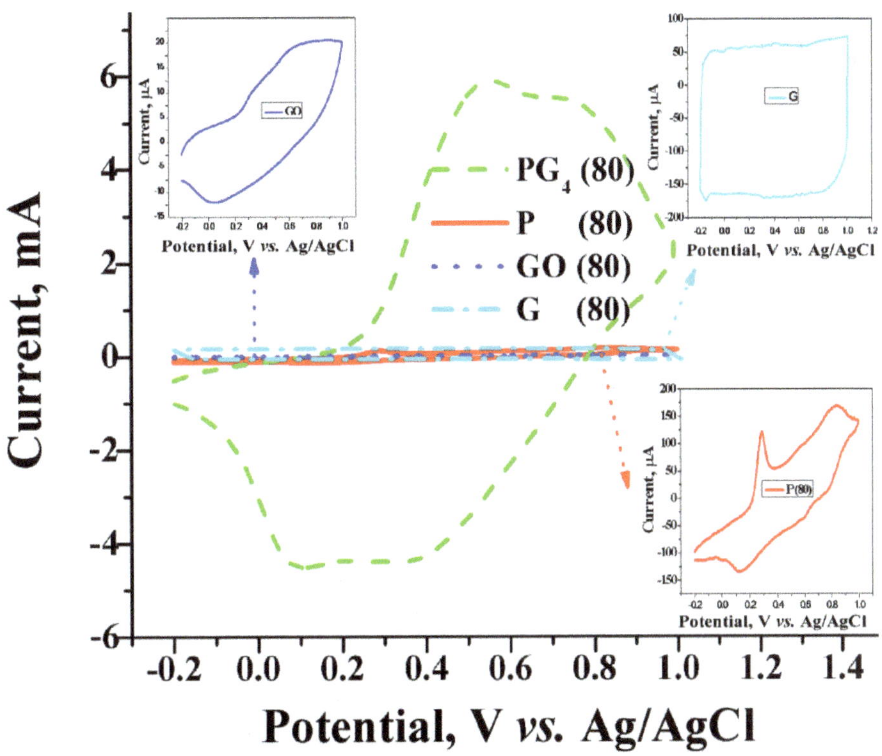

***Figure 7.** Cyclic voltammograms of PG8, polyaniline, graphite oxide and graphene*

The stability of the composite has been checked by repeating the CV plots under the same condition for the same sample for several cycles [43]. Figure 8 shows the average particle size of graphite as received and the nano sized graphite with cumulative contribution in the total number of particles, %. The particle size analysis was done by Fritz particle size analyzer. The average particle size of graphite is 800 µm while the particle size of nano graphite was beyond the range of analyzer. Figure 9 shows that after 100, 200 and 300 CV cycles, at 0.08 V/s scan rate, the current-potential graphs almost superimposes on each other. This reveals the good electrochemical stability of the composite which is essential when it is used as a capacitor.

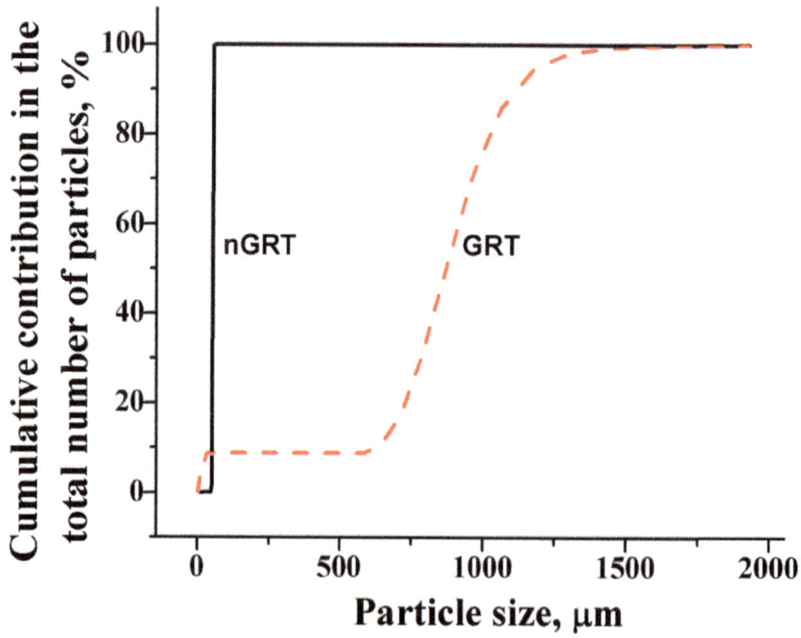

Figure 8. *Particle size analysis of graphite and nano graphite*

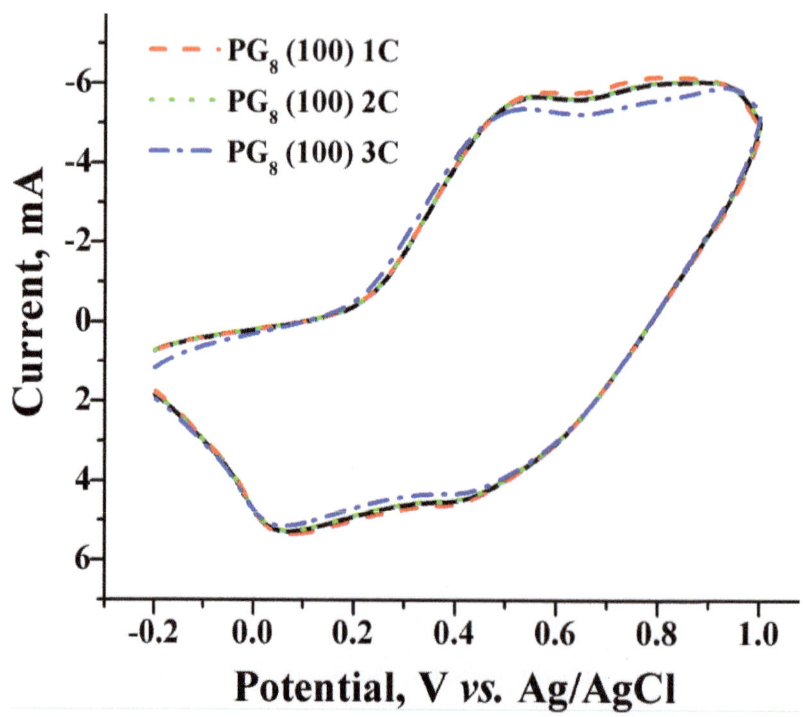

Figure 9. *Cyclic voltammogram of PG8 at 100 mV/s scan rate after 100, 200 and 300 cycles*

Both composition of PANI/graphene composite and size of graphene particles are expected to influence current-voltage characteristics of composites. Figure 10 compares the current potential behavior of polyaniline/graphene composite with variation in graphene weight percent and size of graphene sheet. It shows that with the addition of graphene the peaks in current - potential plots broaden in comparison to PANI. This is due to availability of alternate as well as low energy paths for charge transfer in PANI/graphene system through graphene. The presence of graphene in composite decreases the resistance of system hence there is an increase in the current and broadening of the peak. As shown in Figure 10 the resistance in PnG_4 is almost equal to that of PG_8 composite. This shows that with smaller size of graphene sheet the available surface area is higher and hence it leads to better utilization of PANI. Hence, this infers that with larger surface area the small scale graphene better decreases the resistance of the system and increases the capacitance of composite material by better utilization of PANI in composite. Figure 11 shows that with increase in the graphene weight percent there is no much increase in the current flow relatively to the neat PANI and in PANI/graphene composite as shown in Figure 7 [44]. It could also be seen that at lower scan rates (for example 0.01 V/s) different oxidation and reduction peaks could be distinguished but as the scan rate is increased, the peaks merged. This is due to the time constant effect in CV, since $I=E/R_s\ (e^{-t}/C_dR_s)$, where E is the potential step, R_s being the solution resistance, C_d, the double layer capacitance and t is time [45-46]. Here I decreases as the C_dR_s increases which happens at a low scan rate. As the scan rate is increased C_dR_s doesn't give enough time to the current to decrease hence distinct peaks are not observed.

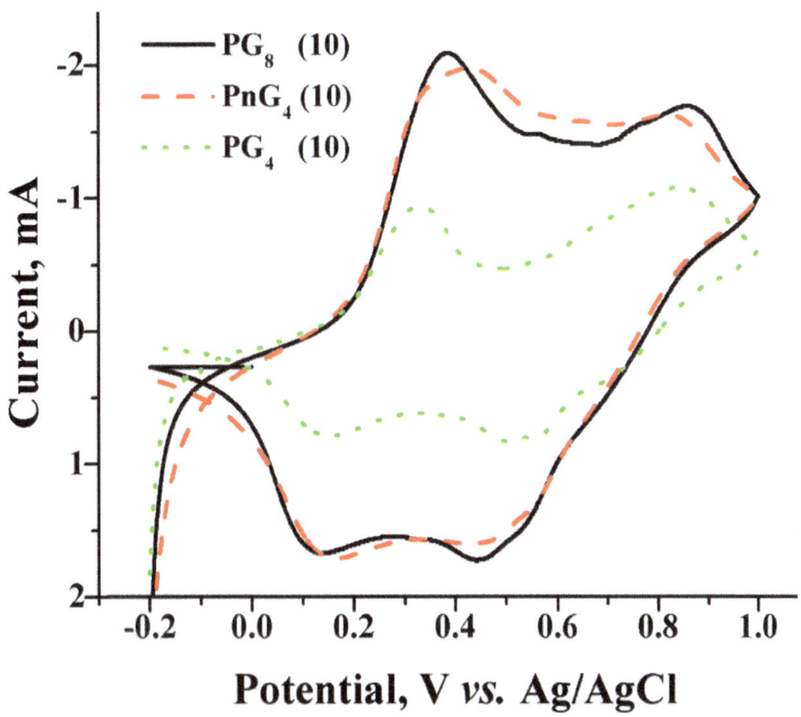

Figure 10. *Cyclic voltammogram of PG4, PG8 and PnG4 at 10 mV/s scan rate*

Conclusions

Different spectroscopic and morphological analysis show that chemical process used in this experimentation leads to successful preparation of graphene sheet from graphite. From CV analysis it has been found that the size of graphene sheet plays an important role in determining the capacitance of the composite. In fact, with the reduction of the size of graphene sheet there is an increase in the capacitance value.

Acknowledgement: *Authors would like to thank CSIR New Delhi for providing the fund to carry out this research work.*

References

[1] R. Ramasubramaniam, J. Chen, H. Liu, *Appl. Phys. Lett.*, **83** (2003) 2928–2930.
[2] M. Rahaman, T. K. Chaki, D. Khastgir, *J. Mater. Sci.*, **46** (2011) 3989–3999.
[3] W. J. Zou, S. S. Mo, S. L. Zhou, T. X. Zhou, N. N. Xia, D. S. Yuan, *J. Electrochem. Sci. Eng.* **1** (2011) 65–71.
[4] N. J. S. Sohi, M. Rahaman, D. Khastgir, *Polym. Compos.*, **32** (2011) 1148–1154.
[5] V. Panwar, R. M. Mehra, *Polym. Eng. Sci.*, **48** (2008) 2178–2187.
[6] N. Syarif, I. A. Tribidasari, W. Wibowo, *J. Electrochem. Sci. Eng.* **3** (2013) 37–45.
[7] X. Luo, D. D. L. Chung, *Compos.: Part B*, **30** (1999) 227–231.
[8] H. Kim, A.A. Abdala, C.W. Macosko, *Macromolecules*, **43** (2010) 6515–6530.
[9] T. Ramanathan, A.A. Abdala, S. Stankovich, D.A. Dikin, M. Herrera-Alonso, R.D. Piner, D.H. Adamson, H.C. Schniepp, X. Chen, R.S. Ruoff, S.T. Nguyen, I. A. Aksay, R.K. Prud'homme, L.C. Brinson, *Nat. Nanotech.*, **3** (2008) 327–331.
[10] Y.W. Zhu, S. Murali, W.W. Cai, X.S. Li, J.W. Suk, J.R. Potts, R.S. Ruoff, *Adv. Mater.*, **22** (2010) 3906–3924.
[11] M.D. Stoller, S.J. Park, Y.W. Zhu, J.H. An, R.S. Ruoff, *Nano Lett.*, **8** (2008) 3498–3502.
[12] C. Lee, X. Wei, J.W. Kysar, J. Hone, *Science*, **321** (2008) 385–388.
[13] X. Du, I. Skachko, A. Barker, E.Y. Andrei, *Nat. Nanotech.*, **3** (2008) 491– 495.
[14] H.B. Zhang, W.G. Zheng, Q. Yan, Y. Yang, J.W. Wang, Z.H. Lu, G.Y. Ji, Z.Z. Yu, *Polymer*, **51** (2010) 1191–1196.
[15] A. Yu, P. Ramesh, X. Sun, E.M. Bekyarova, E. Itkis, R.C. Haddon, *Adv. Mater.*, **20** (2008) 4740–4744.
[16] A.K. Geim, *Science*, **324** (2009) 1530–1534.
[17] D. Li, R. B. Kaner, *Science*, **320** (2008) 1170–1168.
[18] A.G. Green, A.E. Woodhead, *J. Chem. Soc.*, **1910** (1997) 2388–2403.
[19] A. K. Sharma, Y. Sharma, *J. Electrochem. Sci. Eng.* **3** (2013) 47–56.
[20] M. Magioli, B. G. Soares, A. S. Sirqueira, M. Rahaman, D. Khastgir, *J. Appl. Polym. Sci.*, **125** (2012) 1476–1485.
[21] M. Rahaman, L. Nayak, T. K. Chaki, D. Khastgir, *Adv. Sci. Lett.*, **18** (2012) 54–61.
[22] S. Bhadra, D. Khastgir, *Polym. Test.*, **27** (2008) 851–857.
[23] S. Bhadra, S. Chattopadhyay, N.K. Singha, D. Khastgir, *J. Appl. Polym. Sci.*, **108** (2008) 57–64.
[24] M. Rahaman, T. K. Chaki, D. Khastgir, *J. Appl. Polym. Sci.*, **128** (2013) 161–168.
[25] S. Bhadra, D. Khastgir, N.K. Singha, J.H. Lee, *Prog. Polym. Sci.*, **34** (2009) 783–810.
[26] J.C. Chiang, A.G. MacDiarmid, *Synth. Met.*, **13** (1986) 193–205.
[27] W. Hualan, H. Qingli, Y. Xujie, L. Lude, W. Xin, *Electrochem. Commun.*, **11** (2009) 1158–1161.
[28] J.P. Zheng, P.J. Cygan, T.R. Jow, *J. Electrochem. Soc.*, **142** (1995) 2699.
[29] A.J. Burke, *J. Power Sources*, **91** (2000) 37.
[30] B.E. Conway, Electrochemical Supercapacitors, Kluwer-Plenum, New York (1999).
[31] S. Sopčić, M. Kraljić Roković, Z Mandić, *J. Electrochem. Sci. Eng.* **2** (2012) 41–52.
[32] S. Chandra, et al., *Chem. Phys. Lett.*, **519–520** (2012) 59–63.
[33] T. Nakajima, A. Mabuchi, R. Hagiwara, *Carbon*, **26** (1988) 357–361.
[34] L. Anton, H. Heyong, F. Michael, K. Jacek, *J. Phys. Chem. B*, **102** (1998) 4477–4482.

[35] H. Dongxue, C. Ying, *Colloid. Surf.,* **259** (2005) 179–187.
[36] V. Khomenko, E. Frackowiak, F. Beguin, *Electrochimica Acta,* **50** (2005) 2499–2506.
[37] Z. Hao, C. Gaoping, W. Weikun, Y. Keguo, X. Bin, Z. Wenfeng, C. Jie, Y. Yusheng, *Electrochimica Acta*, **54** (2009) 1153–1159.
[38] H. Chi-Chang, L. Jeng-Yan, *Electrochimica Acta*, **47** (2002) 4055–4067.
[39] R. Prakash, *J. Appl. Polym. Sci.,* **83** (2002) 378–385.
[40] E.I. Santiago, E.C. Pereira, L.O.S. Bulhões, *Synth. Met.,* **98** (1998) 87–93.
[41] E. Smela, W. Lu, B.R. Mattes, *Synth. Met.*, **151** (2005) 25–42.
[42] Y. Jun, W. Tong, S. Bo, F. Zhuangjun, Q. Weizhong, Z. Milin, W. Fei, *Carbon*, **48** (2010) 487–493.
[43] Z. Lu, Z. Liang, X. Yuxi, Q. Tengfei, Z. Linjie, S. Gaoquan, *Electrochimica Acta*, **55** (2009) 491–497.
[44] Y. Li, H. Peng, G. Li, Kezheng Chen, *Eur. Polym. J.*, **48** (2012) 1406–1412.
[45] Y.G. Wang, H.Q. Li, Y.Y. Xia, *Adv. Mater.*, **18** (2006) 2619–2623.
[46] J. B. Allen, R.F. Larry, 2nd addition, Electrochemical Methods: Fundamentals and Applications, John Wiley (2001).

13

Electrochemical behaviour of PES ionomer and Pt-free catalyst for PEMFCs

STEFANIA GIORDANO✉, MARIANGELA LONGHI, LEONARDO FORMARO, HERMES FARINA and GIUSEPPE DI SILVESTRO

Dipartimento di Chimica, Università degli Studi di Milano, Via Golgi, 19, 20133, Milano, Italy

✉Corresponding Author

Abstract

Proton Exchange Membrane Fuel Cells (PEMFCs) represent promising technologies to the world economy, with many applications and low environmental impact. A most important aspect concerning their widespread implementation is the cost of polymeric membranes, typically perfluorinated membranes and platinum-based catalytic electrode materials, all of which are necessary to promote electrode reactions, thus increasing fuel cell energy efficiency. In this work, we present some data about non-fluorinated polyetheresulphone (PES) membranes and Pt-free catalysts, as possible substitutes of the above materials. Their electrochemical behaviour in oxygen reduction reaction in acidic media are investigated and compared with available reference materials.

Keywords

ORR, Pt-free catalyst, PES, PEMFC.

Introduction

Polymer electrolyte membrane fuel cells (PEMFCs) have achieved significant progress over the past few decades. They are considered one of the most promising fuel cell technologies for both stationary and mobile applications owing to their high energy efficiency, convenient operation, and environmentally friendly characteristics. The main objective in fuel cell technologies is to develop low cost, high-performance and durable materials [1-6]. At present, platinum is the best cathode catalyst for oxygen reduction reaction (ORR) in PEMFCs; however, because of the scarcity and cost of the metal, there is a strong effort to find alternative metals or alloys with similar activity [7-10]. Promising advances have been made with new composites (non-precious metals/heteroatomic polymers), pyrolysed metal porphyrins (cobalt or iron porphyrins viewed as the most promising precursors) or bio-inspired materials. As they are capable of combining high oxygen-reduction activity with good performance, these materials appear viable alternative catalysts for ORR [11-17]. In PEMFC, catalysts are commonly supported on proton conductive membranes. The most common material of this kind is DuPont's Nafion®, a perfluorinated sulphonic acid polymer with many noteworthy features: good conductivity up to 80°C, which decreases beyond 80°C by dehydration; however, this also has a high cost. Extensive research has been done to produce cheaper membranes to replace Nafion®. New electrolyte membranes obtained by grafting with styrene and sulphonic acid look promising and further development is underway to improve their performance [18-20].

Considering the above mentioned critical aspects, in this work we are looking for new materials for PEMFCs applications, *i.e.* non-fluorinated polymeric membranes and platinum-free catalysts. As possible Nafion substitutes, we synthesised non-fluorinated polyetheresulphones (PES) with different degrees of sulphonation. Polymers were obtained starting from different ratios of sulphonated and non-sulphonated co-monomers. Varying the monomers ratio it was possible to obtain polymers with different values of ion exchange capacity, swelling, hydrodynamic volume and rigidity. PES membranes have been characterised by thermogravimetric analysis (TGA) and differential scanning calorimetry (DSC) in order to study thermal stability and water retention. On the side of Pt-free catalysts, the one presented below was synthesised by a sugar pyrolysis route in the presence of a nitrogen precursor and a non-precious transition metal. Thiamine was chosen as the precursor because, besides nitrogen, it also contains sulphur that may be a useful carbon-doping element [21]. Iron was used as the transition metal. To increase catalyst surface area and to control porosity in the mesoporosity range, a hard template method was adopted using a silica gel in water [22]. Catalyst surface area and porosity were determined by BET and BJH theory, respectively. The electrochemical behaviour of the PES membrane and thiamine catalyst in oxygen reduction (ORR) conditions were investigated and compared to reference materials, *i.e.* Nafion 1100 and EC-20 catalyst (20 % Pt dispersed on Vulcan XC72R). An ORR study was carried on in acidic media by cyclic voltammetry using a rotating disk electrode (RDE).

Experimental

Materials

Glucose, thiamine, Fe(II) acetate, glacial acetic acid, sodium hydroxide, silica (Silica Gel 60 HR), HCl (37 %), H_2SO_4 (95-97 %), ethanol, N-methylpyrrolidone (NMP), dimethyl acetamide (DMAc), and Nafion® 1100 (5 wt. % suspension) were purchased from Aldrich and used as received. $HClO_4$ and K_2CO_3, dimethyl sulphoxide (DMSO) were obtained from Fluka and Carlo Erba, respectively. High purity water from a MilliQ system (Millipore) was used. Nitrogen and oxygen (5.5 and 5 nines respectively) were purchased from Sapio.

Catalyst synthesis

The Pt-based catalyst EC-20 was used as received. Home-made catalyst carbon was obtained by the following procedure. Thiamine was added to a nearly saturated glucose-in-water solution (1.68 mol L^{-1}) in a 1:10 molar ratio. Thiamine dissolution was aided by equimolar glacial acetic acid. Thereafter, acetate iron salt was added (0.96 wt. % on total amount of non-water reactants). Ten millilitres of the solution was stirred with 4.3 g of silica in order to form a gel. Then, the suspension was loaded in a quartz reactor, degassed with nitrogen and inserted in a preheated vertical oven at 600°C for one hour to carbonise the precursors as fast as possible. Silica was removed in 3 M boiling sodium hydroxide followed by repeated carbon washing/filtering. Products were dried in nitrogen (100°C, 24 hours) and finely ground. Materials were heat-activated in a second step at 900 °C (ramping at 6°C min^{-1}, three hours standing) under constant nitrogen flow.

Catalyst specific surface area and porosity

Surface area and porosity were determined using a TriStar® II 3020 apparatus (Micromeritics Instrument Corporation system). Before measurement, samples were outgassed at 150 °C for 4 hours under nitrogen in a FlowPrep 060 degas system. Surface area was determined by low temperature BET (Brunauer, Emmett and Teller) N_2 adsorption. Pore size and pore size distribution were calculated by the Barrett, Joyner, Halenda (BJH) method.

Polyetheresulphones (PES) synthesis

PES copolymers were synthesised by a co-polycondensation method in NMP solution at high temperatures in the presence of K_2CO_3. Different ratios of sulphonated and non-sulphonated co-monomers were used. PES membranes were prepared by casting from DMAc solutions.

PES thermal analysis

Thermal stability and PES water retention were determined by TGA (using a Perkin Elmer TGA 400 system, T = 25-900 °C, 20 °C min^{-1}) and DSC (using a Mettler Toledo DSC 1 system, T = 25-400°C, 20 °C min^{-1}). Data were recorded in N_2. Similar data were recorded for Nafion 1100.

Electrochemical characterization

Before measurement, all glassware was first washed in HCl (37 %), then H_2SO_4 (95-97 %) and finally rinsed with MilliQ water. The working electrode was a rotating disk electrode (RDE-EDI 101 Radiometer) with a glassy carbon tip (cross-sectional area of 0.07 cm^2) embedded in a Teflon sheath. Before use, the carbon tip was gently cleaned with soft sandpaper, polished with diamond powder (Aldrich) and finally degreased with ethanol. The thin film rotating disk electrode method (TF-RDE) was used to immobilise small catalyst amounts onto the polished carbon tip. To improve reproducibility the electrode was initially conditioned by cycling in O_2 saturated solution for 50 min (10 mV s^{-1} without electrode rotation) in the potential range E = –0.275/+0.800 V *vs.* Ag/AgCl. Measurements were thereafter recorded in a range of rotation rates (300–1600 rpm) at 5 mV s^{-1}. When performing measurements on Pt-containing-materials, a Pt counter-electrode was used (0.6 cm^2); a glassy carbon rod (5.5 cm^2) was used instead for Pt-free materials. *I/E* recordings were obtained by means of an Amel 7050 Potentiostat in a standard three-electrode electrochemical cell with an Ag/AgCl external reference electrode in 3 M NaCl. All potentials are reported on the Normal Hydrogen Electrode scale (NHE). Measurements were carried out in 0.1M $HClO_4$ solution at room temperature.

The electrode preparation consisted of two successive deposition steps: the first with a catalyst suspension in water (ink) and the second with an ionomer solution as catalyst binder. In the first step, 10 mg of catalyst was dispersed in 1 mL water and sonicated for 30 min; 7 µL of this mixture was pipetted onto the electrode tip and dried in a bottom-up position under a tungsten lamp. In the second step a few µL of an ionomer solution (see details below) were deposited onto the catalyst layer. This operation was found to be critical, especially in the case of the homemade PES ionomer which, besides being insoluble in water, was also found to be less effective in "gluing" the catalyst layer onto the RDE tip. Because of solubility, organic solvents had to be used for PES; these were also good wetting agents for the RDE Teflon sheath. This feature is unfortunate because it may hamper the accurate determination of the ionomer mass *per* catalyst surface area, depending on the more or less effective wetting by ionomer solutions of the Teflon sheath that surrounds the graphite tip. Therefore, different ionomer deposition procedures were adopted depending on the ionomer used. For Nafion, deposition was straightforward because water is the

Nafion suspension solvent and does not wet Teflon. For PES, several solvents were tested (DMAc, NMP, DMSO). Among them NMP and DMSO were found to be suitable; for DMAc a compromise had to be obtained by mixing DMAc with water in various ratios.

Detailed ionomer deposition conditions were as follows: for Nafion, 7 μL of a Nafion solution in water ($3.5\ 10^{-4}$ g mL^{-1}) was deposited on the electrode tip giving rise to a final Nafion loading of $3.51\ 10^{-5}$ g cm^{-2}; for PES, 4 μL of a PES solution (10^{-2} g mL^{-1}) was used with a final PES electrode loading of $5.7\ 10^{-4}$ g cm^{-2}. The adopted PES solvent and solvent composition will be specified when reporting electrochemical results.

Results and Discussion

Catalyst surface area and porosity

After heat-activation at 900°C, the catalyst surface area was about 800 $m^2\ g^{-1}$, which is lower than that after the first heat treatment at T = 600°C. Fractional pore volume distributions were found to be essentially independent of heating temperature. The pore size distribution shows a maximum (70 %) at about 40-50 nm; however, there was a sizeable presence of micropores (10 %). The high percentage of 40 nm mesopores is presumably useful to ORR behaviour, providing easy diffusional access to the reactive catalyst surface and to the Nafion ionomer units [23, 24].

PES and Nafion thermal analysis

TGA: To increase the PES water storing capability, the virgin polymer was conditioned in HCl and $HClO_4$ (compare the red and green curves in Figure 1). It can be noted that the two acidic conditionings give indistinguishable results. Figure 1 also reports the PES behaviour without acid conditioning (light blue curve). As seen in the figure, acidic conditioning definitely decreases the polymer water retention. Figure 2 reports a comparison between PES and Nafion 1100, with both of them HCl conditioned (red and blue curve, respectively). The most notable feature is the greater PES mass loss in the respect of Nafion up to the crossing point of the curves, at $T \cong 480$ °C. At higher temperatures, the Nafion mass losses become much greater than the PES ones. In a further aspect, the PES mass loss up to $T \cong 100$-120 °C, which is the upper project temperature for PEMFCs operation, is greater than for Nafion. At high temperatures, Nafion behaviour can be interpreted by references to literature data. According to Wilkie *et al.* [25], Almeida and Kawano [26, 27] fluorocarbon polymers exhibit high thermal stability and decompose by a first step (290–400 °C) that may be associated with a polymer desulphonation process, then by a second step (400–470 °C) related to side-chain decomposition and, finally, by a third one (470–560 °C), due to the PTFE backbone decomposition. In comparison, Figure 2 shows similar thermal degradation on PES after the completion of water removal at T = 200-220 °C. At present, no details are available on PES decomposition processes.

DSC: Figure 3 shows some features of the virgin PES (see the black curve) and after acidic conditioning (the red curve). In the black curve, a rather weak endothermic maximum extends over a wide temperature range ($T \cong 50$-170 °C, ΔH = -120 J g^{-1}), followed at higher T by a second sharp endothermal peak that, by comparison with a pure DMAc sample, is due to the removal of residual reaction solvent (DMAc). The sample behaviour is strongly affected by acidic conditioning (red curve). As the most relevant feature, a stronger endothermic process appears in the above-mentioned temperature range ($T \cong 50$-170°C, with a sharp maximum at T = 110 °C, ΔH = -480 J g^{-1}). By comparison with TGA results in Figure 1, the process is due to water removal. By the integration of Figure 3 curves, the water amount released from the acid conditioned samples is approximately four times greater than that from the unconditioned ones. At higher temperatures, the water removal is followed by a second, more diffuse endothermic process ending at $T \cong 200$°C. In comparison, the Nafion behaviour (see Figure 4) is characterised by a sequence of comparatively weaker endothermal (ΔH = -120 J g^{-1}) processes with a first maximum at $T \cong 100$°C and a second, barely distinguishable one at $T \cong 180$°C. By reference again to the TGA results in Figure 1, both of these processes are attributed to water removal. Overall, from a PEMFC operative viewpoint, the water amount retained in PES is favourably greater than in Nafion and is, moreover, more thermally stable.

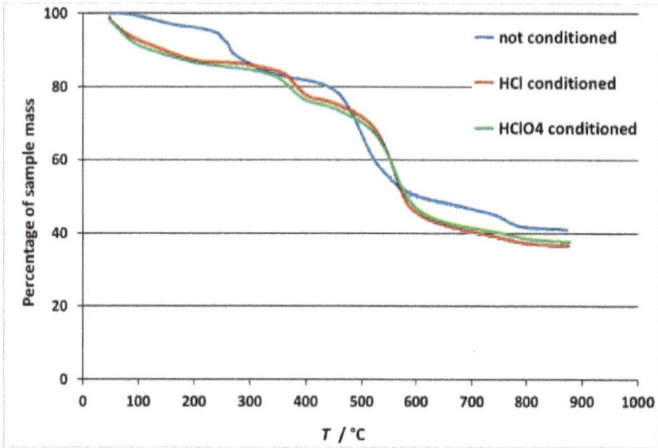

Figure 1. *TGA curves of different PES acidic treatment recorded at 20 °C min^{-1} from 25 to 900 °C in N_2 atmosphere. (a) Light blue curve is relative to non-conditioned PES; (b) red curve for HCl conditioned; (c) green curve for $HClO_4$ conditioned.*

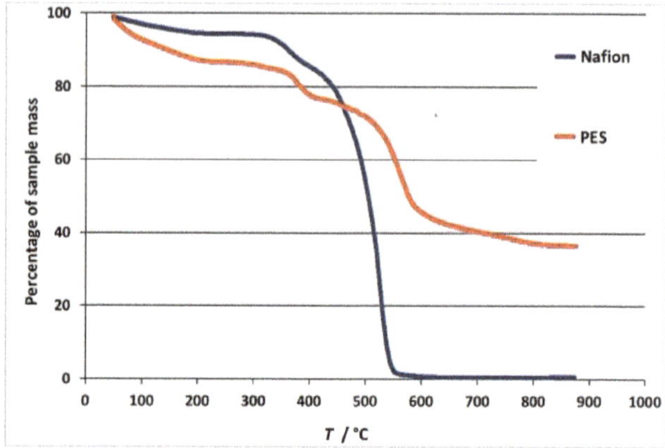

Figure 2. *TGA curves obtained in the same conditions reported in Fig. 1. (a) Red curve is relative to PES membrane conditioned in HCl; (b) blue curve is relative to Nafion membrane conditioned in HCl.*

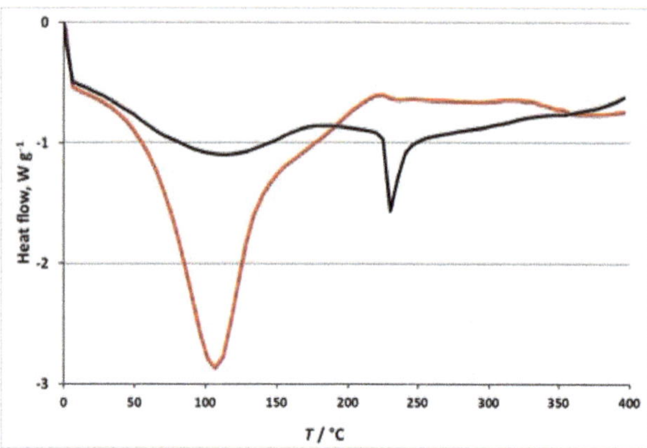

Figure 3. *DSC curves recorded from 25 to 400 °C, heating rate of 20 °C min^{-1} in N_2 atmosphere. (a) Virgin PES is the black curve; (b) PES HCl conditioned is the red one.*

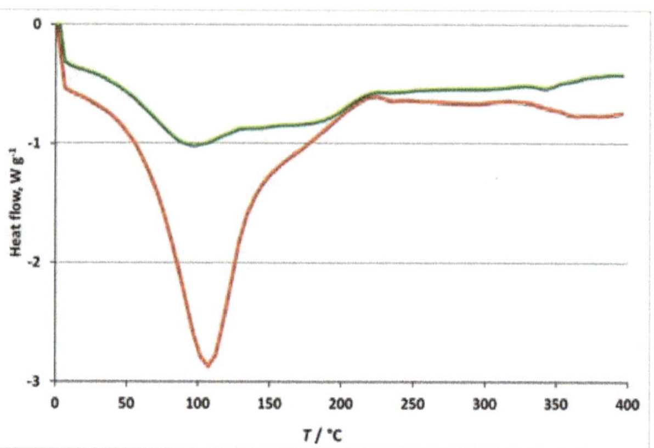

Figure 4. *DSC curves recorded in the same condition of Figure 3. Membranes are HCl conditioned: (a) Nafion is the green curve; (b) PES is the red one.*

Electrochemical characterisation

The following results report net oxygen reduction currents, corrected for background residual currents recorded in N_2. Results from different working electrodes (catalysts, ionomers) are normalised with respect to the geometrical electrode surface area.

Nafion 1100 and EC-20 were used as external references to evaluate the homemade materials, PES and Pt-free catalyst. Relevant reference data in O_2 saturated solution are reported in Figure 5 for many RDE rotation rates. An extended region of limiting current is present and depends on the

RDE rotation rate. The onset potential is about 0.9 V *vs.* NHE, which is very near to the value of massive platinum [28]. By comparison, Figure 6 reports ORR data for EC-20 catalyst with a PES ionomer layer. For the steeper current increase with decreasing electrode potential, ORR results obtained with Nafion are clearly better than PES ones. This difference may be due to the greater loading of PES than Nafion (see the Experimental), or to a greater Nafion acidity/conductivity in respect of PES. Despite many attempts, we were unable to decrease the PES amounts due to increasing mechanical instability of the catalyst layer. As a further main feature, PES causes an apparent slope decrease of j/E plots in the mixed control region ($E \cong 0.9$-0.7 V), and lower limiting current density values at more negative potentials. Figure 7 reports Koutecky-Levich plots for the used Nafion and PES ionomer onto the used Pt catalyst (EC-20). To avoid graphical crowding, the figure is only based on results in the limiting current region (E = 0.535 V *vs.* NHE).

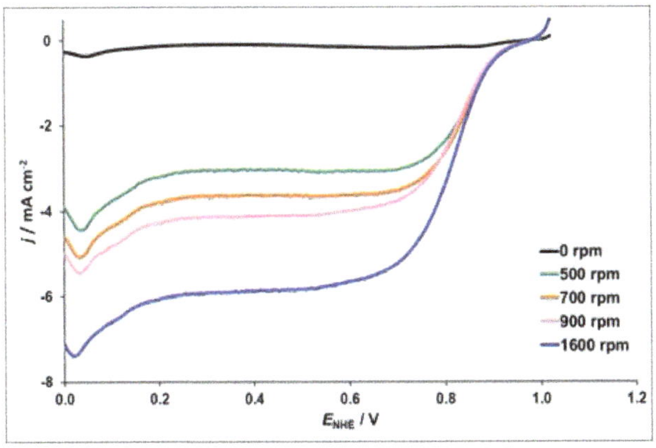

Figure 5. *Voltammetric data for EC-20 with Nafion binder. RDE rotation rates are shown in the figure label. (0.1 M $HClO_4$; v = 5 mV s^{-1}; T = 25 °C).*

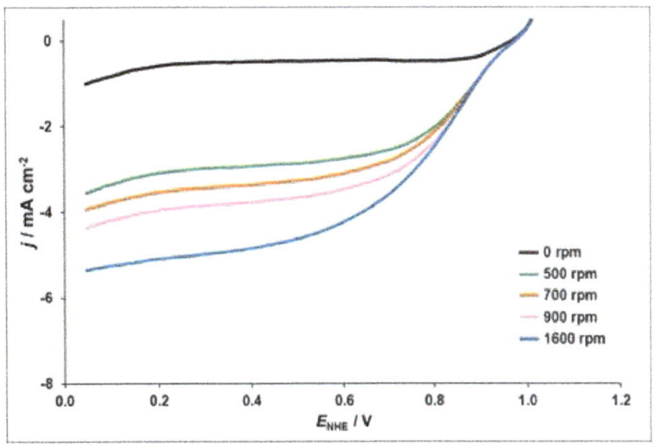

Figure 6. *Voltammetric data obtained for EC-20 with PES binder (DMSO only). RDE rotation rates are shown in the figure label. (0.1 M $HClO_4$; v = 5 mV s^{-1}; T = 25 °C).*

Nafion data are interpolated by a straight line with an almost exactly zero intercept at the axes origin and a slope from which, also taking into account numerical uncertainties, the number of stoichiometric exchanged electrons, n, ranges from 3.6-3.8.

By comparison, PES data give rise to a straight line with a similar slope, which is in accordance with a comparable kinetic hindrance from diffusion in solution. From this slope, n is 3.8; however, there is a greater intercept that, as mentioned above, might be due to the greater PES amount necessary.

Figure 8 shows the Tafel plots of EC-20/Nafion and EC-20/PES electrodes obtained from Koutecky-Levich plots for many potentials of the investigated range. For Nafion results, a first, extended Tafel region, with a slope of ca. -60 mV dec^{-1}, begins at somewhat more cathodic potentials than the ORR onset and ends at $E \cong 0.83$ V. A second, much shorter, Tafel section with a higher slope (ca. -120 mV dec^{-1}) can be specified, although with some difficulty, at more cathodic potentials. These features are in accordance with data from the literature [27-30]. The PES behaviour is characterised by a curved Tafel plot whose slope continuously increases with increasing over-potential.

Figure 9 shows results for EC-20 electrodes with a PES binder layer deposited using various solvents. As mentioned in the Experimental section, many attempts were necessary to balance a good PES solubility in a given solvent with low affinity of the resulting solution for the RDE Teflon sheath. As shown in the figure, results outline a single response in which the behaviour of each electrode closely superimposes on the others. This uniformity in behaviour presumably outlines a range of similar interactions between the PES ionomer, solvent and catalyst.

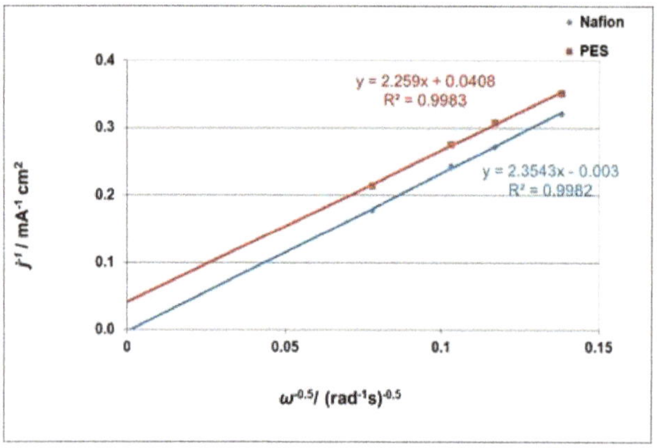

Figure 7. *Koutecky-Levich plots for EC-20 catalyst ORR currents at 0.535 V vs. NHE obtained (a) from Figure 5 for Nafion (blue line); (b) from Figure 6 for PES membrane (red one).*

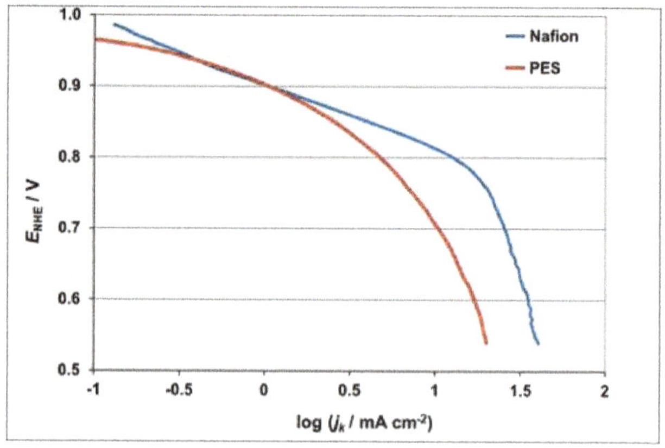

Figure 8. *Tafel plots for EC-20 catalyst obtained (a) from Figure 5 for Nafion (blue curve); (b) from Figure 6 for PES membrane (red one).*

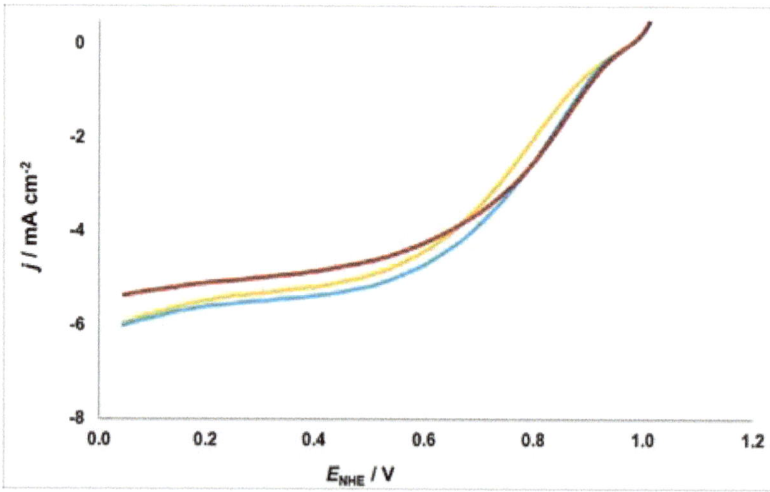

Figure 9. *Voltammetric data obtained for EC-20 with different PES/solvent deposition: (a) yellow curve, DMAc/water 1:2 vol; (b) blue, NMP; (c) red, DMSO. (0.1 M $HClO_4$; v = 5 mV s^{-1}; T = 25 °C; ω = 1600 rpm)*

ORR results for the thiamine sample catalyst with a Nafion binder layer are reported in Figure 10 for various RDE rotation rates. Although not exciting, this behaviour is acceptable overall, with a noteworthy and unfavourable shift of the curves towards more negative potentials in respect of the external reference EC-20.

Figure 11 reports the behaviour of the thiamine sample with a PES, instead of Nafion, binder. A dramatic worsening in behaviour is observed in comparison to Figure 10. This becomes more evident by comparison with a similar result for the PES/EC-20 electrode assembly (the relevant curve of Figure 11 is taken from Figure 6).

In the last figure, PES in itself or the PES/solvent mixture used behaves as a real "killer" for the homemade catalyst, while still affording an acceptable response when used on EC-20, even though less satisfactory than Nafion. This brings to light a complexity of the many interactions that may occur among separate components necessarily involved in building up a final, well behaving PEMFCs electrode catalytic assembly.

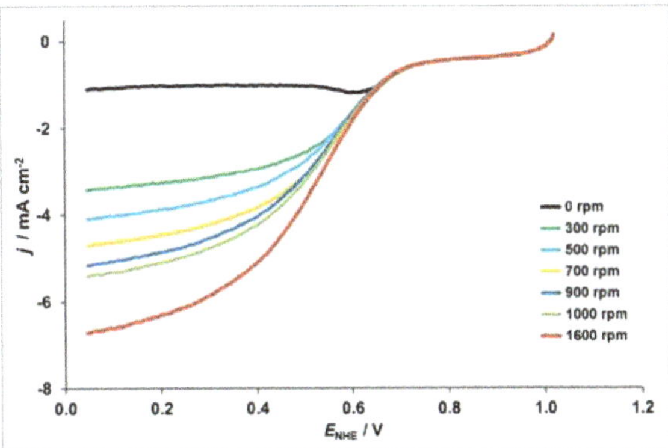

Figure 10. *Voltammetric data obtained for thiamine catalyst with Nafion. RDE rotation rates are shown in the figure label. (0.1 M $HClO_4$; v= 5 mV s^{-1}; T = 25 °C).*

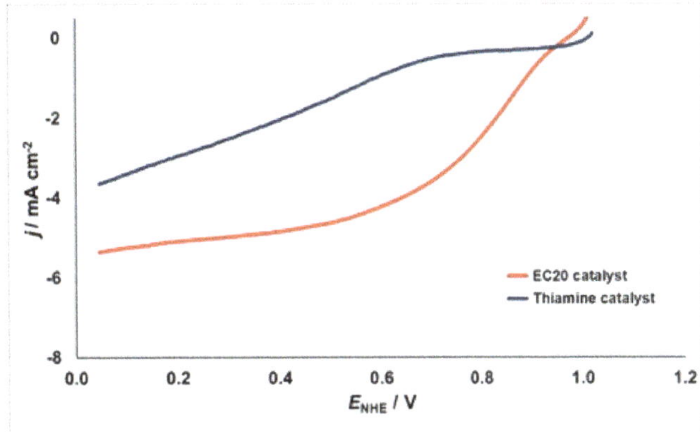

Figure 11. *Voltammetric data obtained for PES membrane (in DMSO only) for (a) EC-20 (red curve) and (b) thiamine catalyst (black curve). (0.1 M $HClO_4$; v = 5 mV s^{-1}; T = 25 °C; ω = 1600 rpm).*

Conclusions

The present homemade catalyst, synthesised by the pyrolysis of sugar thiamine mixtures, is characterised by unsatisfactory ORR features. It is, however, an element of the catalyst family that is currently under investigation as a substitute for Pt.

Similarly, the reported PES sample belongs to a polymer class that is being investigated for favourable water retention and thermal stability. Improvements may likely concern ionic conductivity in PEMFC operational conditions.

It can be also mentioned that the present homemade catalyst and ionomer afford acceptable results when separately tested with one or another reference material. Their ORR response fails completely when used together. This shows that, in view of real PEMFC applications, many mixed interactions have to be taken into account and optimised in detail.

Acknowledgements: *Financial support from Cariplo Foundation (Project 2010-0588 "Non fluorinated polymeric membranes and Platinum-free catalytic systems for Fuel Cells with low cost and high efficiency") are gratefully acknowledged.*

References

[1] X.Z. Yuan, H. Li, S. Zhang, J. Martin, H. Wang, *J. Power Sources* **208** (2012) 96-119
[2] P.J. Ferreira, G.J. la O', Y. Shao-Horn, D. Morgan, R. Makharia, S. Kocha, H.A. Gasteiger, *J. Electrochem. Soc.* **152** (2005) 2256-2271
[3] C.W.B. Bezerra, L. Zhang, H. Liu, K. Lee, A.L.B. Marques, E.P. Marques, H. Wang, J. Zhang, *J. Power Sources* **173** (2007) 891-908
[4] H.A. Gasteiger, S.S. Kocha, B. Sompalli, F.T. Wagner, *Appl. Catal. B: Environ.* **56** (2005) 9-35
[5] K. Sopian, W.R. Wan Daud, *Renew. Energ.* **31** (2006) 719-727
[6] D. Garraín, Y. Lechón, C. de la Rúa, *Smart Grid Renew. Energ.* **2** (2011) 68-74
[7] J.X. Wang, N.M. Markovic, R.R. Adzic, *J. Phys. Chem. B* **108** (2004) 4127-4133
[8] U.A. Paulus, T.J. Schmidt, H.A. Gasteiger, R.J. Behm, *J. Electroanal. Chem.* **495** (2001) 134-145
[9] M.R. Tarasevich, V.A. Bogdanovskaya, L.N. Kuznetsova, A.D. Modestov, B.N. Efremov, A.E. Chalykh, Y.G. Chirkov, N.A. Kapustina, M.R. Ehrenburg, *J. Appl. Electrochem.* **37** (2007) 1503-1513
[10] R. Othman, A.L. Dicks, Z. Zhu, *Int. J. Hydrogen Energy* **37** (2012) 357-372

[11] A. Morozan, B. Jousselme, S. Palacin, *Energy Environ. Sci.* **4** (2011) 1238-1254
[12] G. Liu, X. Li, P. Ganesan, B.N. Popov, *Appl. Catal. B: Environ.* **93** (2009) 156-165
[13] F. Jaouen, E. Proietti, M. Lefèvre, R. Chenitz, J.P. Dodelet, G. Wu, H.T. Chung, C.M. Johnston, P. Zelenay, *Energ. Environ. Sci.* **4** (2011) 114-130
[14] M. Lefèvre, E. Proietti, F. Jaouen, J.P. Dodelet, *Science* **324** (2009) 71-74
[15] M. Annala, S. Lipponen, T. Kallio, J. Seppälä, *J. Appl. Polym. Sci.* **124** (2012) 1511-1519
[16] G. Lalande, R. Côté, D. Guay, J.P. Dodelet, L.T. Weng, P. Bertrand, *Electrochim. Acta* **42** (1997) 1379-1388
[17] F. Jaouen, S. Marcotte, J.P. Dodelet, G. Lindbergh, *J. Phys. Chem. B* **107** (2003) 1376-1386
[18] M.A. Hickner, H. Ghassemi, Y.S. Kim, B.R. Einsla, J.E. McGrath, *Chem. Rev.* **104** (2004) 4587-4612
[19] W. Liu, Y. Xie, J. Liu, X. Jie, J. Gu, Z. Zou, *Int. J. Hydrogen Energy* **37** (2012) 4673-4677
[20] Y. Iwai, T. Yamanishi, *Polym. Degrad. Stabil.* **94** (2009) 679-687
[21] E.J. Biddinger, D.S. Knapke, D. von Deak, U.S. Ozkan, *Appl. Catal. B: Environ.* **96** (2010) 72-82
[22] I. Galbiati, C.L. Bianchi, M. Longhi, A. Carrà, L. Formaro, *Fuel Cells* **10** (2010) 251-258
[23] E. Antolini, *Appl. Catal. B: Environ.* **88** (2009) 1-24
[24] K. Kinoshita, *J. Electrochem. Soc.* **137** (1990) 845-848
[25] C.A. Wilkie, J.R. Thomsen, M.L. Mittleman, *J. Appl. Polym. Sci.* **42** (1991) 901-909
[26] S.H. de Almeida, Y. Kawano, *J. Therm. Anal. Calorim.* **58** (1999) 569-577
[27] L.G. Lage, P.G. Delgado, Y. Kawano, *J. Therm. Anal. Calorim.* **75** (2004) 521-530
[28] K.J.J. Mayrhofer, D. Strmcnik, B.B. Blizanac, V. Stamenkovic, M. Arenz, N.M. Markovic, *Electrochim. Acta* **53** (2008) 3181-3188
[29] F. Gloaguen, F. Andolfatto, R. Durand, P. Ozil, *J. Appl. Electrochem.* **24** (1994) 863-869
[30] N.M. Markovic, H.A. Gasteiger, P.N. Ross, *J. Phys. Chem.* **99** (1995) 3411-3415
[31] D. van der Vliet, D. Strmcnik, C. Wang, V. Stamenkovic, N.M. Markovic, M.T.M. Koper, *J. Electroanal. Chem.* **647** (2010) 29-34

14

Electrochemical determination of an antitumour platinum(IV) complex: trans-[$PtCl_2(OH)_2$(dimethylamine)(isopropylamine)]. Application to biological samples

CARMEN S. HERNANDEZ DOMÍNGUEZ✉ and PEDRO HERNÁNDEZ

Departamento de Química Analítica y Análisis Instrumental, Facultad de Ciencias, Universidad Autónoma de Madrid, Cantoblanco, 28029 Madrid, Spain

✉ Corresponding Author

Abstract

A differential pulse voltammetry (DPV) method has been applied for the first time for determination of trans-Pt[$Cl_2(OH)_2$(dimethylamine)(isopropylamine)]. To this end, all chemical and instrumental variables affecting the determination of trans-Pt[$Cl_2(OH)_2$(dimethylamine)(isopropylamine)] were optimized. From studies of the mechanisms governing the electrochemical response of trans-Pt[$Cl_2(OH)_2$(dimethylamine)(isopropylamine)], it was concluded that it is an electrochemically irreversible system with a reduction under diffusion control, with a reduction potential of -425 mV. Under optimal conditions, the variation in the analytical signal (I_p) with trans-Pt[$Cl_2(OH)_2$(dimethylamine)(isopropylamine)] concentration is linear in the 0.8 μg mL^{-1} to 20 μg mL^{-1} range, with an LOD of 97 ng mL^{-1} and a LOQ of 323 ng mL^{-1}, RSD = 1.58 % and E_r = 0.83 %. The optimized method was applied to the determination of trans-Pt[$Cl_2(OH)_2$(dimethylamine)(isopropylamine)] in biological fluids, human urine and synthetic urine.

Keywords

trans-Pt[$Cl_2(OH)_2$(dimethylamine)(isopropylamine)]; differential pulse voltammetry; biological fluids; human urine.

1. Introduction

The use of *cis*-diamminedichloroplatinum(II), known as cisplatin or *cis*-DDP, in cancer chemotherapy has made a major impact on the observed response rates of some tumour types, such as testicular or ovarian carcinoma [1]. However, cisplatin has two major drawbacks: 1) severe toxicity that includes nephrotoxicity, neurotoxicity and ototoxicity and 2) the acquisition or presence of resistance to the drug [2]. Because tumour resistance to cisplatin limits its efficacy, there is an urgent need to discover new platinum complexes capable of overcoming cisplatin resistance.

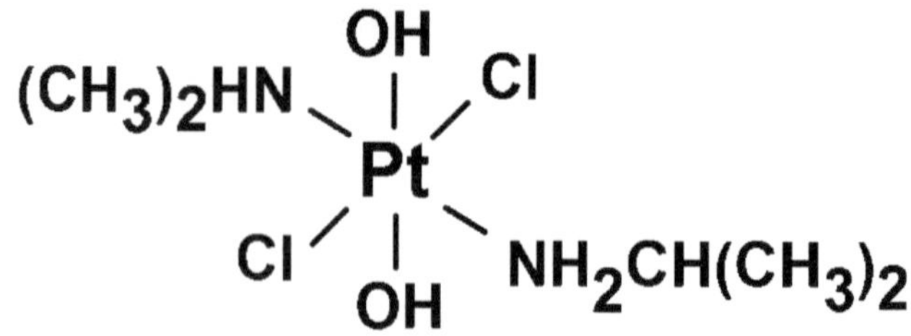

Figure 1. Structure and molecular weight of the platinum complex, trans-Pt[$Cl_2(OH)_2$(dimethylamine)(isopropylamine)] (compound 1), used in this article

Because enhanced removal of cisplatin-DNA adducts has been reported as one of the main causes of cell resistance to cisplatin, there is general consensus that this particular resistance mechanism may be circumvented by platinum complexes that bind to DNA by a different mechanism than cisplatin [3,4]. One platinum compound that possesses DNA-binding properties distinct from those of cisplatin is its *trans* isomer transplatin, *trans*-diamminedichloroplatinum(II), or *trans*-DDP (Fig. 1). In fact, it has been found that *cis*-DDP mainly forms 1,2 intrastrand cross-links on DNA, whereas the main adducts of *trans*-DDP are 1,3 intrastrand cross-links [2]. Unfortunately, early structure-activity relationship studies showed that *trans*-DDP and other *trans*-Pt complexes were inactive as antitumour drugs [5]. However, in 1989, Farrell *et al.* [6] reported the first cytostatic *trans*-Pt(II) complexes. In recent years, several classes of biologically active *trans*-platinum complexes have been reported [7]. Among these unusual classes of platinum drugs, it has recently been found that *trans*-Pt(II)Cl_2 complexed with an asymmetric set of aliphatic amines is able to circumvent cisplatin resistance. It is known that resistance to cisplatin is multifactorial and includes three main mechanisms: decreased cellular accumulation of cisplatin, increased cytoplasmatic detoxification (through increased cytoplasmic levels of glutathione and metallothioneins), and increased DNA repair/tolerance of platinum-DNA adducts[8]. It has also been postulated that alterations in the apoptotic cell death pathway may constitute a fourth mechanism of cisplatin resistance[9].

The group of Carmen Navarro-Ranninger [10] has reported that the *trans*-platinum(II) complex with mixed aliphatic amines, *trans*-Pt[$Cl_2(OH)_2$(dimethylamine)(isopropylamine)] (Fig.1) circumvents cisplatin resistance in cells that overexpress oncogenes. In addition, compound **1** is also able to circumvent resistance to cisplatin in A2780*cis*R ovarian tumour cells, which exhibit resistance through a combination of the three mechanisms mentioned above [11]. Moreover, they observed that circumvention of cisplatin resistance by compound **1** is associated with a higher level of apoptosis induction relative to *cis*-DDP [12,13]. Also of interest was the observation that in A2780*cis*R cells there is a correlation between the DNA interstrand cross-linking efficiency of compound **1** and its ability to induce apoptosis [13].

On these grounds, Navarro-Ranninger *et al.* [10] reported the cytotoxic activity of the *trans*-Pt[$Cl_2(OH)_2$(Dimethylamine)(isopropylamine)] (TPt complex) (Fig. 1) in pairs of cisplatin-sensitive and-resistant human ovarian tumour cell lines. The resistant cell lines showed acquired resistance to cisplatin and were selected with regard to the three major mechanisms of resistance to the drug. In addition, these cell lines have been used previously to identify novel platinum complexes capable of circumventing cisplatin resistance [14-16]. Because cellular and molecular pharmacology studies are essential to understanding the relationships between structure and anticancer properties, Navarro-Ranninger *et al.* [10] have compared the cellular accumulation, DNA binding, interstrand cross-linking efficiency, apoptosis induction, and binding to serum albumin and plasma proteins of the *trans*-Pt(IV) complex (compound **1**) with that of its corresponding *trans*-Pt(II) analog. Finally, the in vivo antitumour activity of compound **1** has been evaluated in mice bearing tumour xenografts. In this article, a new electrochemical method to determine TPt complex in biological fluids is developed.

This paper makes a study of the electrochemical behaviour of this compound by differential pulse voltammetry (DPV) aimed at obtaining the optimal conditions for its determination in biological fluids and establishing its formation constants.

The analysis of biological fluids presents a special relevance because small changes in the concentration of its components are thought to be correlated with several neurological or metabolic disorders [14].

2. Experimental

Reagents: TPt complex [10] was provided by Carmen Navarro-Ranninger. The purity of the analyte was 98 %. This high purity is because the complex of platinum(IV) was a synthetic product, provided by the group of Navarro-Ranninger

Triple distilled mercury was used as working electrode. Buffer solutions prepared with acetic acid, phosphoric acid and borate were adjusted to the desired pH values with sodium hydroxide. Unless otherwise stated, all other reagents were of analytical grade and were used as received. Ultrapure water, obtained from Millipore MilliQ System, was used in the preparation of buffers and solutions and to clean the working electrode of possible substances adsorbed on the gold electrode surface. All solutions were prepared just prior to use, preserved at 4 °C and protected from light.

Apparatus: Electrochemical measurements were performed using a Bioanalytical System (BAS) Epsilon Mercury Stand workstation, an Ag/AgCl/KCl 3M reference electrode, and a platinum wire as counter electrode. For pH adjustment, a pH-meter Methrom C831 was employed as well as a circulation ultrathermostat Frigiterm-10 (P-Selecta, Spain) to control the temperature.

Procedure: TPt complex stock solution (500 μg mL^{-1}) was prepared with ultrapure water and stored at 4±1 °C preserved in aluminium foil covered bottles to avoid photodegradation. Prior to analysis, sample bottles were allowed to equilibrate to room temperature (RT).

Sample preparation: The TPt complex dissolved readily and was further diluted to desired concentrations using 0.5 M acetate buffer solution.

The samples used for the implementation of the method described above included a synthetic urine sample prepared in the laboratory and a sample of real human urine.

The composition of the synthetic urine is presented in Table 1 [17].

Table 1. *Synthetic urine composition*

	NaCl	KCl	$CaCO_3$	$MgCl_2.6H_2O$	H_2SO_4 (98 %)	HCl (conc)	$NH_4H_2PO_4$
c	5.08 g L^{-1}	2.86 g L^{-1}	0.31 g L^{-1}	0.42 g L^{-1}	0.67 mL L^{-1}	8.7 mL L^{-1}	3.09 g L^{-1}

Processing of the human urine sample consisted of:

1) doping of the sample with an aliquot of TPt complex (10 μg mL^{-1}). This value was chosen because, at this time, there are no published references devoted to the use of these compounds in humans;

2) protein precipitation with 0.1 M $HClO_4^-$;
3) digestion of the precipitate for a weekend (48 hours);
4) filtering of the sample. The filters used in the development of this process were cellulose membrane filters with a pore size of 0.45 μm and
5) adjust the pH of the sample to the appropriate value.

3. Results and discussion

The electrochemical study of the TPt complex was performed by DPV and square wave voltammetry in order to determine the best conditions for its analysis.

3. 1. Influence of pH and ionic strength.

In order to select the optimum pH, a pH range between pH 2 and pH 12 was employed. The buffers used were phosphate buffer for pH 2, 3, 6, 7, 8, 10, 11 and 12; acetic acid buffer for pH 4 and 5; and borate buffer for pH 9. The concentration of all the buffers was 0.5 M.

In Figure 2 the variation in peak intensity with respect to pH (Fig. 2a) and the variation in peak potential versus pH (Fig. 2b) are represented. This figure shows that the greatest intensity was reached at pH 4 (Fig. 2a). By representing the variation in peak potential with respect to pH (Fig 2b) three straight segments with clear variations in slope are apparent, indicating that the compound under study has two dissociation constants, which correspond to the values of $K_1 = 10^{-4.2}$, $K_2 = 10^{-7.7}$. This fact can be corroborated by observing the slopes in the E_p *vs.* pH graph (Fig. 2b)

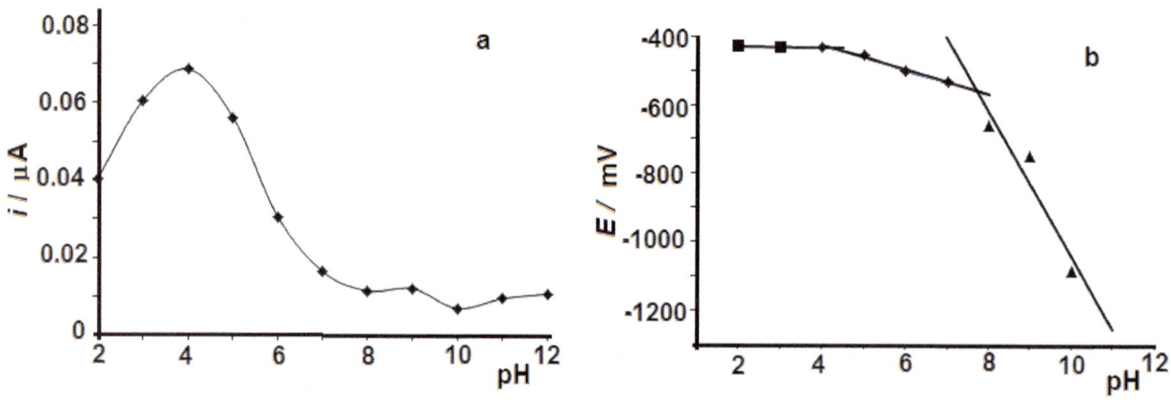

Figure 2. *a) Variation in DPV peak intensity at different pH values. b) Variation in peak potentials. Dissociation constants were calculated from the variations in the slopes. Measurement conditions: ionic strength = 0.05 M; TPt complex concentration = 5 μg mL^{-1}; pulse amplitude = 50 mV; scan rate = 10 mV s^{-1}*

$$E_p = -2\text{pH} - 426;\ r^2 = 1 \quad (1)$$

$$E_p = -35.2\ \text{pH} - 287.4;\ r^2 = 0.995 \quad (2)$$

$$E_p = -211.5\ \text{pH} + 1072.5;\ r^2 = 0.992 \quad (3)$$

From the standpoint of electrochemistry, pH 4 was chosen for further study because it is the value that provided the greater intensity.

The ionic strength necessary to achieve the greatest possible sensitivity in the electrochemical reaction was also studied. For this study we used solutions of acetic acid-acetate buffer (pH 4) in the concentration range between 5×10^{-2} and 2.0 M, and reached a maximum intensity at $5\text{x}10^{-2}$ M. This concentration was used for further studies.

3. 2. Instrumental variables

Once the study of the chemical variables was completed, we proceeded to the optimization of instrumental variables, both by DPV and square wave voltammetry (SWV). In both techniques the values of pH 4 and ionic strength 5×10^{-2} M, determined above, and the analyte concentration of 10 µg mL^{-1} were kept constant.

3. 2. 1. Influence of pulse amplitude

By changing this variable using differential pulse voltammetry, it appeared that the peak current increased linearly according to the equation:

$$i = 0.0015\ \Delta E + 0.0003 \quad (4)$$

using a scan rate of 10 mV s^{-1} throughout the study.

However, at the same time, the width of the half peak increased, so it was necessary to compromise between sensitivity and selectivity. This occurred at a value of pulse amplitude 50 mV.

The peak potential was modified according to the equation:

$$E_p = 0.4338\ \Delta E - 466.88 \quad (5)$$

To calculate the number of electrons exchanged during the reduction a DC voltammetry was carried out, and compared with the findings obtained with DPV (Figure 3).

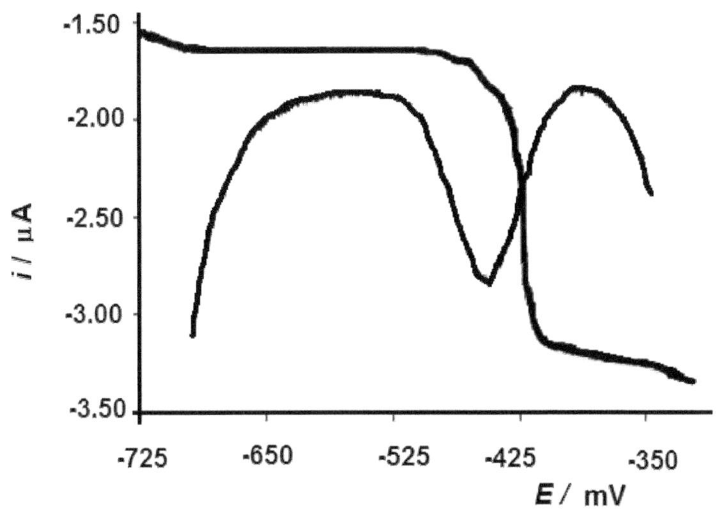

Figure 3. *Comparison of DC and DP polarograms. Measurement conditions: pH 4; ionic strength = 0.05 M; TPt complex concentration = 5 µg mL^{-1}; scan rate = 10 mV s^{-1}*

The number of electrons exchanged was found to be two using the following equation [18]:

$$E_p = E_{1/2} \pm \Delta E/ne \quad (6)$$

where E_p is the peak potential calculated at pH 4 in DPV, $E_{1/2}$ is the potential calculated from DC voltammetry and ΔE is the pulse applied, at a peak potential of -426 mV and -409 mV half-wave.

Therefore, the reduction process can be written as:

$$Pt(IV) + 2e^- \Leftrightarrow Pt(II) \quad (7)$$

The square-wave behaviour is similar to that observed with DPV, the peak current increasing linearly with pulse amplitude according to the equation:

$$I = 0.0035\ \Delta E + 0.32 \quad (8)$$

at a frequency of 25 Hz and a step of 2 mV, which promotes an effective scan rate of 50 mV s^{-1}.

When performing these studies by using square wave voltammetry it is presented the same situation of commitment that in differential pulse voltammetry, in other words, by increasing the amplitude of pulses produces a substantial increase in the half-peak width

The variation in peak potential follows the equation

$$E_p = 0.85\ \Delta E - 457.02 \tag{9}$$

3. 2. 2. Influence of scan rate

As in previous studies, the pH, ionic strength and analyte concentration (10 µg mL^{-1}) were held constant.

When the scan rate was modified at DPV and the pulse amplitude remained constant (ΔE = 50 mV), the variation in peak current was linear with scan rate according the equation:

$$i = 0.0058\ v^{0.5} + 0.32 \tag{10}$$

However, the peak potential does not remain constant, indicating that the electrochemical reduction is an irreversible process controlled by diffusion.

A study of this variable in SWV was carried out by changing the frequency and step, but at a constant pulse amplitude (ΔE = 50 mV). The results were that in modifying the frequency and the step, the peak intensity followed a linear trend fitting the square root of each of the two variables. In addition, the peak potential shifts as well, indicating that the reduction process is irreversible and under diffusion control.

As in previous studies, half-peak width increases, selecting as the most appropriate conditions a frequency of 50 Hz and 2 mV step, representing a scan rate of 100 mV s^{-1}.

The results obtained (Fig. 4) are consistent with those found earlier; the peak intensity fits linearly to the square root of the scan rate by:

$$i = 0.012\ v^{0.5} + 0.12 \tag{11}$$

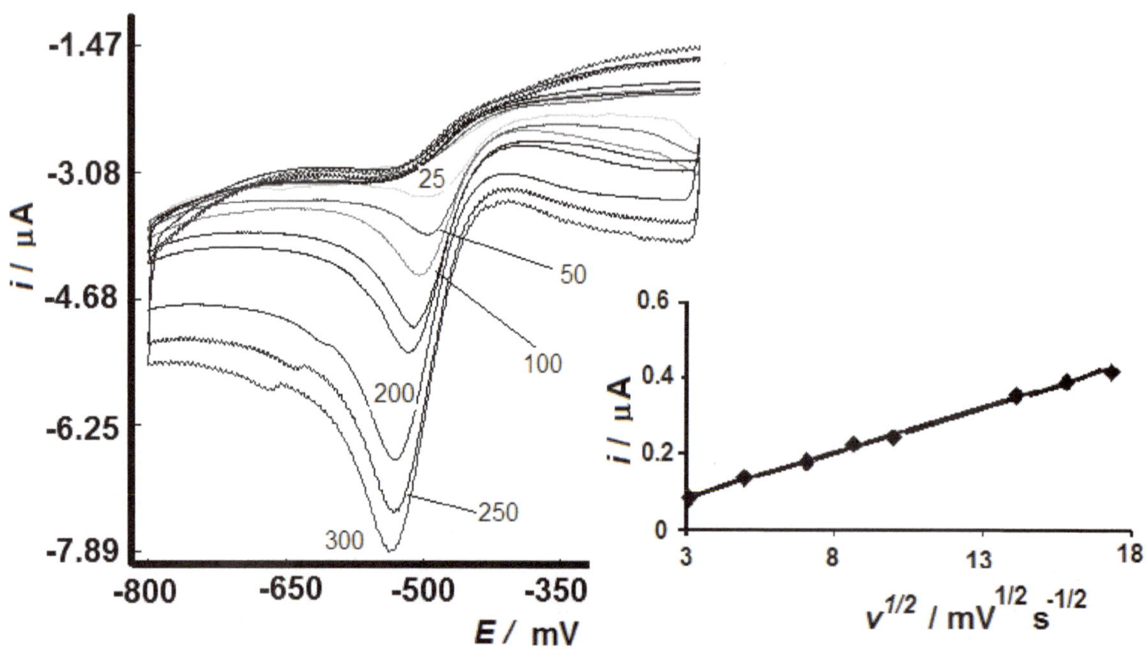

Figure 4. *Cyclic voltammograms obtained at different scan rates. Measurement conditions: pH 4; ionic strength = 0.05 M; TPt complex concentration = 5 µg mL^{-1} pulse amplitude = 50 mV*

and the peak potential also changes linearly as:

$E_p = -2.06\ \Delta E - 487.78$ (12)

This confirms the irreversibility of the process controlled by diffusion.

The reduction process was also studied by cyclic voltammetry in order to contrast the results obtained in earlier techniques. The voltammograms obtained clearly demonstrate that the process is irreversible and under diffusion control.

3. 2. 3. Calibration curve and limit of detection

After choosing the most appropriate measurement conditions for determining the compound under study, the influence of the analyte concentration was studied using the following conditions (Table 2)

Table 2. *Conditions for calibration curve*

	pH	ΔE / mV	v / mV s^{-1}	Ionic strength, M
Differential pulse polarography	4	50	10	0.05
Square wave polarography	4	50	100*	0.05

**Frequency = 50 Hz, Scan increment = 2 mV*

The analytical performance of DPV method developed for TPt complex determination was evaluated. From the voltammograms recorded for increasing amounts of TPt complex using the optimized parameters described above, it was observed that the peak current increased with the analyte concentration over a large range (0.8 µg mL^{-1} to 20 µg mL^{-1}) with good linearity according to $i = 0.015\ c - 0.0053$; $r^2 = 0.998$ with an $E_r = 0.83$ % and an RSD of 1.58 % ($n = 10$ at the 10 µg mL^{-1} level). The minimum detectable amount was 97 ng mL^{-1}, while a concentration of 323 ng mL^{-1} was calculated as the detection limit (Fig. 5).

We proceeded in the same manner when SWV was carried out. In this case, square-wave voltammograms were recorded for increasing amounts of TPt complex under the optimal conditions found for its determination. A linear response was found in the 0.1 µg mL^{-1} to 20 µg mL^{-1} range according to $i = 0.63c - 0.0158$; r^2=0.9962. The sensitivity of the method was inferred from the $LOD_{(3\sigma)}$=105 ng mL^{-1} and $LOQ_{(10\ \sigma)}$=393 ng mL^{-1} values. On the other hand, the RSD and E_r values of 2.10 % and 1.72 %, respectively (n=10 at the 18 µg mL^{-1} level) indicated the accuracy and reproducibility of the proposed method.

3. 3. Analytical application

Prior to using the method for determinations of TPt complex in urine, we conducted a study of the inorganic salts present in a synthetic urine prepared in the laboratory with the composition detailed in Table 1 and doped with 10 µg L^{-1} TPt [17].

To perform this experiment the standard addition method was used, obtaining the equation, $i = 0.0012c - 0.023$ with a correlation coefficient of $r^2 = 0.993$.

Given the suitability of the results, we proceeded to determine the TPt complex in human urine samples. The following procedure was used in pretreating the sample and the DPV.

The sample was doped with an aliquot of 10 µg mL^{-1} TPt complex, then proteins were precipitated with 1 M $HClO_4$. The precipitate was digested for 48 hours, filtered and the filtrate was then brought to pH 4 with a 0.05 M buffer solution and standard measurement by DPV was performed.

Figure 5 shows the voltammograms obtained; the inset shows the reciprocal plot of *I* versus [TPt complex], showing the excellent fit of the experimental data to the equation $i = 0.21c + 1.98$, with a correlation coefficient of r^2=0.997.

When the data obtained were extrapolated into the equation of the calibration curve, the concentration of TPt complex obtained was 9.80±0.05 µg mL^{-1}

The procedure followed in the electrochemical study of the synthetic urine sample was the same as in human urine. The equation obtained when the standard addition was carried out was $i = 0.0034c + 0.0023$, with a correlation coefficient of r^2=0.9903. By extrapolating the obtained data, the concentration of the TPt complex was 9.75 ±0.05 µg mL^{-1}.

The difference between the obtained data for human urine and synthetic urine could be because the matrix of the human urine sample is much more complex than the synthetic urine matrix.

Just as in the calibration section, the differences between the slopes are due to the determination conditions, being that determination of the TPt complex was performed using DPV and SWV, and each of these electrochemical techniques has optimum conditions.

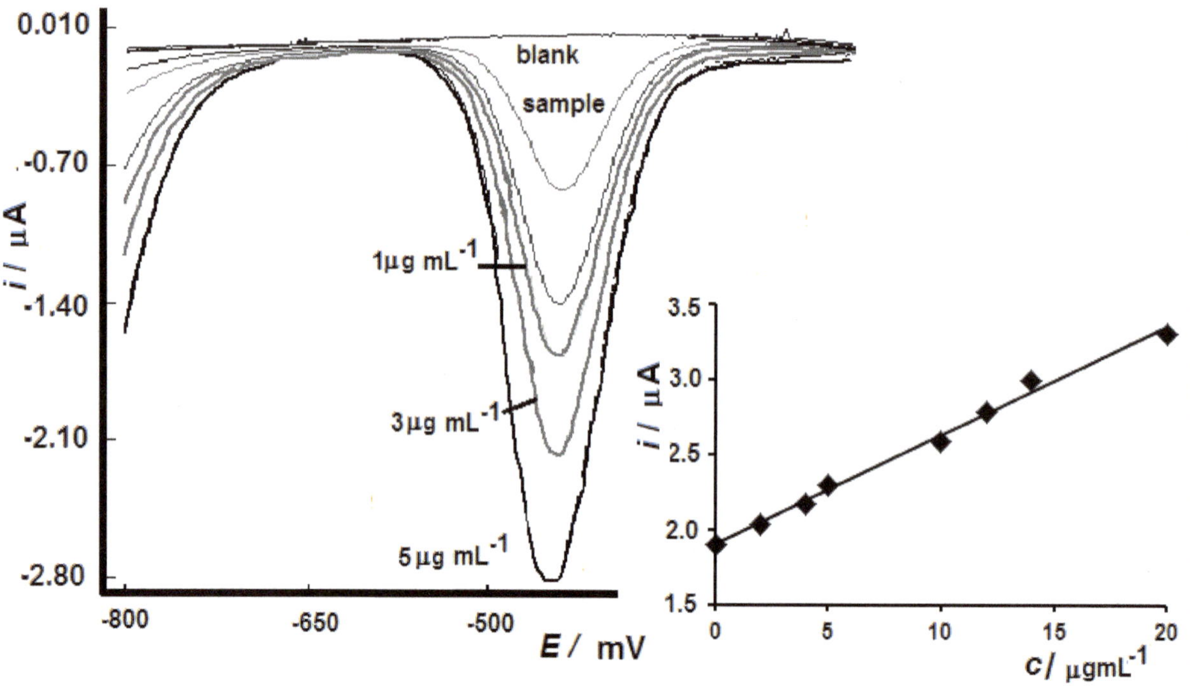

Figure 5. *Calibration curve for DPV in human urine. Measurement conditions: pH 4; ionic strength = 0.05 M; pulse amplitude = 50 mV; scan rate = 10 mVs*$^{-1}$

4. Conclusions

A sensitive differential pulse voltammetry (DPV) method to analyse TPt complex has been applied and the mechanisms governing the electrochemical response have been elucidated. The proposed method can be applied to the determination of the studied Pt(IV)-complex in biological fluids, with the inherent advantages of electrochemical techniques, *i.e.*, greater economy and easy instrumentation with high accuracy and reproducibility, as shown from the obtained results.

***Acknowledgements**: The authors thank Carmen Navarro for providing the platinum complex and MICINN for their economic support.*

References

[1] E. Wong, C.M. Giandomenido, *Chem. Rev.* **99** (1999) 2451-2466.

[2] S.M. Cohen, S.J. Lippard, *Prog. Nucleic Acid Re.* **67** (2001) 93-103

[3] N. Farrell, *Cancer Invest.* **11** (1993) 578-589.

[4] Z.Z. Zdraveski, J.A. Mello, C.K. Farinelli, J.M. Essigmann, M.G. Marinus. *J. Biol. Chem.* **277** (2002) 1255-1260.

[5] T.A. Connors, M.J. Cleare, K.R. Harra, *Cancer Treat. Rep.* **63** (1979) 1499-1502.

[6] N. Farrell, T.T.B. Ha, J.P. Souchard, F.L. Wimmer, S. Cros, N.P. Johnson, *J. Med. Chem.* **32** (1989) 2240-2241.

[7] G. Natile, M. Coluccia, *Coordin. Chem. Rev.* **216** (2001) 383-410.

[8] R.P. Perez. *Eur. J. Cancer* **34** (1998) 1535-1542.

[9] V.M. González, M.A. Fuertes, C. Alonso, J.M. Perez, *Mol. Pharmacol.* **59** (2001) 657-663.

[10] J.M. Perez, L.R. Kelland, E.I. Montero, F.E. Boxal, M.A. Fuertes, C. Alonso, C. Navarro-Ranninger, *Mol. Pharmacol.* **63** (2003) 933-944.

[11] J.M. Perez, E.I. Montero, A.G. Quiroga, M.A. Fuertes, C. Alonso, C. Navarro-Ranninger. *Metal Based Drugs,* **8** (2001) 29-37.

[12] J.M. Perez, E.I. Montero, A.M. González, A. Alvarez-Valdés, C. Alonso, C. Navarro-Ranninger, *J. Inorg. Biochem.* **77** (1999) 37-42.

[13] E.I. Montero, J.M. Perez, A. Schwartz, M.A. Fuertes, J.M. Maligne, C. Alonso, M. Leng, C. Navarro-Ranninger, *ChemBioChem* **3** (2002) 61-67.

[14] L.R. Kelland, G. Abel, M.J. McKeage, M. Jones, P.M. Goddard, M. Valenti, B.A. Murrer, K.R. Harrap. *Cancer Res.* **53** (1993) 2581-2586.

[15] L.R. Kelland, C.F.J Barnard, K.J Mellish, M. Jones, P.M. Goddard, M. Valenti, A. Bryant, B.A. Murrer, K.R. Harrap. *Cancer Res.* **54** (1994) 5618-5622.

[16] L.R. Kelland, C.F.J. Barnard, I.G. Evans B.A. Murrer, B.R.C. Theobald, S.B. Wyer, P.M. Goddard, M. Jones, M. Valenti, *J. Med. Chem.* **38** (1995) 3016-3024.

[17] B.J. Stevens. Clinical Analysis by Atomic Absorption. Department of Pathology. The Royal Children's Hospital. Melbourne, Australia. 1970. Varian Techtron.

[18] J.M. Pingarrón, P. Sánchez Batanero *Química Electroanalítica: Fundamentos y Aplicaciones*, Editorial Síntesis, Madrid, 2003.

15

Electrochemical treatment of leachates from sanitary landfills

ANNABEL FERNANDES⊠, EDITE CATALÃO, LURDES CIRÍACO, MARIA J PACHECO and ANA LOPES

UMTP and Department of Chemistry, University of Beira Interior, 6201-001 Covilhã, Portugal

⊠Corresponding Author

Abstract

The electrochemical treatment of leachate samples from a Portuguese intermunicipal sanitary landfill was carried out using anodic oxidation. The treatment was performed in a pilot plant that possesses an electrochemical cell, with boron-doped diamond electrodes, working in batch mode with recirculation. The influence of the applied current density and the flow rate on the performance of the electrochemical oxidation was investigated. Current density was decreased by steps, during the degradation, in order to study this effect on the efficiency of the process. For the assays run at equal flow rate and initial current intensity, chemical oxygen demand (COD) removal seems to depend mainly on the charge passed and the variation of the current density during the anodic oxidation process can reduce the energetic costs. An increase in the recirculation flow rate leads to an increase in the organic load removal rate and a consequent decrease in the energetic costs, but it decreases the nitrogen removal rate. Also, the bias between dissolved organic carbon and COD removals increases with flow rate, indicating that an increase in recirculation flow rate decreases the mineralization index.

Keywords

Landfill leachate treatment; BDD; anodic oxidation.

Introduction

Leachate generation is an inevitable consequence of the deposition of solid wastes in sanitary landfills. It is the result of rainwater percolation through wastes, that extracts and brings with it several pollutant materials dissolved and in suspension [1]. Sanitary landfill leachate composition is very complex and depends mainly on the type of solid wastes that are deposited, the climatic conditions and the age of the sanitary landfill [2]. Inadequate leachate management involves considerable risks, particularly contamination of water resources, at the surface and groundwater, and soils [1].

A common treatment for sanitary landfill leachates comprises biological reactors with nitrification/denitrification steps, followed by membrane technologies. However, due to variability in the quality and quantity of leachate throughout the life span of the treatment plant, these conventional treatments become ineffective. Thus, it is necessary to implement technologies that can be adjusted to the in situ needs [3]. Electrochemical technologies have shown high efficiency in the elimination of persistent pollutants and several studies have described the application of electrochemical methods in wastewater treatment [4-10].

A promising electrochemical method that can be used in wastewater treatment is the anodic oxidation. Despite several different materials are being used as anodes in the oxidation of persistent pollutants, the best results are obtained with boron-doped diamond (BDD) anodes, due to their unique chemical, electrochemical and structural stabilities that allow their use at high potentials, where most organic pollutants can be oxidized [11-13]. There are already several reports describing the application of electrochemical oxidation with BDD anodes for the treatment of landfill leachates [14-24].

Cabeza and co-workers [14,15] reported the application of electrochemical oxidation process, using a BDD anode, to treat raw leachates and biologically and physicochemically pre-treated leachates from a municipal landfill site. Experimental results showed very high chemical oxygen demand (COD) and ammonia removals, although ammonia removal was slower than that of COD. They also observed that when additional chloride ions were provided, the treatment efficiency increased. Anglada and co-workers [17,20,21] also studied the effect of the applied current density and of the initial concentration of chloride ions, as well as the influence of other operating conditions, such as treatment time and initial pH, in the electrochemical oxidation of landfill leachates, using a BDD anode. They have shown that when high current densities are applied, a change in the mechanism of the organic matter oxidation occurs and that organic matter and ammonia oxidation are highly influenced by the applied current density [17]. Also, they have reported that the concentration of chloride has an effect on the oxidation of ammonia and that chloride ions compete with organic matter to be oxidized at the anode. It was found that some chlorinated organic compounds are formed as a result of organic matter oxidation and their concentration increase continuously with treatment time [21]. Acidic conditions were found to favour the formation of haloacetonitriles and haloketons. A kinetic modelling of the electrochemical removal of ammonium and COD from landfill leachates was proposed in literature [23-24]. Authors found that the use of BDD anodes promotes the generation of hydroxyl radicals, while the high content of chloride induces the simultaneous formation of free chlorine, responsible for the ammonium indirect oxidation and for the formation of undesirable products such as chloramines, chlorate and perchlorate. Chlorine evolution is enhanced at lower COD concentrations. During this process, ammonium removal leads to the formation of nitrogen gas and nitrate as the main oxidation products.

In this work, the influence of the raw leachate dilution on the electrochemical degradation of a biologically pre-treated leachate from a sanitary landfill, using a BDD anode, was assessed and it has shown that mineralization of the organic matter improves with the dilution of the leachate [22]. However, an increase in the dilution greatly increases the energy consumption.

The aim of this work was to study the influence of flow rate and applied current density, carried out with multiple step electro-oxidation, on the performance of the electrochemical oxidation of raw leachate from a sanitary landfill. The energy consumption in the different experimental conditions tested was also assessed.

Experimental

The leachate samples used in this study were collected at a Portuguese intermunicipal sanitary landfill site, in the equalization tank, before any kind of treatment. Samples characterization is presented in Table 1.

Experiments were conducted in a semi-pilot plant operating in batch mode with recirculation, at room temperature and natural pH, without adding background electrolyte. A BDD DiaCell 100 electrochemical cell, with an electrode area of 70 cm^2, and a DiaCell-PS1500 power supply, with automatic polarity reversal, were used. In all assays, automatic polarity reversal occurred every minute. Different current densities, between 50 and 200 mA cm^{-2}, and different flow rates, between 100 and 950 L h^{-1}, were tested in sample volumes of 5, 10 or 15 L. During the degradation process, current density was kept constant or decreased by steps, in order to study this effect on the efficiency of the process. Potential differences between anode and cathode were registered throughout the experiments in order to determine energetic consumptions. All assays were performed in duplicate.

Degradation tests were followed by chemical oxygen demand (COD), dissolved organic carbon (DOC), total nitrogen (TN), total Kjeldahl nitrogen (TKN) and ammonia nitrogen (AN).

Table 1. *Physicochemical characteristics of the raw leachate.*

Parameter	Medium value ±SD*
COD, g L^{-1}	8.9 ± 0.8
BOD_5, g L^{-1}**	1.3 ± 0.3
BOD_5/COD	0.15± 0.05
DOC, g L^{-1}	3.5 ± 0.4
TN, g L^{-1}	2.8 ± 0.2
TKN, g L^{-1}	2.4 ± 0.2
AN, g L^{-1}	2.2 ± 0.3
$c_{Chloride}$ / g L^{-1}	4.5 ± 0.3
$c_{Suspended\ Solids}$ / g L^{-1}	0.7 ± 0.1
$c_{Dissolved\ Solids}$ / g L^{-1}	16.6 ± 0.1
pH	8.3 ± 0.2
Conductivity, mS cm^{-1}	29.1 ± 1.0

*SD - Standard Deviation; **BOD_5 – Biochemical oxygen demand

COD determinations were made using the closed reflux and titrimetric method [25]. DOC and TN were measured in a Shimadzu TOC-V CSH analyser. Before DOC and TN determinations, samples were filtrated through 1.2 μm glass microfiber filters. TKN and AN were determined according to standard procedures using a Kjeldatherm block-digestion-system and a Vapodest 20s distillation system, both from Gerhardt [25].

Results and Discussion

The effect of the applied current density on the rate of electrochemical oxidation was studied by performing the electrodegradation assays at three different current intensities, 4, 7 and 14 A, at a flow rate of 360 L h^{-1}, and using leachate volumes of 15, 5 and 5 L, respectively. Figure 1 presents the results of the normalized COD variation with time and with specific charge passed for these electrodegradation assays. Specific charge was calculated as *It/V*, in C L^{-1}, where *I* is the current intensity, in A, *t* is the time, in s, and *V* is the leachate volume, in L.

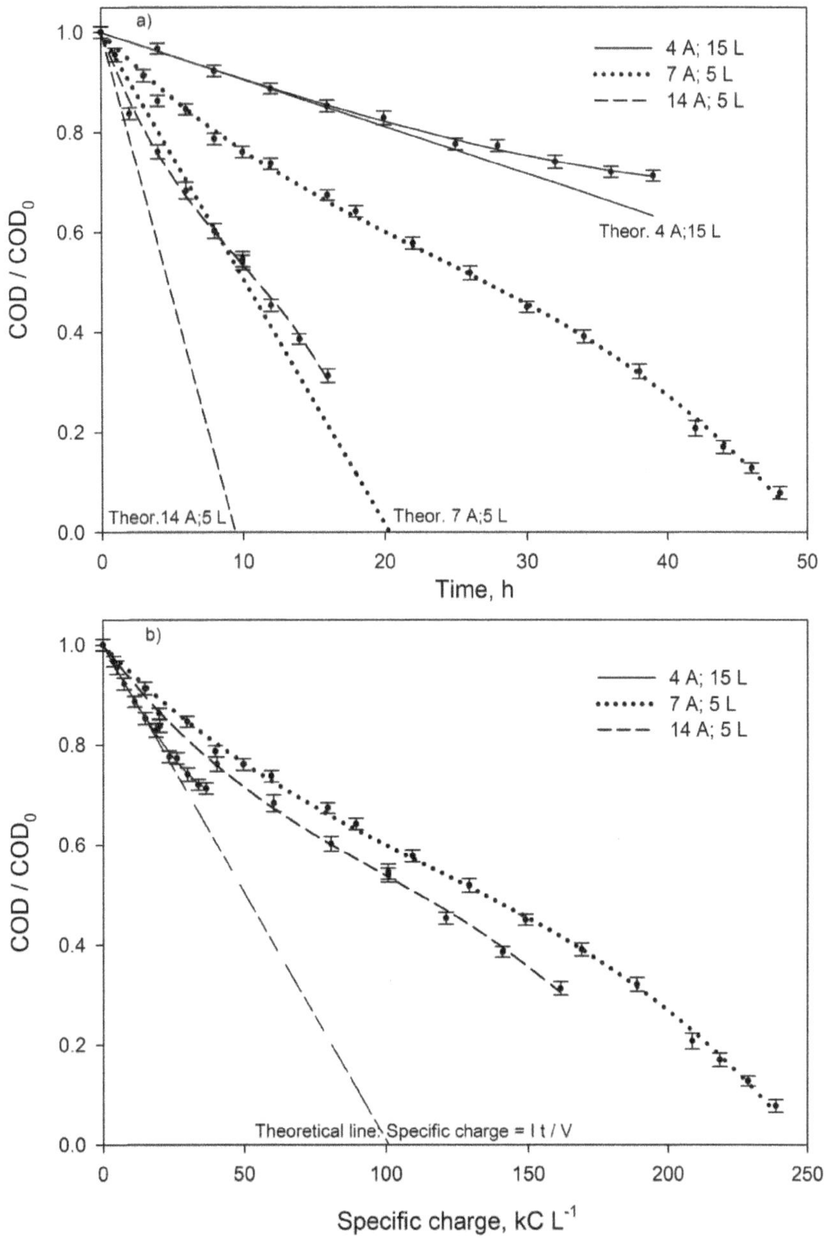

Figure 1. (a) *Normalized COD variation with time for the electrodegradation assays performed at different current intensities, at a flow rate of 360 L h^{-1}.* ***(b)*** *Normalized COD variation with specific charge for the electrodegradation assays performed at different current intensities, at a flow rate of 360 L h^{-1}. Error bars refer to the standard deviation of the COD mean values.*

It can be observed (Fig. 1a) that, for the assays performed with equal leachate volume COD removal rate increases with current density, which points to electrolysis operating under charge transfer control. In fact, for a single-compartment electrolytic reactor similar to the one used in this work, operating at flow rates of 200 and 600 L h^{-1}, mass transport coefficients, k_m, of 1.39×10^{-5} and 1.5×10^{-5} m s^{-1}, for 200 L h^{-1}, and 2.2×10^{-5} m s^{-1}, for 600 L h^{-1}, are presented in literature [26-28]. With these k_m values from literature, limiting currents between 10.4 and 16.5 A were obtained, showing that at least two of the assays presented in Fig. 1 started at current limited control conditions.

According to the model previously proposed in the literature for electrolysis under current limited control [29], *i.e.*, at maximum current efficiency, the trend of COD during electrochemical oxidation can be predicted by Eq. (1), where F is the Faraday constant, 96485 C mol^{-1}, and V is the volume of the samples treated, in m^3. Thus, theoretical slopes of COD *vs.* time, $I/4FV$, can be calculated for each of the assayed current intensities.

$$\mathrm{COD}(t) = \mathrm{COD}_0 - \frac{I}{4\,FV}t \qquad (1)$$

The comparison between these theoretical slopes and the experimental curves (Fig. 1a), for equal recirculation volume, shows that the discrepancy between experimental data and predicted slopes slightly decreases with current density. This can be explained if one assumes that the degradation process happens also by indirect oxidation. The increase in the leachate recirculation volume also seems to contribute to an increase in the efficiency of the process, due to the lower ratio electrode area/treated volume. In fact, when the volume is increased, keeping the same anodic area, the quantity of the compounds that are more easily degraded and that behave ideally augments. Thus, their concentration is kept higher for longer times when the recirculation volume is increased.

The effect of applied current on the trend of the COD with the specific charge consumed during the treatment (Fig. 1b) is less pronounced than the effect on the variation of COD *vs.* time. For equal leachate volume, an increase in current density leads to a more efficient use of the electric charge, since the experimental curve for 14 A is closer to the theoretical prediction. However, since higher current densities imply higher potentials, although the electric charge is more efficient the energetic consumption can be higher. Figure 1b also shows that an increase in the leachate recirculation volume approaches the experimental results to the theoretical prediction.

To try to improve the current efficiency, assays were performed with successive decreases in current density, by steps, during the oxidation process, at different recirculation flow rates. In Figure 2 it can be observed the variation of normalized COD with specific charge for the assays run at constant (14 A) and variable current density (5 h at 14 A + 5 h at 7 A + 10 h at 4 A), at a flow rate of 360 L h^{-1}. COD removal seems to depend only on the charge passed. Variation of normalized COD with time (Fig. 2, inset) shows that during the first five hours, where the applied current density was equal, no difference can be seen in the COD removal rate. But, when the applied current density decreases, in the assays with steps, a decrease in the COD removal rate can be observed.

The influence of the recirculation flow rate in the electrochemical oxidation performance of the assays that were run with current density decreased by steps during the experiment was also studied. Figure 3 shows the normalized COD variation with the specific charge consumed for the assays performed with three or four current density steps at different recirculation flow rates: three steps, 5 h at 14 A + 5 h at 7 A + 10 h at 4 A, flow rates of 160 and 360 L h^{-1}, leachate volume 5 L; four steps, 4 h at 14 A + 4 h at 10.5 A + 4 h at 7 A + 4 h at 4 A, flow rates of 100 and 950 L h^{-1}, leachate volume 10 L.

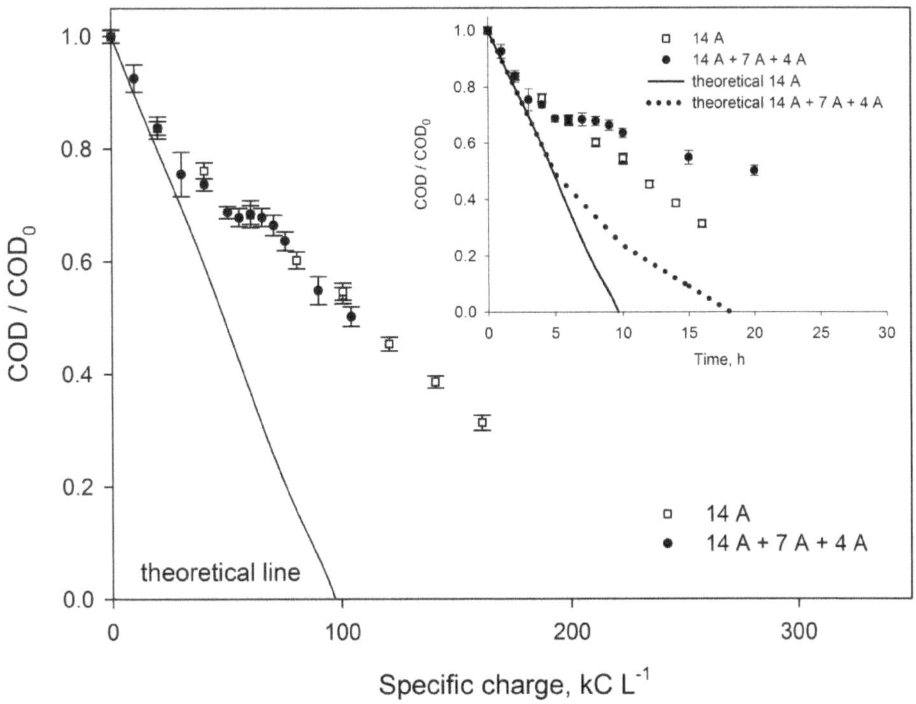

Figure 2. *Normalized COD variation with specific charge passed and with time (inset) for the electrodegradation assays performed at constant and variable current intensity, at a flow rate of 360 L h^{-1}, with a leachate volume of 5 L. Error bars refer to the standard deviation of the COD mean values.*

The theoretical curves at these conditions are also presented. A slight variation in the trend of the COD depletion was observed, pointing to better removals at higher recirculation flow rates. The same behaviour is observed when normalized COD variation with electrolysis time is plotted (Fig. 3, insets). The discrepancy between experimental values and theoretical curves, after the first step of the assays, indicates a high loss in current efficiency that increases when current density is further decreased during the steps process.

Comparing the discrepancy between experimental values and theoretical curves for three and four steps, it can be concluded that charge efficiency is higher when four steps are applied, although this fact must be also related with the higher recirculation volume of leachate used in the four steps experiments.

Nitrogen removal was also assessed. In Figure 4 are plotted the normalized variation with time of ammonia nitrogen and of total nitrogen. Both parameters present similar behaviour of that described for COD in these assays, *i.e.*, a decrease in the applied current density, in the steps assays, leads to a decrease in the nitrogen removal rates. It can be seen that, for the experiments performed at 14 A, an increase in the removal rate is observed after eight hours assay. This fact is consistent with previous reports from other authors [23], which indicate that while BDD anodes promotes the generation of hydroxyl radicals, the high content of chloride induces the simultaneous formation of free chlorine, causing indirect oxidation of ammonium. In fact, this leachate presents high chloride concentration (4.5 g L^{-1}), thus enhancing the chlorine evolution at lower COD concentrations, justifying the increase in the nitrogen removal rate when COD levels are lower. In contrast to what was observed with COD removal, nitrogen removal is higher when the recirculation flow rate is lower. In fact, at higher flowrates COD oxidation is favoured, chlorine evolution, that is a competitive reaction, is delayed as a consequence and thus it influences and slows down the rate of ammonium removal.

In order to analyse the energy consumption, the specific energy consumptions, E_{sp}, in W h/ g COD removed were calculated, by means of Eq. (2):

$$E_{sp} = \frac{UI\,\Delta t}{V\,\Delta \text{COD}} \tag{2}$$

where U is the cell voltage, in V, resulting from the applied current intensity I, in A, Δt is the duration of the electrolysis, in h, V is in m^3 and ΔCOD is the removed COD, in g m^{-3}, during Δt.

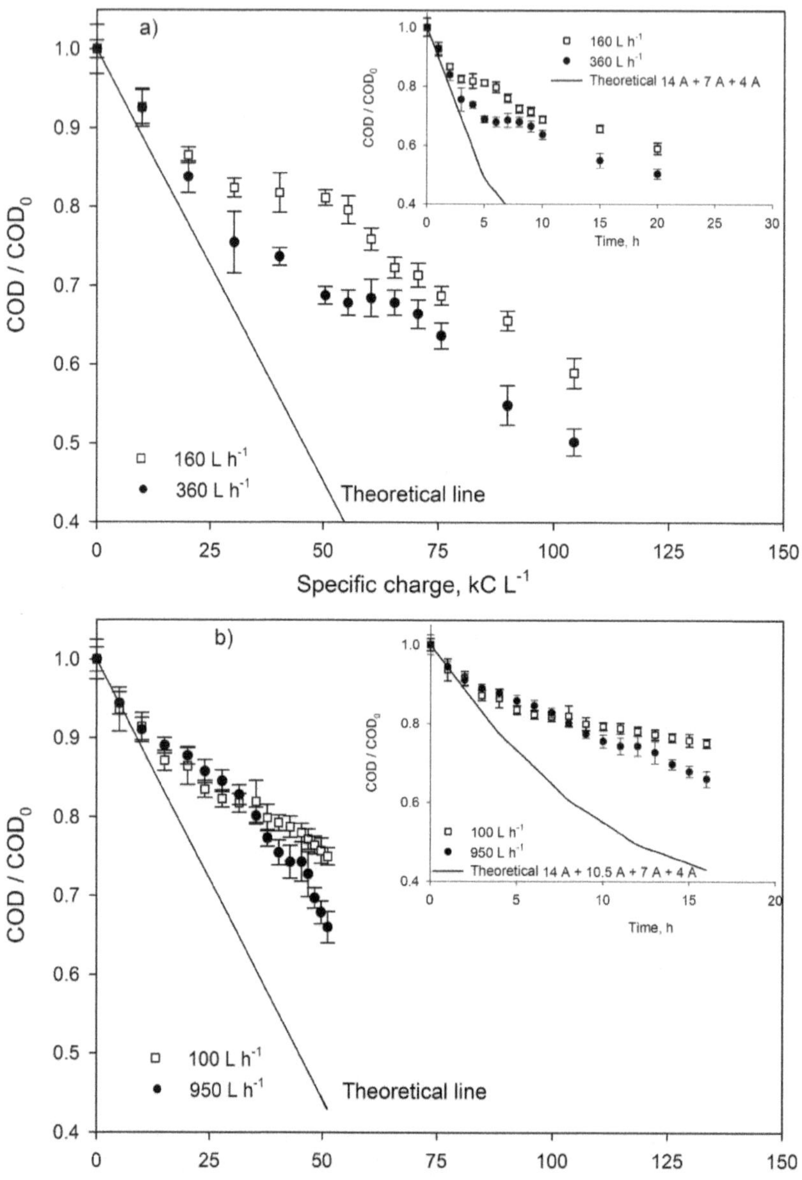

Figure 3. *Normalized COD variation with specific charge and with time (inset) for electrodegradation assays performed with* ***(a)*** *three current density steps, with a leachate volume of 5 L and with* ***(b)*** *four current density steps, with a leachate volume of 10 L. Error bars refer to the standard deviation of COD mean values.*

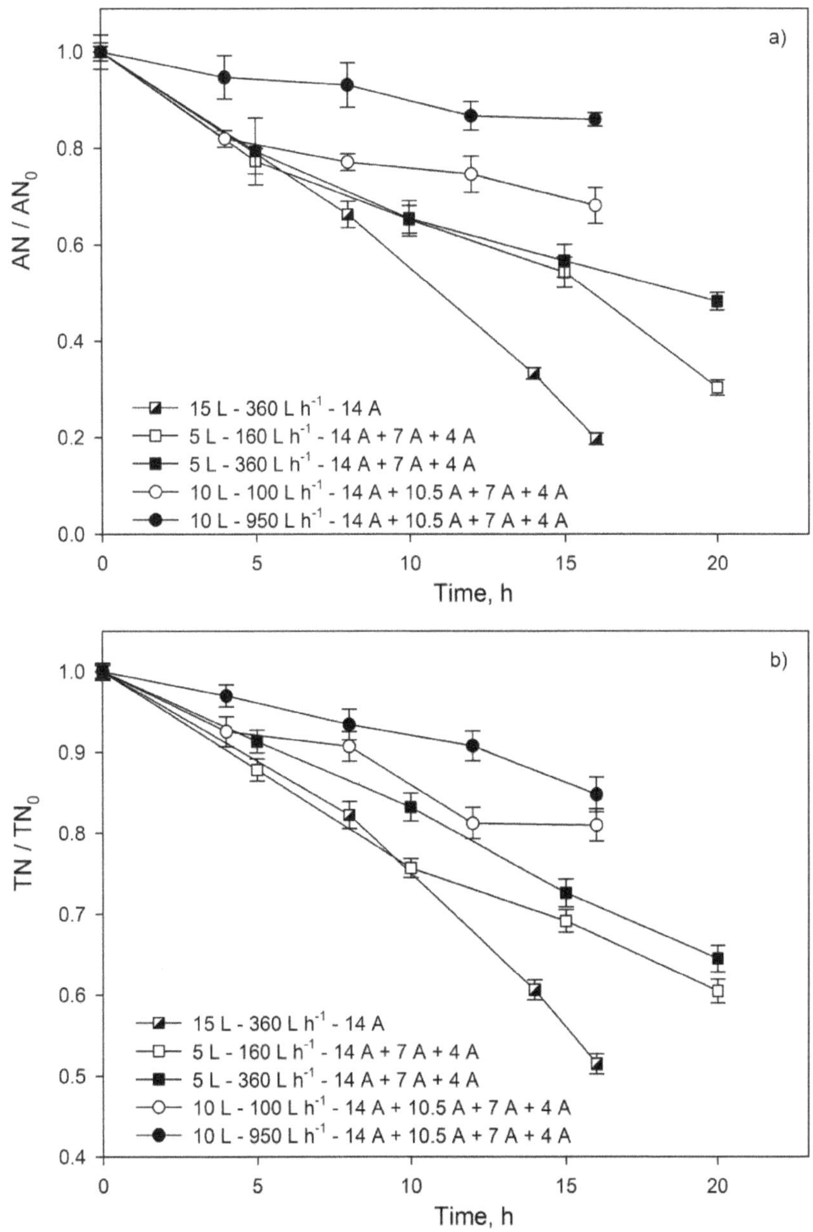

***Figure 4. (a)** Normalized ammonia nitrogen variation with time for the electrodegradation assays performed at constant current density and at three and four current density steps. **(b)** Normalized total nitrogen variation with time for the electrodegradation assays performed at constant current density and at three and four current density steps.*

Figure 5 reports the specific energy consumption as a function of the time for the different assays performed. The specific energy consumption seems to increase with current density (Fig. 5a), which is a consequence of the increase in potential when the current density is increased. When constant current density was imposed (Fig. 5a), there is an increase in the energy consumption during the first part of the assay, followed by a decrease. This behavior must be due to the different types of compounds that are present and that are not degraded simultaneously, being first degraded those that are present in higher concentration and, among them, those who have higher diffusion coefficients. The introduction of steps, although leads to a overall decrease in the energetic consumption, did not present the expected results in terms of specific energy consumption, since it leads to more irregular consumptions rather than lower consumptions

(Fig. 5b). For these assays, an increase in the recirculation flow rate seems to slightly decrease the E_{sp} (Fig. 5c and 5d). On the other hand, the increase in the leachate volume being recirculated really decreases the specific energy consumption, since the values in the yy' axis are much lower in Fig. 5d (10 L) than in Fig. 5c (5 L).

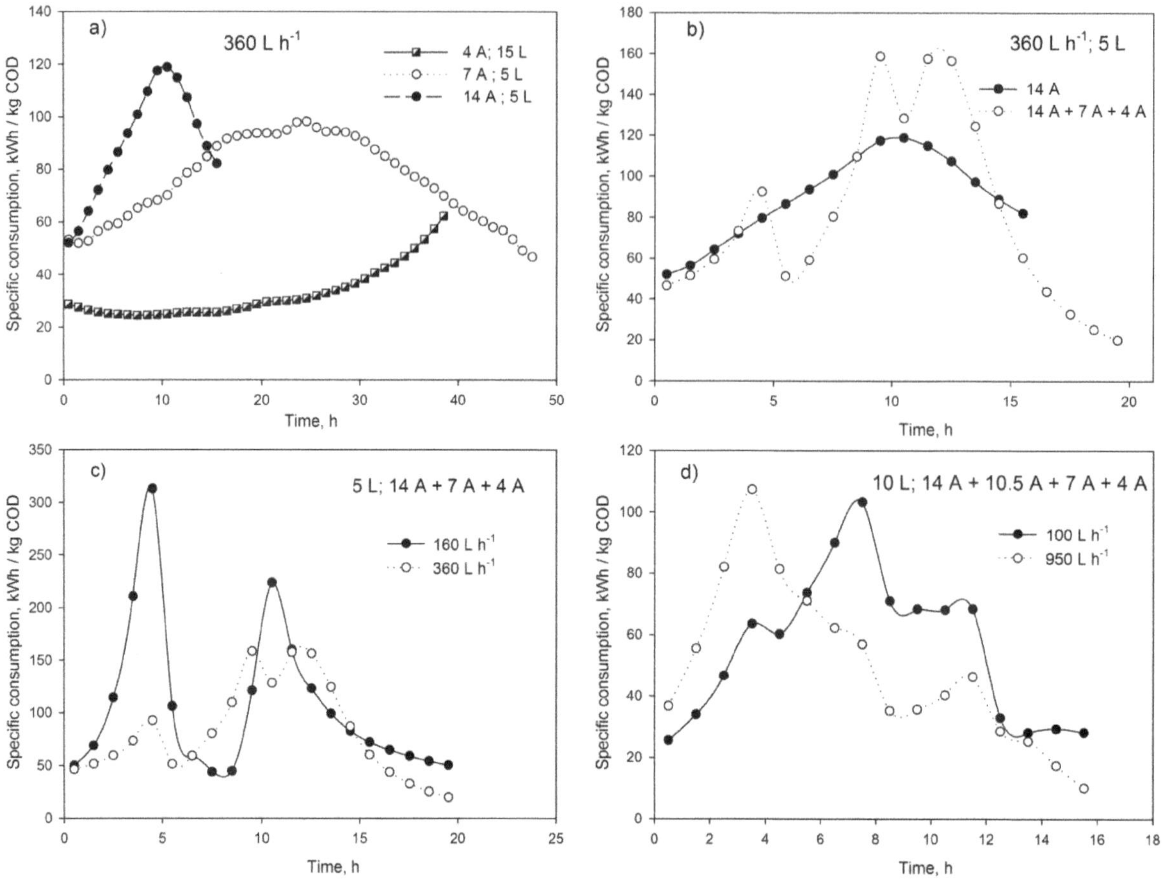

Figure 5. *Evolution of specific energy consumption with time for **(a)** electrodegradation assays performed at different current densities, at a flow rate of 360 L h^{-1} **(b)** electrodegradation assays performed at constant and variable current density, at a flow rate of 360 L h^{-1} **(c)** electrodegradation assays performed with three current density steps at different recirculation flow rates, with a leachate volume of 5 L **(d)** electrodegradation assays performed with four current density steps at different recirculation flow rates, with a leachate volume of 10 L.*

The removals in COD, DOC, TN, TKN and AN for all assays performed with current density decreased by steps, as well as the medium specific energy consumption, are presented in Table 2. This table includes also the results obtained in the assay performed at constant current intensity of 14 A and 360 L h^{-1} recirculation flow rate, in order to allow comparison between assays performed with and without reduction in the current intensity during the assay. The apparent discrepancy between absolute and percentage values presented in Table 2 is due to the variation of the experimental determinations of those parameters for the different assays, due to the complexity and heterogeneity of the leachate suspension. Data reported confirm the previous analysis, showing that for both multiple step designs, with 3 or 4 current density steps, and for a wide range of recirculation flow rate, from 100 to 950 L h^{-1}, an increase in the recirculation flow rate increases COD removal rate and decreases nitrogen removal rate (TN, TKN and AN). Also, it can be seen that DOC removals are always lower than COD removals and these differences increase with flow rate, indicating that a decrease in the flow rate increases the mineralization index. Regarding the energy consumption, an increase in the recirculation flow rate leads to a decrease in the medium energy consumption, mainly because COD removal rate increases with recirculation flow rate.

Table 2. *COD, DOC, TN, TKN and AN removals and medium specific energy consumption for assays performed with one, three and four current density steps at different recirculation flow rates.*

Parameter		Experimental conditions				
		16 h (14 A) V = 5 L; t = 16 h	5 h (14 A) + 5 h (7 A) + 10 h (4 A) V = 5 L; t = 20 h		4 h (14 A) + 4 h (10.5 A) + 4 h (7 A) + 4 h (4 A) V = 10 L; t = 16 h	
		360 L h^{-1}	160 L h^{-1}	360 L h^{-1}	100 L h^{-1}	950 L h^{-1}
COD Removal	g L^{-1}	5.42	3.11	4.09	2.56	2.57
	%	69	41	50	25	34
DOC Removal	g L^{-1}	1.35	0.50	0.94	0.61	0.18
	%	44	19	30	15	6
TN Removal	g L^{-1}	1.23	1.04	0.93	0.59	0.36
	%	48	39	35	19	15
TKN Removal	g L^{-1}	1.72	1.11	1.06	0.83	0.33
	%	72	53	45	35	15
AN Removal	g L^{-1}	1.66	0.99	0.95	0.78	0.24
	%	80	60	45	32	14
$\overline{E_{sp}}$ / kW h (kg COD)$^{-1}$		90.1	106.0	80.9	55.7	49.5

Conclusions

The anodic oxidation was used to treat leachate from an intermunicipal sanitary landfill and the following conclusions can be drawn:

- Organic load removal rate increases with applied current density. This happens mainly because, due to the high organic load content, the electrochemical processes are under current control most of the assay period.
- An increase in the recirculation flow rate leads to an increase in the organic load removal rate. However, it decreases the nitrogen removal.
- By reducing the current density along the anodic oxidation process it is possible to reduce energetic costs. Similar results can be obtained by increasing the recirculation flow rate.
- DOC removals are always lower than COD removals and these differences increase with flow rate. Thus, a decrease in flow rate seems to increase the mineralization index.

Thus, although huge variations can be found in the composition of leachates from sanitary landfills, the anodic oxidation, performed with a BDD anode, can be an alternative/complement to treat this kind of wastewaters. Also, the variation found in the medium specific energy consumption shows that it is possible to optimize the process in order to reduce energy costs.

Acknowledgements: *Financial support from FEDER, Programa Operacional Factores de Competitividade – COMPETE, and FCT, for the projects PTDC/AAC- AMB/103112/2008 and PEst-OE/CTM/UI0195/2011 of the MT&P Unit and for A. Fernandes grant to SFRH/BD/81368/2011.*

References

[1] T. Eggen, M. Moeder, A. Arukwe, *Science of The Total Environment* **408** (2010) 5147-5157
[2] C.B. Öman, C. Junestedt, *Waste Management* **28** (2008) 1876-1891
[3] A.A. Abbas, G. Jingsong, L.Z. Ping, P.Y. Ya, W.S. Al-Rebaki, *Journal of Applied Sciences Research* **5** (2009) 534-545
[4] G. Chen, *Separation and Purification Technology* **38** (2004) 11-41
[5] C.A. Martínez-Huitle, S. Ferro, *Chemical Society Reviews* **35** (2006) 1324-1340

[6] C. Comninellis, A. Kapalka, S. Malato, S.A. Parsons, I. Poulios, D. Mantzavinos, *Journal of Chemical Technology & Biotechnology* **83** (2008) 769-776

[7] A. Anglada, A. Urtiaga, I. Ortiz, *Journal of Chemical Technology & Biotechnology* **84** (2009) 1747-1755

[8] C.A. Martínez-Huitle, E. Brillas, *Applied Catalysis B: Environmental* **87** (2009) 105-114

[9] M. Panizza, G. Cerisola, *Chemical Reviews* **109** (2009) 6541-6569

[10] D. Ghernaout, M.W. Naceura, A. Aouabeda, *Desalination* **270** (2011) 9-22

[11] M. Panizza, G. Cerisola, *Electrochimica Acta* **51** (2005) 191-199

[12] J.H.T. Luong, K.B. Male, J.D. Glennon, *Analyst* **134** (2009) 1965-1979

[13] M. Panizza, M. Delucchi, I. Sirés, *Journal of Applied Electrochemistry* **40** (2010) 1721-1727

[14] A. Cabeza, A. Urtiaga, M.J. Rivero, I. Ortiz, *Journal of Hazardous Materials* **144** (2007) 715-719

[15] A. Cabeza, A. Urtiaga, I. Ortiz, *Industrial & Engineering Chemistry Research* **46** (2007) 1439-1446

[16] Y. Deng, J.D. Englehardt, *Waste Management* **27** (2007) 380-388

[17] A. Anglada, A. Urtiaga, I. Ortiz, *Environmental Science & Technology* **43** (2009) 2035-2040

[18] A. Urtiaga, A. Rueda, Á. Anglada, I. Ortiz, *Journal of Hazardous Materials* **166** (2009) 1530-1534

[19] G. Zhao, Y. Pang, L. Liu, J. Gao, B. Lv, *Journal of Hazardous Materials* **179** (2010) 1078-1083

[20] A. Anglada, A.M. Urtiaga, I. Ortiz, *Journal of Hazardous Materials* **181** (2010) 729-735

[21] A. Anglada, A. Urtiaga, I. Ortiz, D. Mantzavinos, E. Diamadopoulos, *Water Research* **45** (2011) 828-838

[22] A. Fernandes, M.J. Pacheco, L. Ciríaco, A. Lopes, *Journal of Hazardous Materials* **199-200** (2012) 82-87

[23] A. Urtiaga, I. Ortiz, A. Anglada, D. Mantzavinos, E. Diamadopoulos, *Journal of Applied Electrochemistry* **42** (2012) 779–786

[24] G. Pérez, J. Saiz, R. Ibañez, A.M. Urtiaga, I. Ortiz, *Water Research* **46** (2012) 2579–2590

[25] A. Eaton, L. Clesceri, E. Rice, A. Greenberg, M.A. Franson, *Standard Methods for Examination of Water and Wastewater, twenty-first ed.*, American Public Health Association, Washington, DC, 2005

[26] R. Bellagamba, P.A. Michaud, Ch. Comninellis, N. Vatistas, *Electrochemistry Communications* **4** (2002) 171-176

[27] L. Gherardini, P.A. Michaud, M. Panizza, Ch. Comninellis, N. Vatistas, *Journal of the Electrochemical Society* **148** (2001) D78–D82

[28] E. Chatzisymeon, N. Xekoukoulotakis, E. Diamadopoulos, A. Katsaounis, D. Mantzavinos, *Water Research* **43** (2009) 3999-4009

[29] M. Pannizza, P.A. Michaud, G. Cerisola, C. Comninellis, *Journal of Electroanalytical Chemistry* **507** (2001) 206–214

16

Formation of the honeycomb-like electrodes by the regime of pulsating overpotential in the second range

NEBOJŠA D. NIKOLIĆ✉, GORAN BRANKOVIĆ* and MIOMIR G. PAVLOVIĆ

ICTM-Institute of Electrochemistry, University of Belgrade, Njegoševa 12, 11000 Belgrade, Serbia

**Institute for Multidisciplinary Research, University of Belgrade, Kneza Višeslava 1a, Belgrade, Serbia*

✉Corresponding Author: E-mail: nnikolic@tmf.bg.ac.rs

Abstract

In this study the honeycomb-like copper structures electrodeposited by the regime of pulsating overpotential in the second range were analyzed by the technique of scanning electron microscopy. The overpotential amplitude of 1000 mV, deposition pulse of 1 s, and pause durations of 1, 5, 10 and 15 s were selected for the production of this type of structures. The size of holes which remained upon detachment of hydrogen bubbles do not depend on the length of pause duration. On the other hand, the change in morphology of electrodeposited copper around holes from cauliflower-like agglomerates of copper grains to degenerated dendrites is observed when pause duration was increased. Effects of the application of the regime of pulsating overpotential in the second range on the formation of the honeycomb-like structures were less pronounced than the effects attained by the application of the same regime in the millisecond range. However, they were more pronounced than those attained by electrodeposition in the regime of constant potential.

Keywords

Copper; electrodeposition; pulsating overpotential; hydrogen; scanning electron microscope.

Introduction

Open porous structures with an extremely large surface area are of high technological significance, because they are very suitable as electrodes in many electrochemical devices, such as fuel cells, batteries and sensors [1,2], and in catalysis [3]. Electrodeposition technique was shown to be very favorable way for the production of this type of electrodes [1,2,4]. Open porous copper electrodes, denoted as both 3-D foam [1,2,4] or honeycomb-like electrodes [5-13], are formed by electrodeposition at high current densities and overpotentials, when, parallel to electrodeposition process, the hydrogen evolution reaction occurs. The main characteristics of these electrodes are holes or pores formed upon detachment of hydrogen bubbles, surrounded by the agglomerates of metal grains or dendritic particles. This way of the preparation of porous electrodes is denoted as gas bubble dynamic template method, since the hydrogen bubbles function as a dynamic template for the formation of this type of deposits. The specific surface area of these structures is determined by the number and size of the holes, as well as by the width of the walls in-between [4]. Aside from copper, which is the most studied system [1,2,4-15], open porous structures of some other technologically important metals, such as tin [1], nickel [16], silver [17,18], gold [19], lead [20], and copper-tin alloys [2], were also investigated.

The conditions of the formation of open porous copper electrodes by constant regimes of electrolysis (the potentiostatic and galvanostatic ones) are well examined and systematized [1,13]. Two ways have been proposed for the increase in specific surface area and the improvement of micro- and nano-structural characteristics of the open porous structures: a) the addition of additives to the electroplating baths [4,21-23] and b) the application of periodically changing regimes of electrolysis, such as those of pulsating overpotential (PO) [24-27], pulsating current (PC) [28-30] and reversing current (RC) [31].

All benefits in the formation of the honeycomb-like structures by the electrochemical deposition processes at periodically changing rate were reported for the cases of electrodepositions in the millisecond range. However, the application of the periodically changing regimes of electrolysis on the formation of open porous structures in the second range of pulsating is completely unknown. Therefore, the aim of this paper is to analyze the formation of the honeycomb-like structures by the regime of pulsating overpotential (PO) in the second range.

Experimental

Copper was electrodeposited from 0.15 M $CuSO_4$ in 0.50 M H_2SO_4, in an open glass cylindrical cell of 100 mL at the room temperature. Doubly distilled water and analytical grade chemicals were used for the preparation of the solution for electrodeposition of copper.

Copper electrodeposition was performed by the square-wave pulsating overpotential (PO) technique. In all experiments, the overpotential amplitude of 1000 mV was used. Pause durations of 1, 5, 10 and 15 s were applied, while the applied deposition pulse was 1 s.

The working electrodes used for electrodeposition of copper were cylindrical copper wires. The diameter of these copper electrodes was 0.08 cm, while their length was 2 cm. The counter electrode was copper foil of 0.80 dm^2 surface area placed tightly to the walls of the cell, whereas the reference electrode was copper wire, the tip of which was positioned at the distance of 0.2 cm from the surface of the working electrode. The working electrode was always situated in the centre of cell.

The preparation of working electrodes consisted of alkaline degreasing at 70°C followed by an acid etching (20 % H_2SO_4) at 50°C. After each of these preparation phases, working electrodes were rinsed by distilled water. This pretreatment of the electrodes was important in order to obtain the same state of initial electrode surface prior to each experiment.

Copper was electrodeposited by consuming 10 mAh cm^{-2} of electricity.

The obtained copper deposits were examined by a scanning electron microscope TESCAN Digital Microscopy.

Results and Discussion

Figure 1 shows the honeycomb-like structures obtained by the regime of the pulsating overpotential (PO) in the second range. For the formation of the shown honeycomb-like structures, the following parameters of the PO regime were applied: the overpotential amplitude of 1000 mV, the deposition pulse of 1 s, and the pause durations of 1 (Fig. 1a), 5 (Fig. 1b), 10 (Fig. 1c) and 15 s (Fig. 1d). It can be noticed from Fig. 1 that there is no significant difference in the number of holes formed upon detachment of hydrogen bubbles from the electrode surface. The number of the formed holes was estimated to be 125 ± 5 holes per mm^2 surface area of the copper electrode.

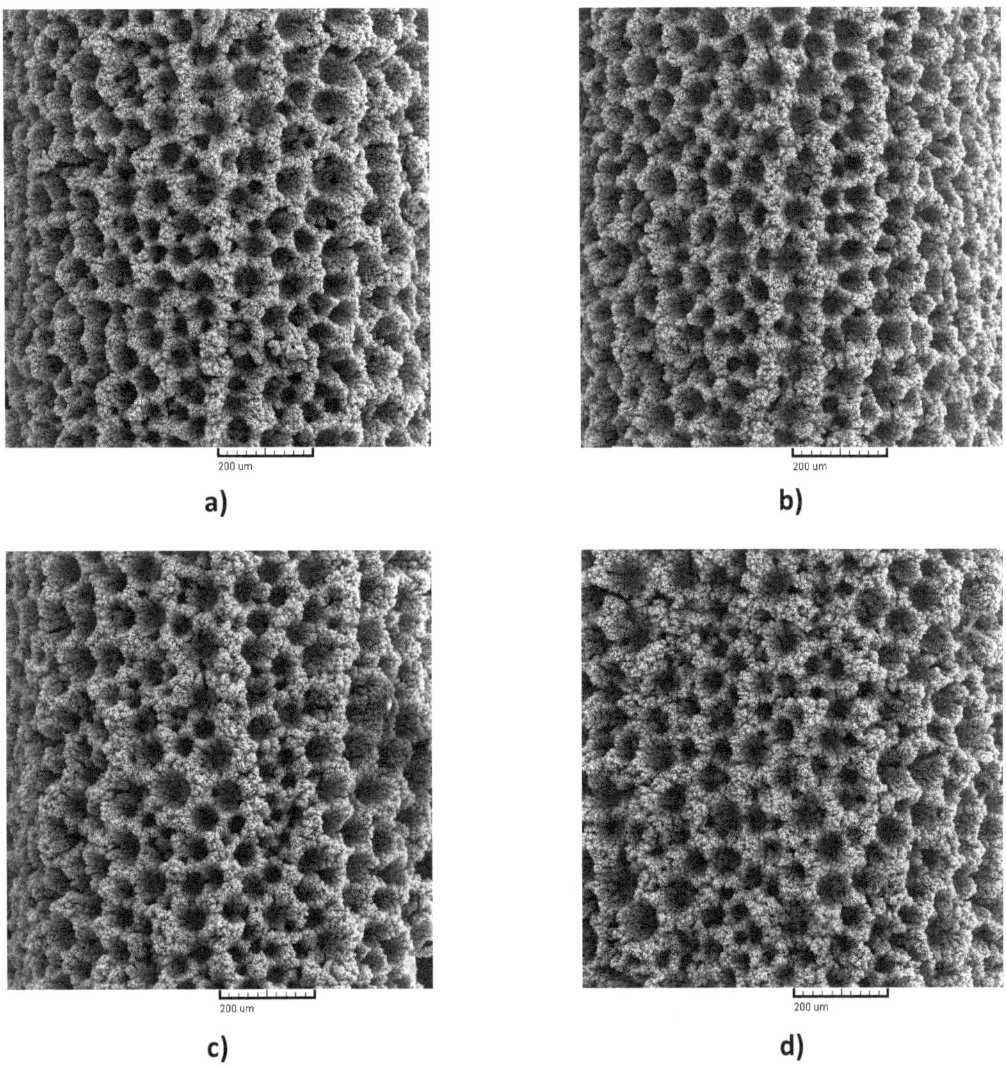

Figure 1. *The honeycomb-like copper structures obtained by the regime of pulsating overpotential (PO) in the second range with pause durations of:* ***a)*** *1,* ***b)*** *5,* ***c)*** *10 and* ***d)*** *15 s. The overpotential amplitude: 1000 mV. Deposition pulse: 1 s.*

The detailed analysis of these honeycomb-like electrodes showed negligible presence of holes obtained by the coalescence of closely formed hydrogen bubbles. The analysis of the holes at higher magnifications showed that their size does not depend on the pause duration. The typical hole, formed by detachment of a hydrogen bubble, which was obtained with pause duration of 5 s is shown in Fig. 2. The average hole size was estimated to be 80 ± 2.0 μm.

Figure 2. *The typical hole obtained by the PO regime with pause duration of 5 s. The overpotential amplitude: 1000 mV. Deposition pulse: 1 s.*

Morphologies of copper deposits formed around holes are shown in Fig. 3. It can be seen in Fig. 3 that macro-structure of the obtained deposits strongly depends on the pause duration. The cauliflower-like agglomerates of copper grains were formed with pause durations of 1 (Fig. 3a) and 5 s. The increase in pause duration to 10 and 15 s led to the formation of deposits with dendritic characteristics (Figs. 3b – 3d). Dendritic character is more pronounced for the copper deposit obtained with pause duration of 15 s (Figs. 3c and 3d) than with the one that lasted 10 s (Fig. 3b). These morphological forms can be denoted as degenerate dendrites. Contrary to the strong effect of the pause duration on the macro-structure of the honeycomb-like electrodes, their micro-structure was only negligibly affected. Both degenerate dendrites and cauliflower-like agglomerates of copper grains consist of small copper grains agglomerates. The size of grains of both, the cauliflower-like agglomerates of grains and the degenerate dendrites, approached the nano-size dimensions.

Generally, the two types of pores (or holes) were formed by hydrogen evolution reaction during electrodeposition process [32]. The first type relates to macro-pores (or holes) formed upon detachment of hydrogen bubbles. These pores originate from the hydrogen bubbles formed in the initial stage of electrodeposition process. They are responsible for the creation of the honeycomb-like structure and they, upon detachement from the electrode surface, contribute neither to the stirring of the solution nor to the change in hydrodynamic conditions in the near-electrode layer.

The origin of the second type of pores is related to the hydrogen bubbles generated at the peaks of agglomerates of copper grains during the growth process (the current density distribution effect). The agglomerates of copper grains were formed at the electrode surface in the initial stage of electrodeposition with hydrogen bubbles. Some of these freshly formed hydrogen bubbles will coalesce with hydrogen bubbles formed in the initial stage of electrodeposition leading to the growth of the former as the electrolysis proceeds (the increase in the size of macro-pores). Simultaneously, some of newly formed hydrogen bubbles will not coalesce with primarily formed hydrogen bubbles because they are situated between freshly formed copper nuclei, which represent a barrier for the coalescence with the hydrogen bubbles formed in the initial stage of electrodeposition, or for development of larger hydrogen bubbles. These newly formed hydrogen bubbles will detach from the electrode surface very fast, thus forming micro- and nano-pores through the interior of the deposit. "The stream of hydrogen" is formed of these detached small hydrogen bubbles, which causes the stirring of the solution and the change in hydrodynamic conditions in the near-electrode layer. The channel structure through the interior of the deposit was formed of this micro- and nano-pores.

Figure 3. *Morphology of electrodeposited copper formed around holes by the PO regime with pause durations of:* ***a)*** *1,* ***b)*** *5,* ***c)*** *10 and* ***d)*** *15 s.*

The following conveniences in the formation of the honeycomb-like electrodes were achieved by the application of the PO regimes in the millisecond range in relation to the constant potentiostatic regime [24-27]:

a) the improvement of the current density distribution. Due to both, suppression of coalescence of closely formed hydrogen bubbles and the prevention of hydrogen bubbles to remain captive in the interior of deposit, the increase in the number of holes formed by detachment of hydrogen bubbles was observed;
b) the decrease in the amount of evolved hydrogen contributing to the effective stirring of the solution in the near-electrode layer. The remarkable change in surface morphology around holes from cauliflower-like agglomerates of metal grains to branchy dendrites was observed due to the decrease of this amount of evolved hydrogen.

Both of these conveniences result from the dissolution process caused by the anodic current density during the pause. It is clear that the condition (a) is fulfilled by the application of the PO regime in the second range (the increase of number of holes formed from detached hydrogen bubbles), while condition (b) is partially fulfilled (the mild change of surface morphology around holes). The possible reason for the smaller influence of the regime in the second range on the morphology of the honeycomb-like structures lies in the fact that the current during the pause is negligibly small in comparison with the current during deposition pulse. On the other hand, deposition pulses in the second range are sufficiently long for the current response during these pulses to approach the response of the constant potentiostatic regime. This increased difference between cathodic and anodic currents prevents the negligibly small anodic current during the pause to make significant effect on the dissolution process and hence on the amount of evolved hydrogen and surface morphology around the holes.

It is very clear from the previous consideration that the specific surface area of the honeycomb-like electrodes obtained in the second range is larger than the one obtained in the constant potential mode. The number of the holes obtained by the potentiostatic electrodeposition at an overpotential of 1000 mV (on the same type of working electrodes) was 45 per mm^2 of electrode surface, while the average hole size was about 115 μm [24].

It is necessary to note that during the formation of the honeycomb-like structures by the PO regime in both millisecond and second range with pause durations longer than deposition pulse, the parameters of the PO regime affect neither the number nor the size of the holes formed by detachment of hydrogen bubbles.

The concept of "effective overpotential" can be proposed to explain the formation of honeycomb-like structures by the PO regime in the range of seconds. This concept was originally proposed to explain the change in surface morphology during copper electrodeposition in the constant potentiostatic regime at high overpotentials, when, parallel to copper electrodeposition, the vigorous hydrogen evolution occurs [5,13]. According to this concept, when hydrogen evolution is vigorous enough, then electrodeposition process occurs at an overpotential which is effectively lower than the specified one, and this overpotential is denoted as "effective" in a deposition process. From the morphological point of view, this means that morphologies of metal deposits become similar to those obtained at some lower overpotentials at which hydrogen evolution does not occur or it is very slow. This concept can be adapted for the case of electrodeposition by the PO regime and denoted as "effective overpotential amplitude" concept. The appearance of degenerate dendrites clearly indicates the increase in the degree of diffusion control with the increase in pause duration.

Anyhow, morphological characteristics of the honeycomb-like structures obtained by the PO regime in the second range were between those obtained by the constant potential regime and by the PO regime in the millisecond range. Although the improvement of micro- and nano-structural characteristics of the honeycomb-like electrodes achieved by the application of the PO regime in the second range was smaller than the one achieved by the application of the same regime in the millisecond range, it is clear that the useful effects of the application of this electrolysis regime are maintained in relation to those attained by the electrodeposition in the constant potential mode.

Conclusions

Electrodeposition of copper in the potential range of the hydrogen co-deposition was examined by applying the regime of pulsating overpotential (PO) deposition in the second range. The obtained morphologies of electrodeposited copper were characterized by the scanning electron microscope (SEM). The parameters of square-wave PO were selected to enable formation of the honeycomb-like structures. The increase of pause duration affected neither the number nor the size of the holes formed by detachment of hydrogen bubbles. Simultaneously, the appearance of degenerate dendrites, instead of cauliflower-like agglomerates of copper grains, is observed when the pause duration was increased. The effect the PO regime in the second range on micro- and nano-structural characteristics of the honeycomb-like structures was less pronounced than that attained by the application of this regime in the millisecond range. However, the effect is more pronounced than the one observed by the electrodeposition in the constant potential mode.

Acknowledgements: *The work was supported by the Ministry of Education and Science of the Republic of Serbia under the research project: "Electrochemical synthesis and characterization of nanostructured functional materials for application in new technologies" (No. 172046).*

References

[1] H.-C. Shin, J. Dong, M. Liu, *Adv. Mater.* **15** (2003) 1610-1614.
[2] H.-C. Shin, M. Liu, *Adv. Funct. Mater.* **15** (2005) 582-586.
[3] J. Yin, J. Jia, L. Zhu, *Int. J. Hydrogen Energy* **33** (2008) 7444-7447.
[4] H.-C. Shin, M. Liu, *Chem. Mater.* **16** (2004) 5460-5464.
[5] N.D. Nikolić, K.I. Popov, Lj.J. Pavlović, M.G. Pavlović, *J. Electroanal. Chem.* **588** (2006) 88-98.
[6] N.D. Nikolić, K.I. Popov, Lj.J. Pavlović, M.G. Pavlović, *Surf. Coat. Technol.* **201** (2006) 560-566.
[7] N.D. Nikolić, K.I. Popov, Lj.J. Pavlović, M.G. Pavlović, *J. Solid State Electrochem.* **11** (2007) 667-675.
[8] N.D. Nikolić, Lj.J. Pavlović, M.G. Pavlović, K.I. Popov, *Electrochim. Acta* **52** (2007) 8096-8104.
[9] N.D. Nikolić, K.I. Popov, Lj.J. Pavlović, M.G. Pavlović, *Sensors* **7** (2007) 1-15.
[10] N.D. Nikolić, Lj.J. Pavlović, M.G. Pavlović, K.I. Popov, *J. Serb. Chem. Soc.* **72** (2007) 1369-1381.
[11] N.D. Nikolić, Lj.J. Pavlović, S.B. Krstić, M.G. Pavlović, K.I. Popov, *Chem. Eng. Sci.* **63** (2008) 2824-2828.
[12] N.D. Nikolić, G. Branković, M.G. Pavlović, K.I. Popov, *J. Electroanal. Chem.* **621** (2008) 13-21.
[13] N.D. Nikolić, K.I. Popov, *"Hydrogen Co-deposition Effects on the Structure of Electrodeposited Copper"* in *Electrodeposition: Theory and Practice, Series: Modern Aspects of Electrochemistry*, S.S. Djokić, Ed., Vol. 48, Springer (2010) pp. 1 – 70.
[14] S. Li, R. Furberg, M.S. Toprak, B. Palm, M. Muhammed, *Adv. Funct. Mater.* **18** (2008) 2215-2220.
[15] R. Kim, D. Han, D. Nam, J. Kim, H. Kwon, *J. Electrochem. Soc.* **157** (2010) D269-D273.
[16] C.A. Marozzi, A.C. Chialvo, *Electrochim. Acta* **45** (2000) 2111-2120.
[17] S. Cherevko, X. Xing, C-H. Chung, *Electrochem. Commun.* **12** (2010) 467-470.
[18] S. Cherevko, C-H. Chung, *Electrochim. Acta* **55** (2010) 6383-6390.
[19] S. Cherevko, C.-H. Chung, *Electrochem. Commun.* **13** (2011) 16-19.
[20] S. Cherevko, X. Xing, C-H. Chung, *Appl. Surf. Sci.* **257** (2011) 8054–8061.
[21] Y. Li, W.-Z. Jia, Y.-Y. Song, X.H. Xia, *Chem. Mater.* **19** (2007) 5758-5764.
[22] J. Kim, R. Kim, H. Kwon, *Electrochem. Commun.* **10** (2008) 1148-1151.
[23] D. Nam, R. Kim, D. Han, J. Kim, H. Kwon, *Electrochim. Acta* **56** (2011) 9397-9405.
[24] N.D. Nikolić, G. Branković, M.G. Pavlović, K.I. Popov, *Electrochem. Commun.* **11** (2009) 421-424.
[25] N.D. Nikolić, G. Branković, V.M. Maksimović, M.G. Pavlović, K.I. Popov, *J. Solid State Electrochem.* **14** (2010) 331-338.

[26] N.D. Nikolić, G. Branković, V.M. Maksimović, M.G. Pavlović, K.I. Popov, *J. Electroanal. Chem.* **635** (2009) 111-119.

[27] N.D. Nikolić, V.Maksimović, M.G. Pavlović, K.I. Popov, *J. Serb. Chem. Soc.* **74** (2009) 689-696.

[28] N.D. Nikolić , G. Branković, *Electrochem. Commun.* **12** (2010) 740-744.

[29] N.D. Nikolić, G. Branković, K.I. Popov, *Mater. Chem. Phys.* **125** (2011) 587-594.

[30] N.D. Nikolić, G. Branković, V. Maksimović, *J. Solid State Electrochem.* **16** (2012) 321-328.

[31] N.D. Nikolić, G. Branković, V.M. Maksimović, *J. Electroanal. Chem.* **661** (2011) 309-316.

[32] N.D. Nikolić, G. Branković, M.G. Pavlović, *Powder Technol.* In press.

Electrochemical treatment of trypan blue synthetic wastewater and its degradation pathway

ANANTHA N. SUBBA RAO and VENKATESHA T. VENKATARANGAIAH✉

Department of P.G. Studies and Research in Chemistry, School of Chemical Sciences, Kuvempu University, Shankaraghatta-577451, Shimoga, Karnataka, India

✉Corresponding Authors

Abstract

The trypan blue (TB) dye synthetic wastewater was treated in presence of chloride ions by electrochemical method. The effect of current density, pH, initial concentration of dye and supporting electrolyte on color and COD removal were investigated. The UV-Vis absorption intensity, chemical oxygen demand (COD), cyclic voltammetry (CV), Fourier transform-infrared spectroscopy (FT-IR), gas chromatography – mass spectrometry (GC-MS) analysis were conducted to investigate the kinetics and degradation pathway of TB dye.

Keywords

Cataract, Teratogenic, Carcinogenic, Niagara blue, Diamine blue, Decolorization, COD removal, Active anode, Oxidants, Indirect oxidation.

Introduction

Dyes find extensive applications in industries like textile, leather, pulp and paper, printing, photographs [1]. The wastewater from these industries is therefore loaded with the dyes. It is aesthetically unbearable and can turn out to be fatal if this wastewater is allowed to combine with the natural water resources. The presence of dyes or organic molecules increases the biological oxygen demand (BOD), chemical oxygen demand (COD) and total organic carbon (TOC) content of natural water, which especially poses challenges to the survival of aquatic life besides causing serious health hazards leading to the imbalance in ecology [2-4].

Chemical, physical, biological, and advanced oxidation processes are in practice to treat the wastewater. The electrochemical method is efficient, less time consuming and it does not require the addition of any chemical reagent. In this method, the organic pollutant is oxidized either by direct electron exchange with the anode (direct oxidation) or by the electrochemically generated oxidants (indirect oxidation). Direct oxidation is possible if the organic pollutant is oxidizable within the overpotential of oxygen (OER) and hydrogen evolution reactions. Indirect oxidation is brought about by the oxidants generated during electrolysis, such as hydroxyl radical ($^{\bullet}OH$). Hydroxyl radical ($^{\bullet}OH$, H^+/H_2O) is a powerful oxidizing agent with standard reduction potential of 2.7 V *vs*. standard hydrogen electrode. Active chlorine species (oxidants) like Cl_2, Cl_3^-, ClO_2^-, ClO_3^-, HOCl, ClO^- are generated in presence of chloride electrolyte and these species can oxidize the organics into simple biodegradable molecules [5,6].

The electrochemical degradation of methyl violet 2B, acridine, eosin yellow [7], alphazurine dye [8], reactive blue 81, reactive red 2 [9], reactive blue-4, reactive orange-16 [10], reactive red 120 [11], reactive brilliant red K-2BP [12], methyl green [13], methyl red [14] and other dyes has been successfully achieved in presence of suitable electrolyte and operating conditions. Furthermore, the results from these works infer that the extent of degradation of an organic pollutant not only depends on the operating conditions employed, but also on the structural, chemical and electrochemical properties of the pollutant itself [7]. The nature of the electrolyte and its concentration, electrode material, pH, current density, wastewater flow rate (stirring rate) and the mechanism of oxidation also significantly influence the extent of degradation of organics. Complete incineration of organics with high efficiency can be achieved by electrochemical method under optimum conditions.

The molecules *o*-tolidine, benzidine and *o*-dianisidine are carcinogenic in nature [16,17]. Trypan blue (TB) (Congo blue 3B / Niagara blue 3B / Diamine blue 14) is an o-tolidine based diazo dye, well known for its ability to selectively color the dead cells blue. It is also used in the dyeing of textiles, leather and paper [18].

***Table 1:** Properties and structure of Trapan blue*

Properties	Chemical structure
Molecular formula: $C_{34}H_{24}N_6Na_4O_{14}S_4$	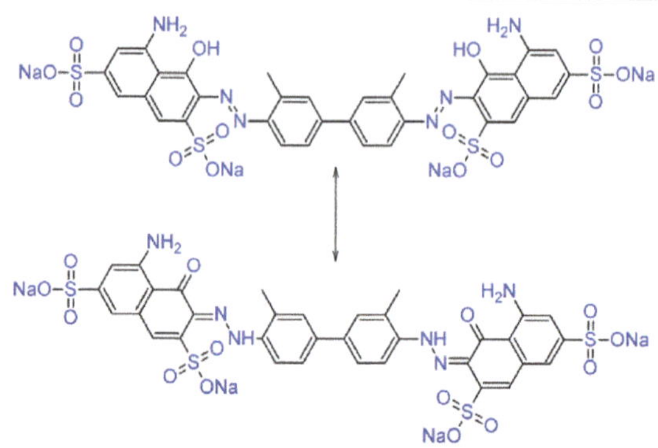
Molecular weight: 960.8	
Color Index number: 23850	
λ_{max} / nm = 590 [19]	

TB dye (Table I) emerged as one of the most frequently used staining agent in cataract, as well as other anterior segment surgeries in the late 1990s [20]. However, there are reports which unveil the complications in cataract surgery with the usage of TB. C Lüke *et al.* [21] reported the toxic effects of TB after a short duration of retinal exposure and recommended minimizing the usage of TB in intraocular applications [20,22]. TB carries associated risk at higher concentrations and longer duration of exposure [20-22]. Long term presence of 0.2 % TB can cause considerable damage to the retina [23]. International Agency for Research on Cancer provides sufficient evidences for the carcinogenic and teratogenic effects of TB in animals and possibly in humans [24,25].

The photocatalytic degradation of TB is reported in the literature [15,26] but there are no work solely focused on the degradation of TB by electrochemical method [15]. In the present work, the electrochemical oxidation of TB with Pt as anode material has been investigated and probable degradation pathway of TB has been proposed.

Materials and Chemicals

Chemicals

Commercially available trypan blue (TB) dye was (purchased from HiMedia, Mumbai, India) used for the studies. Analytical grade NaCl, Na_2SO_4, Na_2CO_3 and $NaNO_3$ and their aqueous solutions were (HiMedia Laboratories Pvt. Ltd, Bangalore, India) used as electrolytes. pH of each electrolyte was adjusted by H_2SO_4 and NaOH. The dil. HNO_3 was used for the pretreatment of electrodes (HiMedia Laboratories Pvt. Ltd, Bangalore, India).

Experimental

A single compartment glass container was used as electrolytic cell (6.5 cm diameter x 10 cm height, covered with Teflon® lid provided with gaps to insert the electrodes). Two Pt foils (dimension 0.5 cm × 0.5 cm, thickness 0.018 cm, supplied by Systronics India Ltd. Bangalore, India) of surface area, 0.54 cm^2 were used as anode and cathode which were separated by 0.5 cm in the electrolytic cell. The potentials of working electrode (anode) were recorded relative to saturated calomel electrode (SCE). A DC source with galvanostatic mode (Model PS-618 chemilink systems, Navi Mumbai, India) was used for power supply. The synthetic wastewater of TB dye was prepared with ultrapure water obtained from Elix 3 Milli-pore system (resistivity, > 18 MΩ cm at 25 °C). A magnetic stirrer (550 rotations per minute) was used for the agitation of wastewater during electrolysis.

Electrolysis

The solution with the composition of 50 mg L^{-1} TB dye solution and NaCl, 0.2 % was prepared in Millipore water and was subjected to electrolysis under galvanostatic condition for 60 minutes. The Pt electrodes were dipped in dilute HNO_3, sonicated for 1 minute and thoroughly washed with Millipore water. All experiments were carried out at ambient temperature. To monitor the progress of the degradation process, electrolyzed samples were collected at appropriate time intervals and subjected to analysis. The effect of current density, pH, nature of electrolyte and its concentration on COD and color removal efficiency was examined. The optimum condition for the maximum color and COD removal efficiency was determined.

Analysis

Cyclic voltammetry (CV)

The cyclic voltammetric measurements were performed at room temperature with conventional three electrode system connected to software controlled electrochemical work station (CH Instruments 660C, USA). The working electrode was a Pt disk and counter electrode was a Pt foil. The working electrode surface was polished with 0.05 µm alumina, washed with methanol, dipped in 10 % HNO_3, sonicated for 2 min and then washed thoroughly with Millipore water before use. The open circuit potential was recorded in 0.2 % NaCl solution and the potential window from +1.1 V to -1.1 V was selected for CV scans. CV for blank solution (0.2 % NaCl) and TB dye solution (50 mg L^{-1} + NaCl 0.2 %) were recorded at the scan rate of 100 mV s^{-1}.

Spectroscopic and chemical analysis

The extent of degradation was monitored by evaluating color and COD removal. UV-Vis absorbance spectra of samples were recorded by Ocean Optics UV-Vis HR 4000 spectrophotometer (UV-Vis-NIR light source, DT-MINI-2-GS and Jaz detector). The percentage color removal was calculated by using the relation (1) with reference to the absorption intensity at λ_{max} of 590 nm.

$$\text{Color removal, \%} = \frac{[Abs_0 - Abs_t]}{Abs_0} \times 100 \quad (1)$$

Where Abs_0 and Abs_t are the absorbance at 590 nm at time 0 and t minutes of electrolysis respectively.

The COD ($g(O_2)\ L^{-1}$) estimation was carried out by open reflux method and its values used to monitor the progress in degradation. The percentage COD removal was calculated using the relation (2).

$$\text{COD removal, \%} = \frac{[\text{COD}_0 - \text{COD}_t]}{\text{COD}_0} \times 100 \quad (2)$$

Where COD_0 and COD_t are the COD at time 0 and t minutes of electrolysis respectively.

The percentage average current efficiency (ACE) was calculated using the following relation (3) [6].

$$\text{ACE, \%} = \frac{[\text{COD}_0 - \text{COD}_t]FV}{8It} \times 100 \quad (3)$$

Where F is the Faraday's constant (96,485 C mol^{-1}), COD in (g O_2 L^{-1}), I is the current (A), V is the volume (L) of electrolyte, t – time (s), 8 – gram equivalent mass of oxygen (g $equiv^{-1}$).

To determine the degradation pathway, the test samples at selected time intervals (15 and 40 min) were subjected to freeze drying using Lyophilizer (Delvac, India) at -50 ± 5 °C under reduced pressure of 0.12 mbar. Freeze drying eliminates the chances of decomposition of intermediates and only the volatile compounds are removed. The solid residue thus obtained was subjected to UV-Vis spectroscopy, Fourier Transform Infrared Spectroscopy FT-IR (Alpha, Brucker) and gas chromatography mass spectrometry (GC-MS) analyses. The solid residue was finely mixed with KBr, pressed into a disk and subjected to FT-IR analysis. The FT-IR spectrums recorded are the average 16 scans from wave number 4000 cm^{-1} to 500 cm^{-1}. GC-MS analysis was carried out in order to detect the end products of TB degradation. The samples were analyzed using GC-MS system (Trace GC Ultra and Trace DSQ THERMO-Electron Corporation, Austin, TX, USA) with a HP-5MS, 5 % phenyl methyl silox column (30 m × 0.25 mm, film thickness 0.25 µm). A temperature gradient program at 10° C min^{-1} was applied between 40 °C and 250 °C.

Results and Discussion

Cyclic Voltammetry Studies

The cyclic voltammograms are shown in Fig. 1. A peak at -0.23 V was observed during cathodic sweep of blank solution. The hydrogen evolution begins approximately at -0.80 V. The voltammogram obtained for the dye solution showed the same pattern as the blank and no new peak was observed during the anodic sweep. This indicates that the TB dye does not undergo direct electrochemical reaction on Pt in the potential range from -1.1 to +1.1 V. The decrease in anodic peak (at -0.80 V *vs.* SCE) current can be attributed to the adsorption of dye on the surface of working electrode.

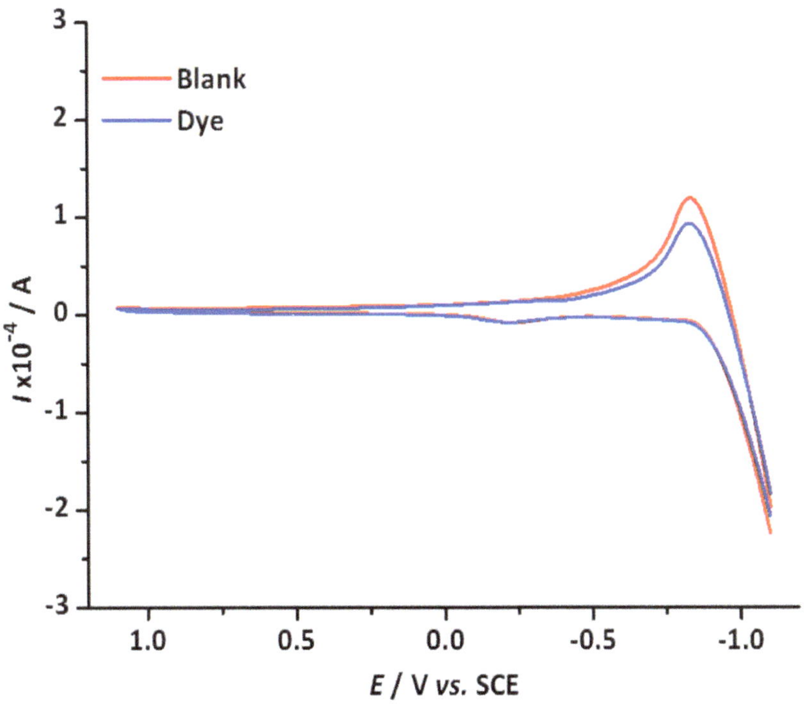

Figure 1. *CV of 0.2 % NaCl and 50 mg L^{-1} TB + 0.2 % NaCl solutions; scan rate 100 mV s^{-1}*

Effect of Supporting Electrolyte

To determine the effect of different supporting electrolytes, preliminary investigations on TB degradation were carried out separately in presence of chloride (Cl^-), sulfate (SO_4^{2-}), nitrate (NO_3^-) and carbonate (CO_3^{2-}) ions. In all the cases 34.2 mM solution of electrolyte was was used. On electrolysis of the solution for 60 min, showed a 100 %, 42 %, 27 %, 15 % color removal in presence of Cl^-, CO_3^{2-}, NO_3^-, SO_4^{2-} ions respectively. Similar trend was observed by Fornazari *et al.* [27] in the electrochemical degradation of phenol-formaldehyde mixture [27,28]. The presence of Cl^- ions is not only essential for the color removal but also its *in situ* generated oxidants cleave the TB bonds. Therefore, NaCl was selected as the supporting electrolyte for the present studies. Fig. 2A and Fig. 2B depict the effect of different electrolytes on absorption intensity and color removal of dye.

The Pt is an active anode and during electrolysis the generation of hydroxyl radical is limited [6]. In presence of CO_3^{2-}, NO_3^-, SO_4^{2-} ions, the oxidation of TB is possible only either by the hydroxyl radicals generated on Pt anode or by direct electron transfer with the anode. But Pt anode does not show any electrochemical activity for the oxidation of TB (Fig. 1). On Pt, in presence of oxygen and above 0.8 V there exists Pt and PtO mixture [29]. Slight decolorization of TB in presence CO_3^{2-}, NO_3^-, SO_4^{2-} ions is perhaps due to oxidation by the hydroxyl radicals generated on Pt and by PtO/Pt couple. Below are the possible reactions (4) – (7) on the surface of the Pt anode [19,30].

$$Pt + H_2O \rightarrow Pt\text{-}OH_{ads} + H^+ + e^- \quad (4)$$

$$Pt\text{-}OH_{ads} \rightarrow PtO + H^+ + e^- \quad (5)$$

$$xPtO + R_{aq} \rightarrow ROx + xPt \quad (6)$$

$$2PtO \rightarrow 2Pt + O_2 \quad (7)$$

Decoloration is the clear evidence for the cleavage of the chromophoric (–N=N-) group of TB dye. Since decolorization was very slow in presence of CO_3^{2-}, NO_3^-, SO_4^{2-} electrolytes, hence the COD removal cannot be more than color removal and is very small. Therefore COD was not measured in these cases.

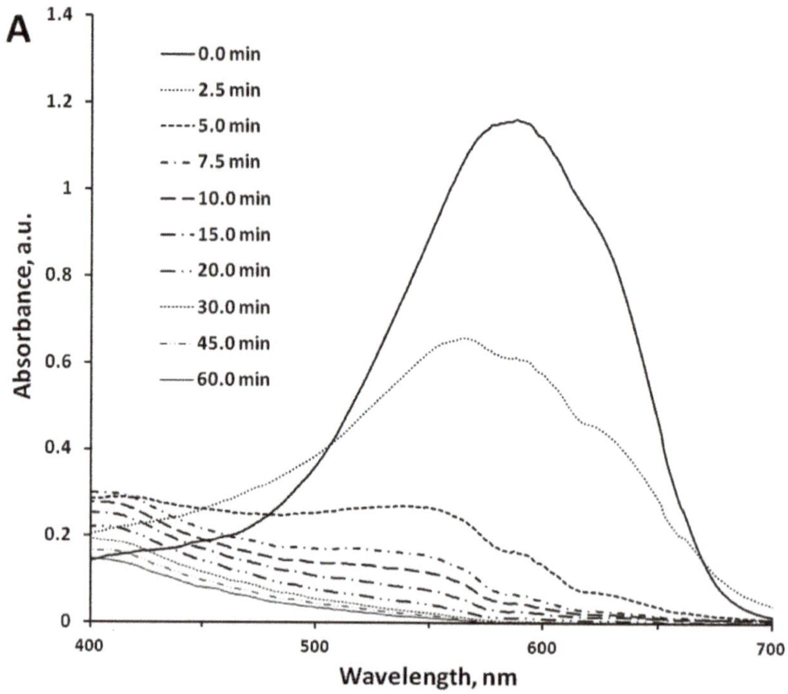

Figure 2A. *The variation of absorption with electrolysis time. Conditions; 0.2% NaCl + 50 mg L^{-1} TB, current density – 93 mA cm-^{2}, λ_{max} = 590 nm*

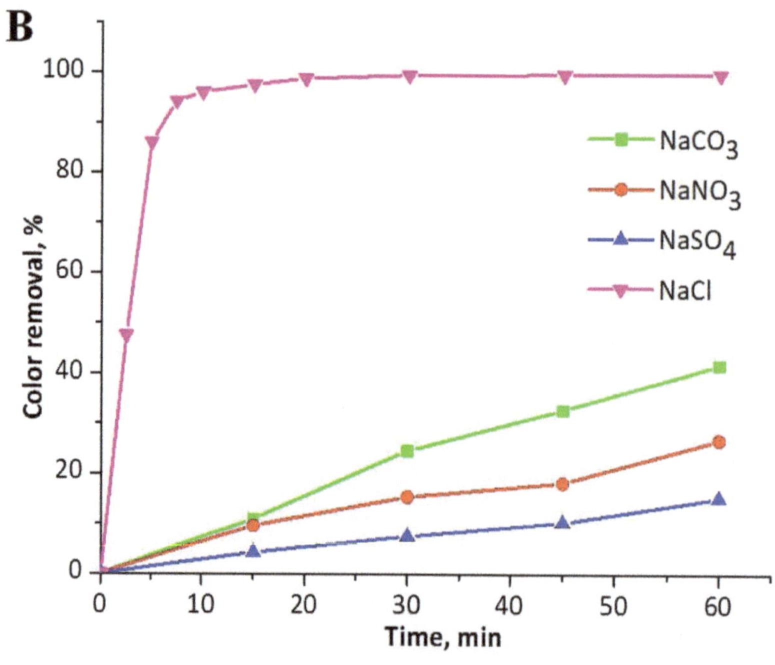

Figure 2B. *Effect of supporting electrolytes on color removal*

Decoloration is the clear evidence for the cleavage of the chromophoric (–N=N-) group of TB dye. Since decolorization was very slow in presence of CO_3^{2-}, NO_3^-, SO_4^{2-} electrolytes, hence the

COD removal cannot be more than color removal and is very small. Therefore COD was not measured in these cases.

Effect of NaCl Concentration

The effect of NaCl concentration on the rate of color and COD removal was investigated with NaCl concentration ranging from 0.05 % to 3.0 %. Fig. 3A and 3B, show the effect of NaCl on color and COD removal respectively.

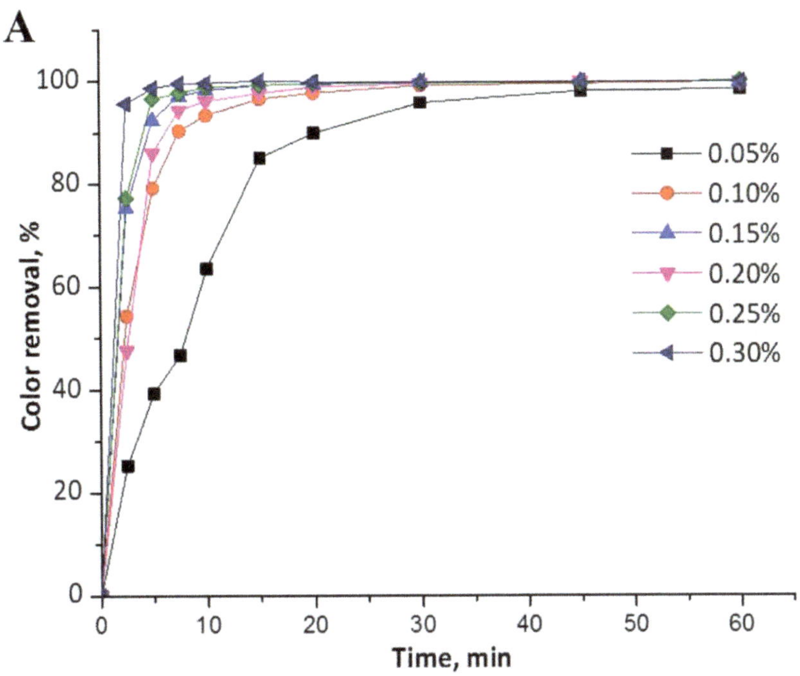

Figure 3A. *Effect of NaCl concentration on color removal*

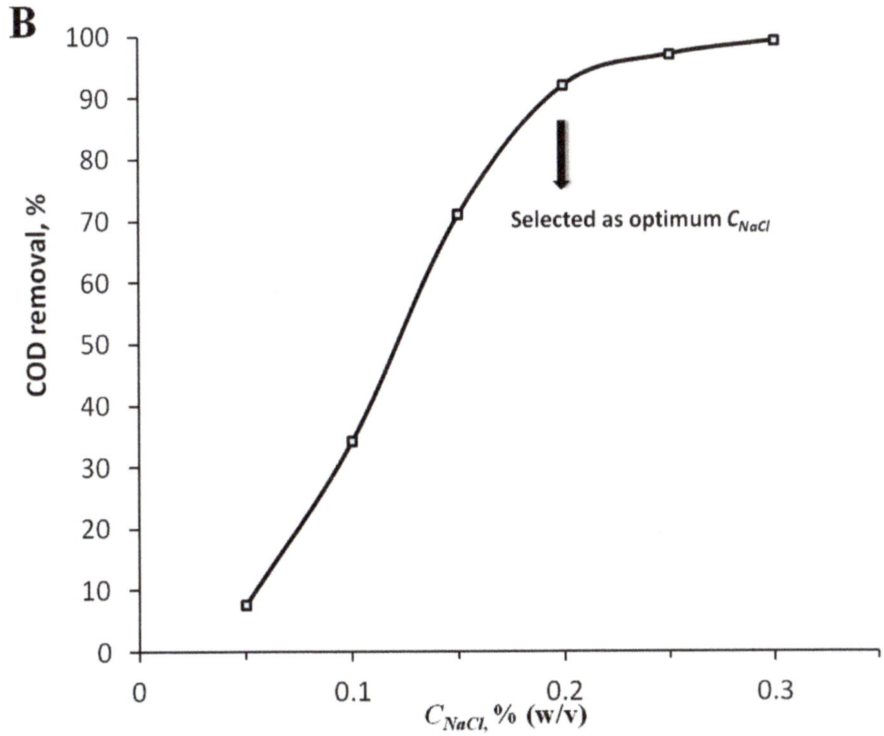

Figure 3B. *Effect of NaCl concentration on COD removal (electrolysis time 60 min)*

The amount of oxidants on the anode surface increases with the applied current density. But, at the given current density, the amount of Cl^- ions is the limiting factor. If Cl^- ion concentration in the solution is less, the current supplied would be lost in oxygen evolution, which is otherwise could be used to generate active chlorine species. The optimum Cl^- ion concentration can facilitate high current efficiency and thus less energy consumption. At higher concentrations of Cl^- ions, the applied current density is efficiently utilized to generate more oxidants [6]. At the same time, the concentration of Cl^- ions should not exceed the concentration level commonly found in any real wastewater sample. So, the effect of Cl^- ion concentration was tested within 0.3 % NaCl.

Complete decolorization of TB solution occurred in 30 min of electrolysis with all concentrations of NaCl from 0.1 % to 0.3 %, except with 0.05 % NaCl, in which 98 % color removal was achieved. The rate of decoloration increased with increase in Cl^- ion concentration. Concomitant COD removal was observed with decolorization. The COD removal after 60 min electrolysis increased largely with NaCl concentration from 0.05 % to 0.2 %. After this concentration, only slight improvement in the COD removal was noticed. Therefore, 0.2 % NaCl was chosen as the optimum concentration for further studies.

Effect of Current Density

The efficiency and energy consumption for the degradation process is dependent on the applied current density and electrolysis time. Higher current density always leads to high energy consumption and low current efficiency. The degradation of TB in the present case takes place by indirect oxidation brought about by the oxidants generated during electrolysis. Higher current density enhances the rate of electrode reactions and hence rate of generation of oxidants also elevates. A range of current density from 37 to 111 mA cm^{-2} was applied to determine its effect on color and COD removal. The rate of color and COD removal increased with applied current density (Fig. 4A and 4B). The COD removal increased as current density increased from 37 to 55 mA cm^{-2}, beyond which slow and steady raise in COD was noticed. In contrast, the current efficiency decreased with increse in applied current density.

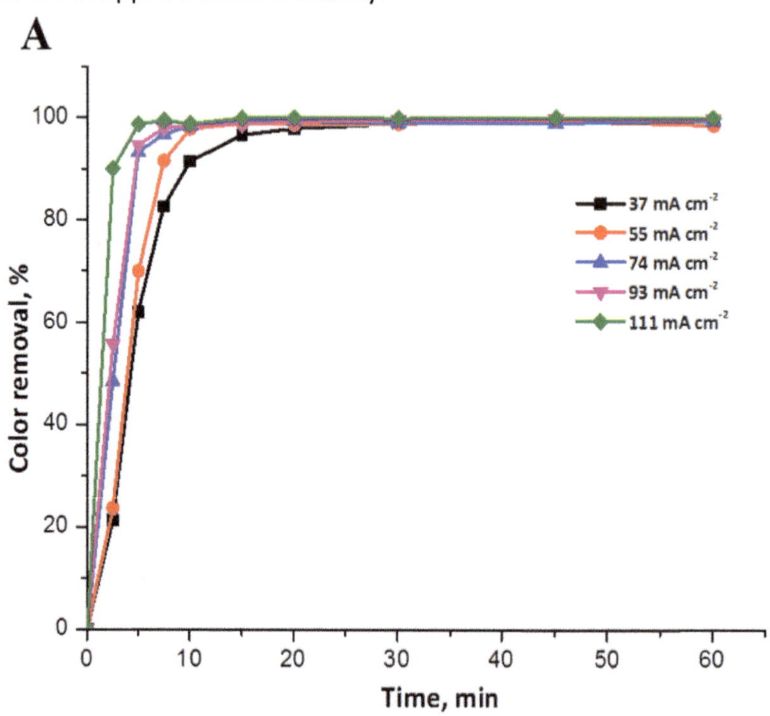

Figure 4A. *Effect of current density on color removal*

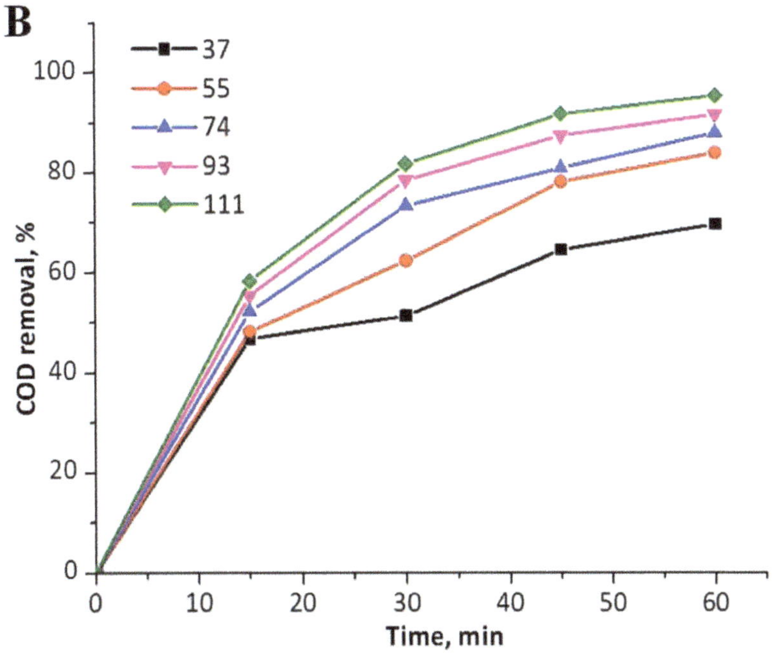

Figure 4B. *Effect of current density on COD removal; conditions - 0.2 % NaCl, current density- 93 mA cm^{-2}, dye – 50 mg L^{-1}; electrolysis time – 60 min*

The color removal was 100 % in all cases and the COD removal was 70 % to 95 % with 25 % to 11 % current efficiency. This implies that with increase in current density, though the production of oxidants increased, the undesired hydrogen and oxygen evolution reactions also increased alongside and consumed major part of the supplied current. A charge consumption of 3.60 kC L^{-1} was recorded for 92 % of COD removal under the current density of 93 mA cm^{-2} (60 min) whereas, 2.88 kC L^{-1} charge was consumed for the same COD removal under the current density of 37 mA cm^{-2} (120 min). This indicates that higher COD removal can be achieved with less charge (and hence energy) consumption, but requires longer time duration.

Effect of pH

pH influences the chemical form of the oxidant present in the electrolyte. The electrolysis was carried out at four different pH. The electrochemical reaction in presence of chloride ions can be represented as in reactions (8) – (11);

$2Cl^- \rightarrow Cl_{2(aq)} + 2e^-$ (Standard potential of Cl_2 (E^o = 1.40 V *vs.* SHE) (8)

$Cl_{2(aq)} + H_2O \rightarrow HOCl + H^+ + Cl^-$ (E^o of HOCl = 1.49 V *vs.* SHE) (9)

$HOCl \leftrightarrow H^+ + OCl^-$ (E^o of ClO^- = 0.89 V *vs.* SHE) (10)

$TB + OCl^- \rightarrow Intermediates \rightarrow CO_2 + H_2O + Cl^-$ (11)

The pH variation influences the equilibrium in reaction (10) and the HOCl predominates in the pH range 3-8 but above pH 8, ClO^- is the major candidate [26,31]. The oxidation by HOCl and $Cl_{2(aq)}$ [5,6] is therefore expected to be more favorable and faster at near neutral pH. Furthermore, reactions in reaction (12) and (13) are expected to take place on the active anode such as Pt. The MOx(HOCl) mediates the oxidation unselectively and hence higher COD removal efficiency was noticed [26].

$MOx(\bullet OH) + Cl^- \rightarrow MOx(HOCl) + e^-$ (12)

$MOx(HOCl) + TB \rightarrow Intermediates \rightarrow MOx + CO_2 + H_2O + Cl^- + H^+$ (13)

The effect of pH on color and COD removal with electrolysis time is shown in Figs. 5A and 5C, respectively. Fig. 5B illustrates the slow decrease in the absorption intensity of the peak in the visible region with increase in pH (11.8). It is evident that color and COD removal is above 90 % in pH 3.1 and 6.2. As the pH increased to 9.0, the rate of color and COD removal reduced to 82 % and became 67 % in pH 11.8.

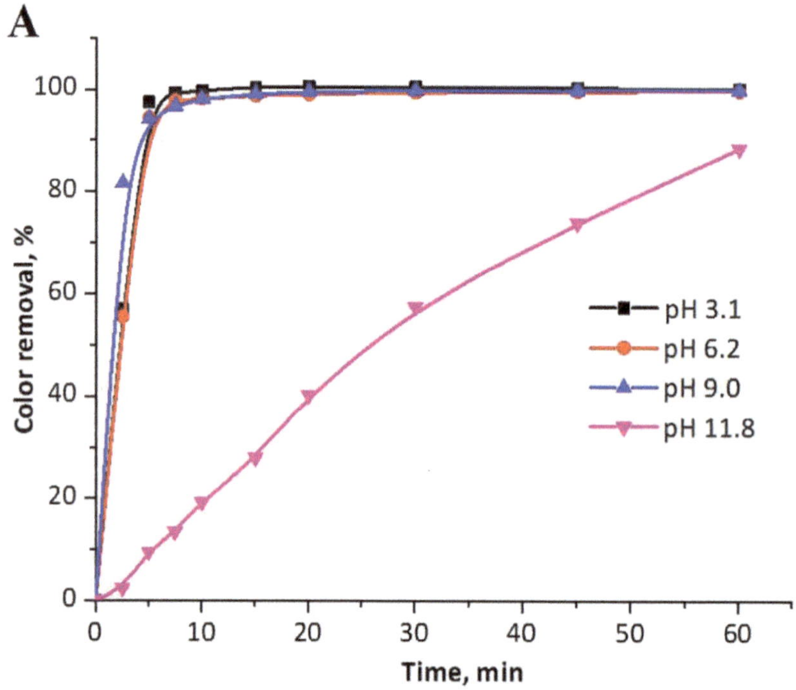

Figure 5A. *Effect of pH on color removal*

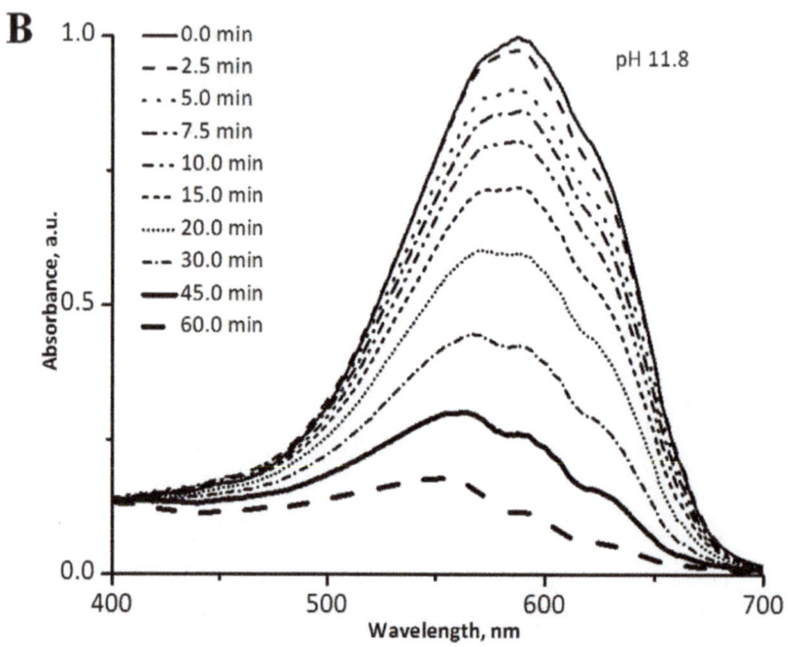

Figure 5B. *Variation in the intensity of peak at 590 nm in pH 11.8*

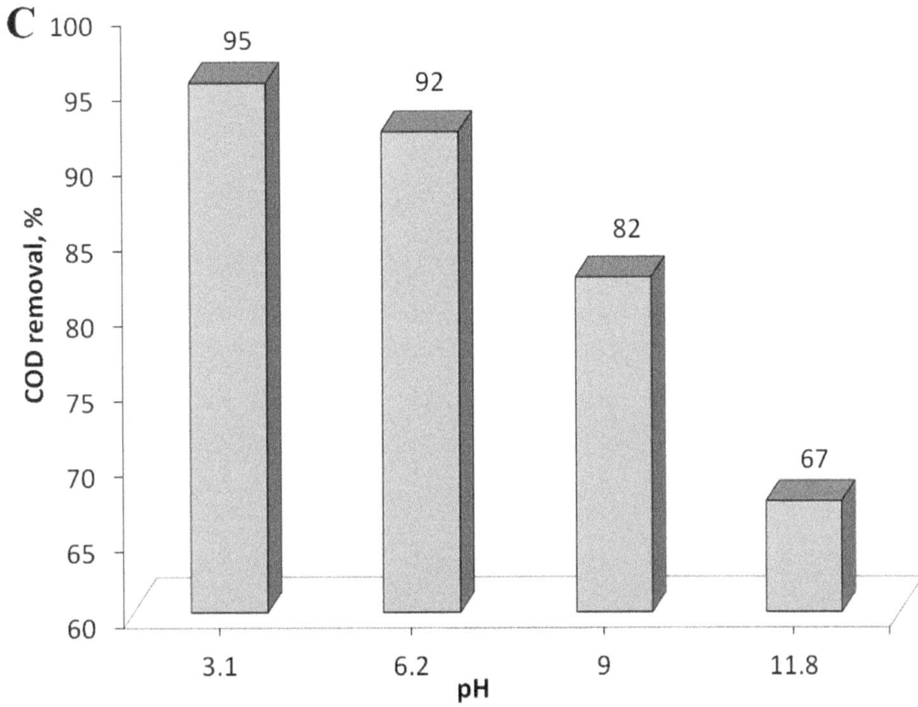

Figure 5C. *Effect of pH on COD removal; conditions c_{NaCl}, 0.2 % (w/v), current density – 93 mA cm^{-2}, [TB] – 50 mg L^{-1}, electrolysis time – 60 min*

Complete decoloration was achieved in all pH except with pH 11.8, for which 88 % color removal was achieved. The difference in COD removal in pH 3.1 and pH 6.2 was only 3 % and in both the cases it was above 90 %. But COD removal decreased above pH 9.0. Hence, for further studies, pH 6.2 was selected.

Effect of Initial Dye Concentration

At the optimized operating conditions of electrolysis process, only fixed amount of oxidants are generated at the anode. The energy would be simply lost if the generation of oxidants is more than that required to oxidize the available organics in the wastewater. Sufficient concentration of organics should be available for the efficient utilization of generated oxidants. The TB solutions of different concentrations ranging from 25 to 100 mg L^{-1} were prepared and electrolysed under optimized operating conditions (93 mA cm^{-2}, pH 6.3 and 0.2 % NaCl concentration). The color and COD removal decreased gradually with increase in initial concentration of TB. Fig. 6 shows the trend of color and COD removal with time for different initial concentrations of TB dye. A 100 % decolorization was achieved within 30 min for all initial concentrations of TB. But 100, 91, 75 and 67 % COD removal was attained with 25, 50, 75 and 100 mg L^{-1} initial TB concentrations respectively after 60 min electrolysis.

At initial concentration of 25 mg L^{-1}, the in situ generated oxidants were efficienty oxidized the TB completely. But, the dye concentration was less than the amount of oxidants available in the electrolysis cell (dye concentration is the limiting factor). When the TB dye concentration was increased to 100 mg L^{-1}, the amount of oxidants was less than the equivalent value, not sufficient to degrade the entire amount of TB (oxidant concentration is the limiting factor). Under the currrent density - 93 mA cm^{-2} and 60 min of electrolysis, the COD removal was 92 % for the initial concentraiton of 50 mg L^{-1} chosen as the optimum condition.

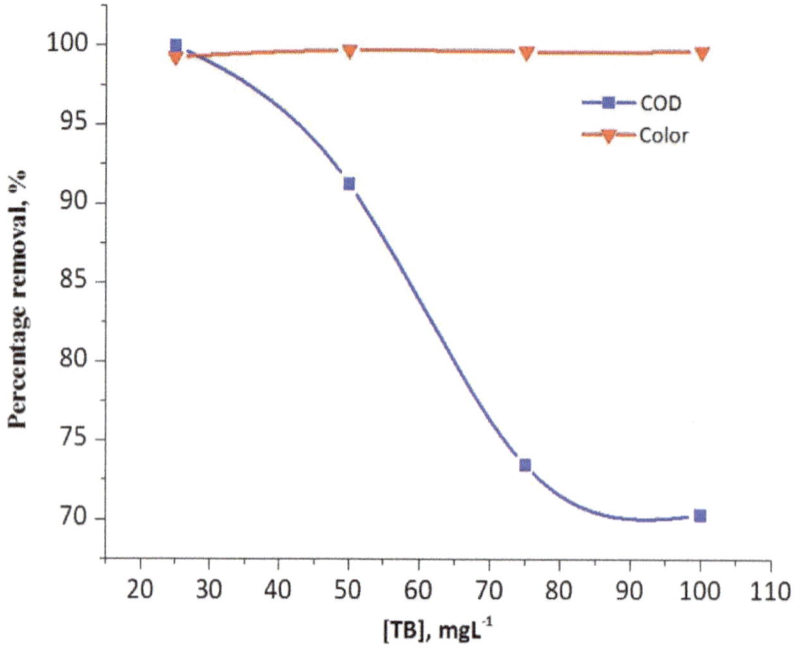

Figure 6. *Effect of initial concentration of TB on COD and color removal.*

Kinetics of TB Degradation

During the electrolysis, the concentration of TB in synthetic wastewater decreased exponentially for all applied current densities but the rate of oxidation increased with increase in current density, indicating that the degradation of TB followed the first order kinetics [32]. If we consider that the oxidants generated on electrolysis spontaneously react with TB, then from the steady state approximation, the concentration of oxidants (Ox) can be estimated [32] and the rate of oxidation of TB can be expressed as in Eq. (13) and (14);

Table 2. *Degradation kinetics data*

Parameters	j / mA cm^{-2}*					Initial concentration of TB, mg L^{-1} #			
	37	55	74	93	111	25	50	75	100
k_{app} / min^{-1}	0.022	0.032	0.037	0.044	0.053	0.058	0.043	0.026	0.021
Regression	0.867	0.978	0.967	0.974	0.993	0.974	0.986	0.92	0.841

* TB concentration - 50 mgL^{-1}; # Current density - 93 mA cm^{-2}

$$\frac{d[\mathrm{COD}]}{dt} = -k\,[\mathrm{Ox}][\mathrm{COD}] \quad (13)$$

$$= -k_{\mathrm{app}}\,[\mathrm{COD}] \quad (14)$$

On integrating the above equation;

$$Ln\frac{[\mathrm{COD}]_0}{[\mathrm{COD}]_t} = k_{app}\,t \quad (15)$$

where [Ox] is the concentration of oxidants, k_{app} is the apparent pseudo first-order rate constant ([Ox] is considered as very high), COD_0 and COD_t are the COD at time 0 and *t* minutes of electrolysis respectively.

The slope gives k_{app} which are tabulated in Table II. The first order rate constant steadily increased with increase in applied current density and decreases with increase in initial concentration of TB dye (Fig. 7A and 7B). The regression value indicates the deviation from the first order reaction rate equation for current density 37 mA cm^{-2} and TB concentration 100 mg L^{-1} can be attributed to the fact that at lower applied current density and high initial concentration of TB, the active concentration of oxidants is the limiting factor. Moreover, the area of the anode used in all these experiments was only 0.57 cm^2.

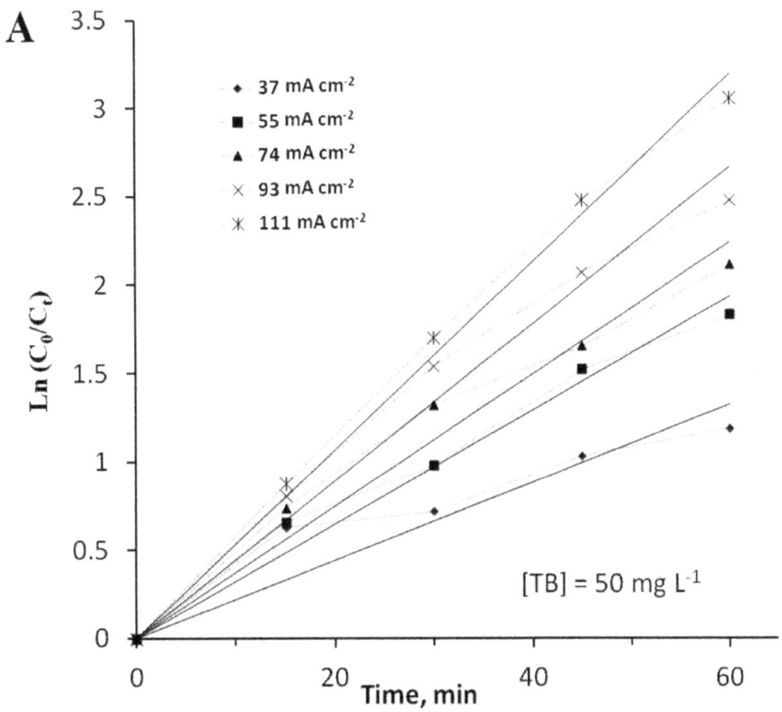

Figure 7A. *Plot of ln (C_o/C_t) vs. time for different applied current densities*

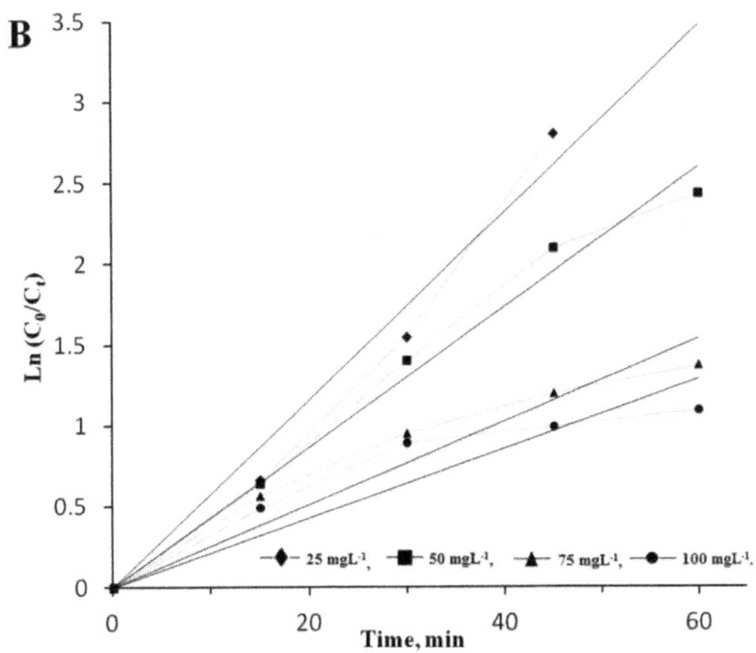

Figure 7B. *Plot of ln(C_o/C_t) vs time for different initial dye concentrations*

The oxidants generated on the surface of such a small area of Pt anode under the current density of 37 mA cm^{-2} might not be so high to satisfy the steady state approximation for first order kinetics. Similarly, oxidants generated under the current density of 93 mA cm^{-2} and 100 mg L^{-1} initial concentration of TB dye, a similar deviation was observed.

Spectroscopic Analysis

UV-vis spectroscopic analysis

UV-Vis spectrum of TB dye showed three major peaks at 235, 318 and 590 nm. The peak at 590 nm corresponds to the chromophoric azo group. On electrolysis (50 mgL^{-1} dye, 93 $mAcm^{-2}$), the intense blue color of the solution turned in to wine red in the first 10 min, then transformed slowly into pale yellow color, which persisted till the end. The intensity of the peak at 590 nm rapidly decreased to zero absorbance on electrolysis, which can be ascribed to the cleavage of azo group.

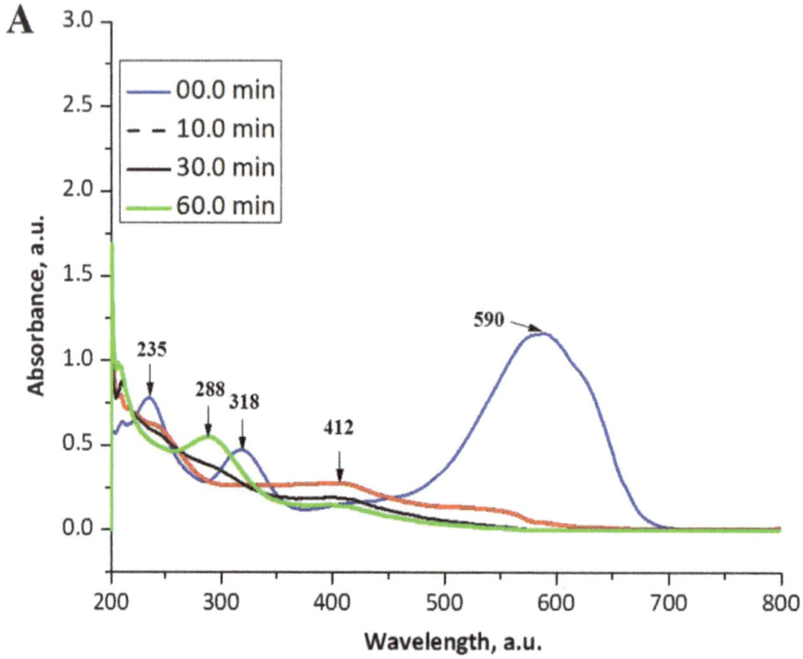

Figure 8A. *UV-Visible absorbance of TB at different time intervals of electrolysis*

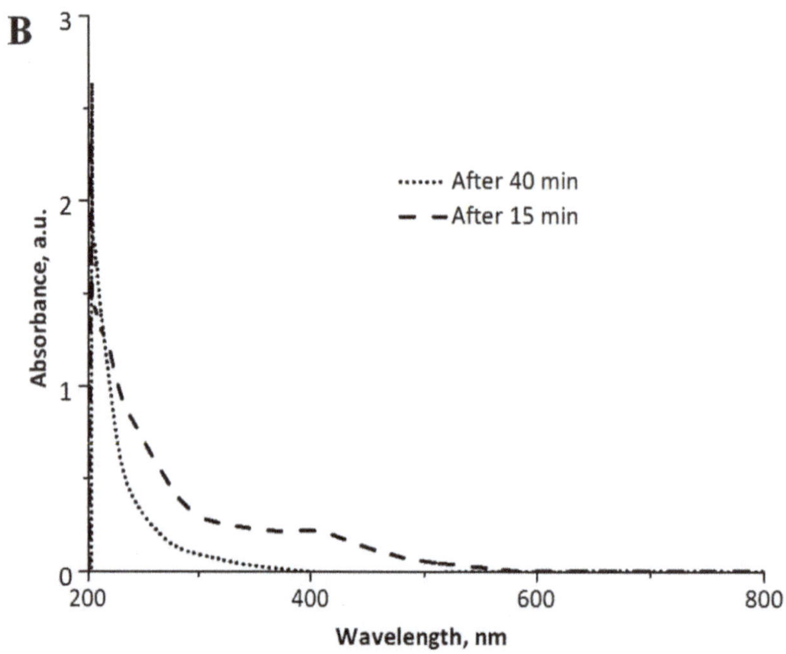

Figure 8B. *Absorbance of residue obtained by the evaporation of electrolyzed TB solution*

For the first 10 min of electrolysis, the peak at 585 nm showed blue shift and gradually reached zero absorbance. The blue shift can be attributed to the detachment of auxochromes from the parent dye structure during electrolysis. A new peak at 407 nm was observed, which can be ascribed to the formation of a fragment from the parent dye. This peak diminishes very slowly. The pale yellow color is due to the absorption at 412 nm, (absorption color – violet (412 nm), perceived color – yellow) which reduced slowly with time (Fig. 8A). Peak at 318 nm reduced with time, reached minimum value and again increased with a blue shift at 288 nm. This peak can be assigned to the low molecular weight oxidized fragments with conjugated π bonds in which n→π* and π→π* transitions are allowed (-C=O, -C=C-, -C=N-, -(CO)NH-). Similar effect was observed by Hanene et al. [34] in the degradation of Amido black (AB) on boron doped diamond electrode.

The electrolyzed dye solution was evaporated at lower temperature with reduced pressure. UV-Vis spectrum of the solid thus obtained was taken by dissolving it in water. Interestingly, the peak at 288 nm was absent. This observation validates our suspicion that compounds leading to this peak were low molecular volatile compounds (Fig. 8B). After 40 min electrolysis, the UV-Vis spectrum of the residue obtained by evaporation does not show any peak, except a peak in 200 to 220 nm range. This is due to the small amount of aromatic compounds resistant to indirect oxidation remained in the solution, which also accounts for the residual COD after 60 min.

FT-IR analysis

The FT-IR spectrum of TB dye (Fig. 9A) exhibits a characteristic band at 1579 cm^{-1} corresponding to azo group [34, 35]. The bands at 1492 cm^{-1} and 1423 cm^{-1} correspond to C=C aromatic ring stretching in combination with C=N [34] and methyl bending respectively. The keto form of the TB is dominant (Table I) [36]. The peak at 1617 cm^{-1} is due to *α,β*-unsaturated C=O stretch. The peaks at 1339, 1188 and 1128 cm^{-1} are due to aromatic amine C-N and $-SO_3^-$ (S=O) stretch [35]. The peak obtained at 1042 cm^{-1} is due to C-O. The broad doublet peak at 674 cm^{-1} and 644 cm^{-1} indicates the aromaticity. In Fig. 9B and 9C, the peaks at 1579, 1492, 1423 cm^{-1} and 1339, 1128, 1042, 674 cm^{-1} disappeared completely indicating the breakage of corresponding groups.

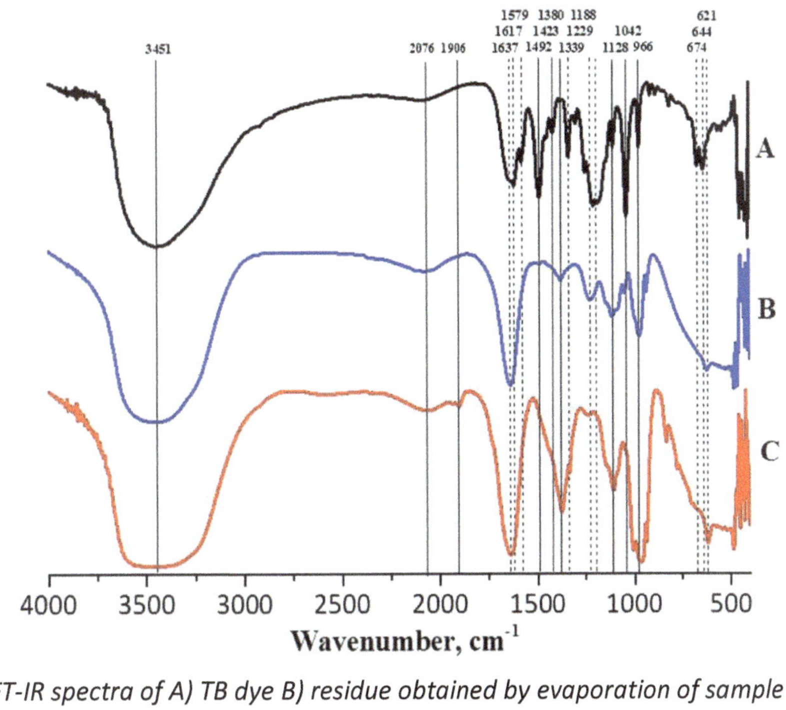

Figure 9. *FT-IR spectra of A) TB dye B) residue obtained by evaporation of sample electrolyzed for 15 min C) electrolyzed for 40 min*

The peak at 1229 cm^{-1} disappeared in Fig. 9C, indicating that group corresponding to this frequency is susceptible to oxidation by the electrochemically generated oxidants [37]. The peak at 1119 cm^{-1} shifted to 1108 cm^{-1}. While peak 674 cm^{-1} disappeared completely, 644 cm^{-1} shifted to 621 cm^{-1}. In Fig. 9C the peak at 1637 cm^{-1} is due to the carbonyl stretch of carboxylate salts, 1380 cm^{-1} is due to C-N of aliphatic amines. This suggests the formation of carboxylic acids and amines. 1008 cm^{-1} is devoted to C-O stretch and peak at 966 cm^{-1} is due to the formation of ClO_3^- and OCl^- during electrolysis [34,35]. These changes in the FT-IR spectrum are the clear evidence for the degradation of TB into simpler molecules, like aromatic and aliphatic amines and carboxylic acids.

Identification of intermediate products and degradation pathway

The intermediates of TB degradation were identified by gas chromatography mass spectrometry (GC-MS). On the basis of the intermediates and end products as depicted by GC-MS, a possible degradation pathway is proposed in Fig. 10. GC-MS revealed that the parent TB molecule was completely disintegrated into simpler molecules.

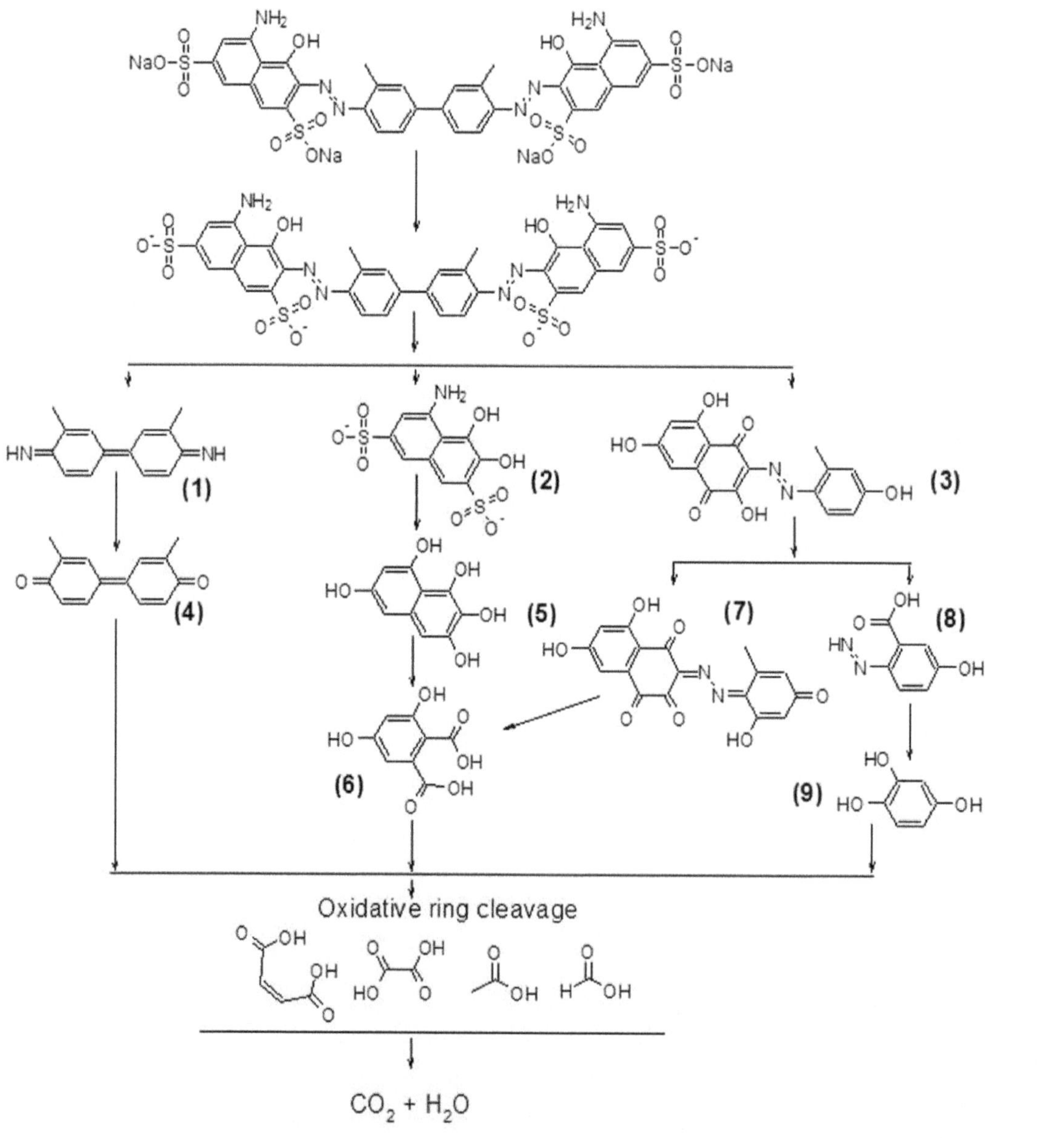

Figure 10. *A possible degradation pathway of TB*

Decomposition of TB was initiated with the cleavage of azo bonds resulting in the formation of three different intermediates **1** (3,3'-dimethyl-1,1'-bi(cyclohexa-2,5-dien-1-ylidene)-4,4'-diimine), **2** (5-amino-3,4-dihydroxynaphthalene-2,7-disulfonic acid) and **3** (2,5,7-trihydroxy-3-[(4-hydroxy-2-methylphenyl)diazenyl]naphthalene-1,4-dione). Intermediate **3** resulted from the cleavage of one of the azo bonds in the parent molecule was found in very small concentration. The absorption at 412 nm in UV-Vis spectroscopy can be attributed to this intermediate. The sulfonic group in intermediate **2** renders it water soluble. A foamy layer was formed on the surface of the electrolytic bath solution during electrolysis gradually disappeared with time. This can be attributed to the formation of intermediates **1** and **3** at the beginning of electrolysis. Further oxidation of intermediates **2** and **3** produced aromatic acids and phenols, **5** (naphthalene-1,2,3,6,8-pentol), **6** (3,5-dihydroxybenzene-1,2-dicarboxylic acid), **9** (benzene-1,2,4-triol). Aliphatic acids were generated on the oxidative ring opening of **4**, **6** and **9**, which decomposed into CO_2 to accomplish the degradation process.

Conclusion

The TB dye synthetic wastewater was successfully treated in presence of Cl^- with Pt foil as anode material. The removal of TB was dependent on the applied current density, pH, type of supporting electrolyte and concentration of TB dye and NaCl. The degradation of TB was achieved only in presence of NaCl as supporting electrolyte suggesting that the active chlorine species (Cl_2, ClO^-, HOCl) generated during the electrolysis played crucial role in the oxidation of TB. The rate of removal of TB increased with increase in current density and NaCl concentration and decreased with rise in initial concentration of TB and pH. Under the optimum conditions (current density – 93 mA cm^{-2}, pH - 6.2, NaCl concentration - 0.2 %, initial TB concentration – 50 mgL^{-1}) maximum COD and color removal achieved after 60 min electrolysis was 92 % and 100 % respectively. The first order rate constant for COD removal at these conditions was 0.044 min^{-1}. The UV-Vis spectroscopy, FT-IR and GC-MS analysis confirmed the complete degradation of TB into simple aliphatic carboxylic acid and CO_2 suggesting that the treatment of wastewater containing TB dye as pollutant can be successfully accomplished by electrochemical method.

Acknowledgement: *We thank department of chemistry, Kuvempu University, Shankaraghatta and UGC, New Delhi for their financial support.*

References

[1] H. Zollinger, Synthesis, Properties of Organic Dyes and Pigments. In: Color Chemistry. New York, USA: VCH Publishers; (1987) pp-92-102

[2] M. B. Ibrahim, N. Poonam, S. Datel, M. Roger, *Bioresour Technol.* **58** (1996) 217-227

[3] E. Forgacs, T. Cserháti, G. Oros, *Environ Inter.* **30** (2004) 953971

[4] W. Przystaś, E. Zabłocka-Godlewska, E. Grabińska-Sota, *Water, Air, Soil Pollut.* **223** (2012) 1581-1592

[5] S. Song, J. Fan, Z. He, L. Zhan, Z. Liu, J. Chen, X. Xu, *Electrochim. Acta* **55** (2010) 3606-3613

[6] C. A. Martinez-Huitle, E. Brillas, *Appl Catal B: Environ* **87** (2009) 105-145

[7] F. Zaviska, P. Drogui, J. F. Blais, G. Mercier, *J. Appl. Electrochem.* **39** (2009) 2397-2408

[8] J. L. Nava, M. A. Quiroz, C. A. Martínez-Huitle, *J. Mex. Chem. Soc.* **52** (2008) 249-255

[9] E. Kusmierek, E. Chrzescijanska, M. Szadkowska-Nicze, J. Kaluzna-Czaplinska, *J. Appl. Electrochem.* **41** (2011) 51–62.

[10] R. G. da Silva, S. A. Neto, A. R. de Andrade, *J. Braz. Chem. Soc.* **22** (2011) 126-133

[11] T. Panakoulias, P. Kalatzis, D. Kalderis, A. Katsaounis, *J. Appl. Electrochem.* **40** (2010) 1759-1765

[12] L. Jiancheng, Y. Jie, L. Weishan, H. Qiming, X. Hongkang, *J. Electrochem. Sci. Eng.* **2** (2012) 171-183

[13] D. A. Carvalho, J. H. B. Rocha, N. S. Fernandes, D. S. da Silva, C. A. Martínez-Huitle, *Latin Am. Appl. Res.* **41** (2011) 127-133

[14] M. Panizza, G. Cerisola, *Ind. Eng. Chem. Res.* **47** (2008) 6816-6820

[15] J. Marugán, M. J. López-Muñoz, R. van Grieken, J. Aguado, *Ind. Eng. Chem. Res.* **46** (2007) 7605-7610

[16] Health Hazard Alert - Benzidine-, o-Tolidine-, and o-Dianisidine- Based Dyes. NIOSH Report DHEW (NIOSH) Publication No. 81-106, December 1980.

[17] H. B. Mansour, D. Corroler, D. Barillier, K. Ghedira, L. Chekir, R. Mosrati, *Food Chem. Toxicol.* **45** (2007) 1670-1677

[18] Toxicology data network, National Library of Medicine, Hazardous Substance Data Bank (HSDB), U.S department of Health and Human services. http://toxnet.nlm.nih.gov/cgi-bin/sis/search/f?./temp/~s8y7RI:1, August, 2013

[19] P. Jiang, J. Zhou, A. Zhang, Y. Zhong, *J. Environ. Sci.* **4** (2010) 500-506

[20] V. Jhanji, E. Chan, S. Das, H. Zhang, R. B. Vajpayee, *Eye* **25** (2011) 1113-1120

[21] C. Lüke, M. Lüke, T. S. Dietlein, A. Hueber, J. Jordan, W. Sickel, B. Kirchhof, *Br. J. Ophthalmol.* **89** (2005) 1188-1191

[22] B. T. van Dooren, W. H. Beekhuis, E. Pels, *Arch. Ophthalmol.* **122** (2004) 736-742

[23] M. Veckeneer, K. van Overdam, J. Monzer, K. Kobuch, W. van Marle, H. Spekreijse, J. van Meurs, *Graefe's Arch. Clin. Exp. Ophthalmol.* **239** (2001) 698-704

[24] M. M. Turbow, *Nature,* **206** (1965) 637.

[25] C. H. Waddington, M. M. Perry, *J. Embryol. Exp. Morph.* **4** (1956) 110-119

[26] S. A. Elfeky, A. S. A. Al-Sherbini, *Kinet and Catal.* **52** (2011) 391-396

[27] A. L. T. Fornazari, G. R. P. Malpass, D. W. Miwa, A. J. Motheo, *Water, Air, Soil Pollut.* **223** (2012) 4895-4904

[28] M. Gaber, N. A. Ghalwa, A. M. Khedr, M. F. Salem, *J. Chem.* **2013** (2013) 1-9

[29] C. Song, J. Zhang, PEM fuel cell electrocatalysts and catalyst layers, fundamentals and applications, Spinger-Verlag London Limited, Vancouver, Canada, 2008, p. 89-134 http://www.springer.com/978-1-84800-935-6

[30] R. B. Moghaddam, P. G. Pickup, *Electrochim. Acta* **97** (2013) 326-332

[31] G. R. P. Malpass, D. W. Miwa, R. L. Santos, E. M. Vieira, A. J. Motheo, *Environ. Chem. Lett.* **10** (2012) 177-182

[32] M. Panizza, G. Cerisola, *Chemosphere* **77** (2009) 1060-1064

[33] H. Akrout, L. Bousselmi, *Arab. J. Geosci.* (2012) 5033-5041

[34] H. Z. Zhao, Y. Sun, L. N. Xu, J. R. Ni, *Chemosphere* **78** (2010) 46-*51*

[35] S. Raghu, C. W. Lee, S. Chellammal, S. Palanichamy, C. A. Basha, *J. Hazard. Mater.* **171** (2009) 748-754

[36] J. P. Graham, M. A. Rauf, S.Hisaindee, M. Nawaz, *J. Molecul. Struc.* **1040** (2013) 1-8

[37] S. S. Vaghela, A. D. Jethva, B. B. Mehta, S. P. Dave, S. Adimurthy, G. Ramachandraiah, *Environ. Sci. Technol,* **39** (2005) 2848-2855

18

Determination of nevirapine in the presence of cucurbit(7)uril with a gold electrode

ANA M. ESTEVA, ELÍAS BLANCO*,✉, JUAN J. PIÑA, ABEL I. BALBIN, CARMEN QUINTANA*, PEDRO HERNÁNDEZ*

Departamento de Química Analítica, Facultad de Química, Universidad de La Habana, La Habana 10400, Cuba
**Departamento de Química Analítica y Análisis Instrumental, Facultad de Ciencias, Universidad Autónoma de Madrid, Cantoblanco 28049, Madrid, Spain*

✉Corresponding Author

Abstract
The electrochemical oxidation of nevirapine, an anti-HIV drug, at a gold electrode was studied by voltammetric techniques. Nevirapine showed a signal that interfered with a working electrode wave. This interference was solved by the use of cucurbit(7)uril allowing nevirapine to be determined in tablets (80.4 % recovery, presence of stavudine and lamivudine) and urine (98.4 %).

Keywords
Antiretroviral, Voltammetry, Tablets, Urine

Introduction

Nevirapine (NEV, Figure 1) is a non-nucleoside reverse transcriptase inhibitor (NNRTI) of HIV-1 that causes acquired immunodeficiency syndrome (AIDS). The drug directly bounds to and blocks the activities of RNA and DNA polymerases, both dependent, which caused breakdown of the enzyme catalytic site. NEV activity was not competitive with the reverse transcriptase enzyme or with nucleoside triphosphates. Reverse transcriptase (RT) of HIV-2 and DNA polymerases of eukaryotic cells (eg. human DNA polymerases alpha, beta, gamma and sigma) were not inhibited by nevirapine. The in-vitro antiviral activity was determined in peripheral blood mononuclear cells (PBMC), monocyte-derived macrophages and a lymphoblastoid cell line. The values of the 50 % inhibitory concentrations (IC_{50}) were in the range of 10 to 100 μM against laboratory and clinical isolates of HIV-1. In cell cultures, nevirapine demonstrated additive to synergistic action against

HIV-1 in combination regimens with zidovudine, didanosine, stavudine, lamivudine, saquinavir and indinavir [1].

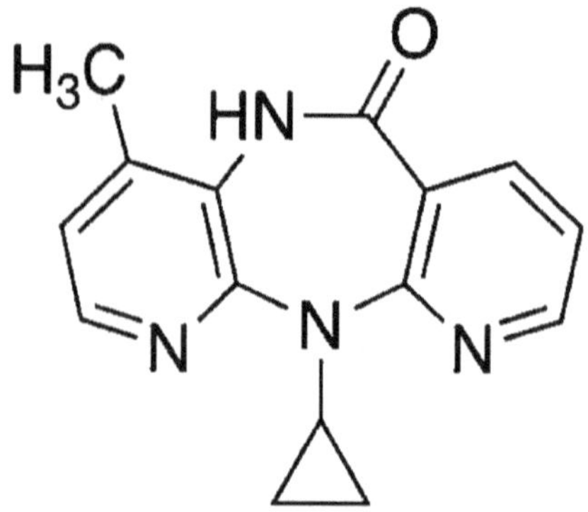

Figure 1. *Nevirapine; 11-cyclopropyl–5,11–dihydro–4–methyl–6H di-pyrido[3,2–b:2',3'-e]-[1,4]diazepin–6–one.*

Different analytical techniques were used to detect NEV, including high performance liquid chromatography (HPLC) [2], matrix-assisted laser desorption/ionization-time of flight mass spectrometry (MALDI-TOF) [3], and capillary electrophoresis [4]. These techniques require expensive equipment, costly reagents for sample preparation and analysis and quite some time.

Electroanalytical methods are an accurate and cheap alternative which offer very low detection limits for electroactive molecules. Different drugs were determined by this technique achieving very low detection limits [5]. Some articles have recently been published about the electroanalytical determination of NEV by means of different working electrodes [6-8].

The family of compounds of cucurbit(*n*)urils (CB(*n*)) are polymeric macrocycles obtained by the condensation reaction of glicoluryl and formaldehyde in acid conditions and have *n* units of glicoluryl bridged by methylene groups. They bind molecules by hydrophobic and ion-dipole interactions (but not exclusively) due to the cavity portals delineated by a rim of carbonylic oxygens. The hydrophobic cavity allowed the inclusion of different molecules depending on the CB(*n*) homologue and the size of the guest [9,10].

We developed a method for the analysis of NEV using CB(7) and gold electrode. Low detection limits were obtained. The method was applied to biological fluids (urine) and a pharmaceutical formulation (which also contained lamivudine and stavudine) and it was demonstrated that the methodology had fewer steps than other ones.

Experimental

Reagents

NEV was provided by the Center for State Control of Drugs (CECMED-Cuba). Aqueous solutions of the analyte were prepared at a 2 mg mL^{-1} concentration in acid medium (pH < 3). Diluted solutions were prepared in supporting electrolyte just before use. CB(7) was supplied by Sigma-Aldrich Chemical Co. All reagents were of analytical grade (> 98 %) and were provided by Scharlau. Ultrapure water was produced by a Milli-Ro and Milli-Q system (Millipore, Waters). Solutions of these compounds were stored at 4 °C and protected from light. Britton-Robinson buffer solutions (BR, mixture of boric, acetic and phosphoric acids) were used as supporting electrolytes, prepared at a final concentration of 0.04 M and the buffer pH was adjusted with 0.1 M NaOH.

NEV was determined in a tablet sample whose nominal content was 250 mg per tablet together with 40 mg of stavudine and 150 mg of lamivudine. A tablet was dissolved in methanol and filtered through a cellulose membrane of 0.45 µm pore size to get a 2.5 mg mL^{-1} NEV solution which was stored at 4 °C.

Apparatus

Electrochemical measurements were performed by means of a µAutolab III potentiostat made by Eco-Chemie in a three electrodes cell: an Au working electrode (2.01 mm^2 geometric area) provided by BAS, a coiled platinum wire as counter electrode and an Ag/AgCl (3 M KCl) reference electrode (all potentials in this paper were referred against it). The pH was controlled by means of a Methrom 827 pH meter with combined glass and an Ag/AgCl/ (3 M KCl) electrode.

Procedure

Activation and regeneration of the gold electrode surface was carried out by successive scanning in 0.1 M sulphuric acid between 0.0 V and 1.5 V at 100 mV s^{-1} by cyclic voltammetry (CV). An ultrasonic bath was used to clean the electrode surface when required and prior to the described activation procedure. Differential pulse voltammetry (DPV) was the chosen technique for the analyte determination in solutions of a NEV:CB(7) ratio of 1:2, the measurements started at 0.4 V and the chosen scan rate and pulse amplitude were 25 mV s^{-1} and 25 mV, respectively.

Results and Discussion

Our studies were performed with NEV at a concentration of 100 µM (26.6 µg mL^{-1}) by CV at gold electrode in 0.04 M BR buffer at pH 2. The analyte showed irreversible redox behaviour and a reduction wave was observed at 0.66 V. As it can be seen in Figure 2, a broad and intense signal at 1.14 V was seen in the anodic scan (green line) but close to the gold oxidation wave at 1.3 V (black line). It shifted to potential values lower than 1.1 V between pH 2 and 6 until disappearance at higher pH. Above that pH, NEV was not electroactive.

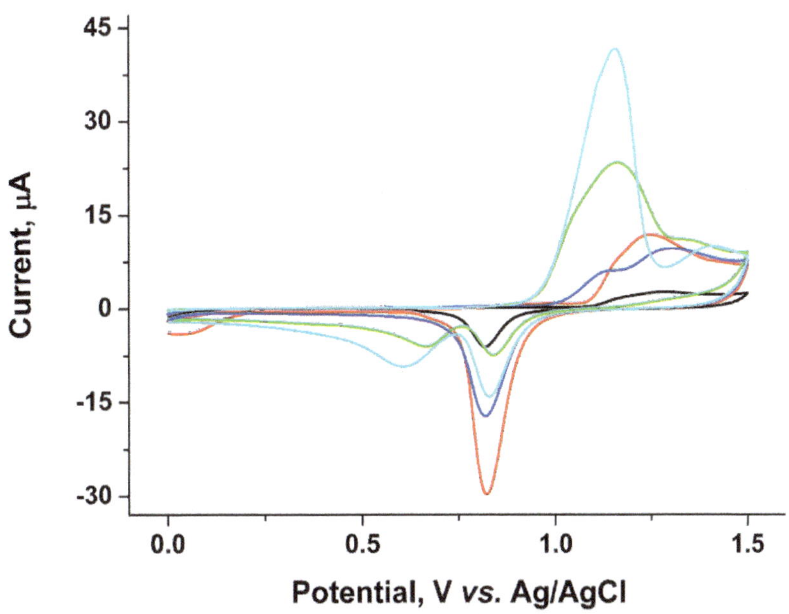

***Figure 2.** Cyclic voltammograms of NEV at 100 mV s^{-1} in 0.04 M BR buffer at pH 2. Black line, supporting electrolyte; red line, 20 µM CB(7); green line, 100 µM NEV; blue line, 10 µM NEV; cyan line, 10 µM NEV and 20 µM CB(7).*

An increase in the sweep rate (V_b) between 10 and 400 mV s^{-1} entailed a variation of the intensity and potential of the signals when a 266 μg mL^{-1} analyte solution in 0.04 M BR at pH 2 was analysed. When the logarithm of the anodic peak current was represented versus the logarithm of the scan rate in Figure 3 (black points), a straight line was obtained with a 0.53 slope value (close to 0.5) so the oxidation could happen after diffusion of the analyte to the electrode surface. However, it was showed in Figure 2 (green line) that NEV was oxidized by means of at least two processes and its shape was not a diffusion-like one so the 0.53 slope value was a chance. The dependence between the logarithm of the cathodic peak current of the NEV oxidation product and the logarithm of the scan rate was studied (Figure 3, red points) and a slope close to 1 was found so the reduction of that product could be concomitant with an adsorption process. Nevertheless, that wave was overlapped with the gold oxide reduction wave and both processes were connected.

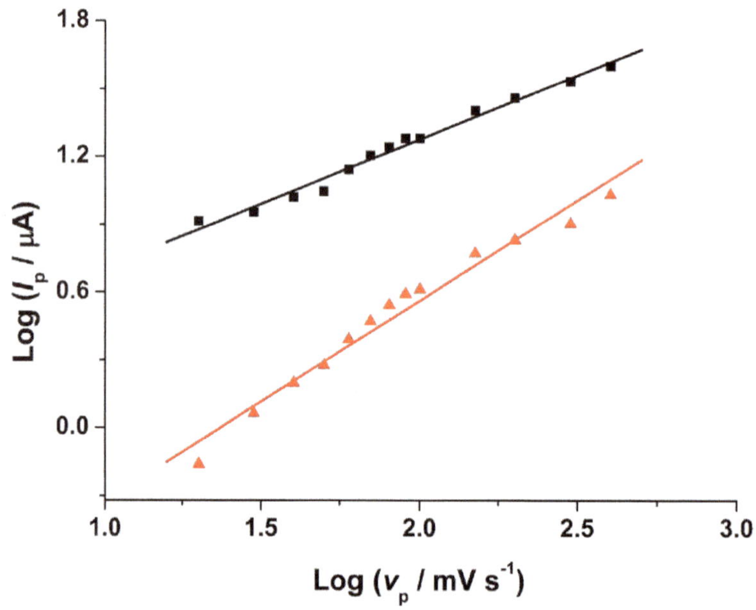

Figure 3. *Effect of the change of the scan rate on the oxidation (black points and line) and reduction peak currents (red points and line).*

As before stated, the NEV oxidation wave was at a potential very close to the gold oxides formation one so the measurements analysis could be complicated or even impossible if the analyte concentrations of the sample solution were low, as in the case of a 10 μM NEV (blue line, Figure 2). When CB(7) was added to solutions of this low NEV concentration (10 μM NEV and 20 μM CB(7), cyan line, Figure 2), the anodic signals were more separated, the NEV wave was narrower and a huge increase in the peak current was observed when they were compared to the signal of solutions of the same analyte concentration and no added CB(7) (blue line, Figure 2). Voltammetric measurements of blank solutions of CB(7) did not show any signal but the same waves observed when the cell just contained supporting electrolyte, as it can be seen in Figure 2, red and black lines, respectively.

If a 1:2 NEV:CB(7) molar ratio was kept constant, the separation of NEV and gold waves in the anodic scan observed at pH 2 in Figure 2 (cyan line) continued up to neutral pH. In these conditions, the NEV cathodic signal was not observed at pH higher than 2. The peak potential (E_p) depended on the medium pH so it can be concluded that the anodic reaction was coupled to an acid-base one. In this case, this dependence followed a straight line whose equation was

E_p / V = 1.19 - 0.021 pH (R^2 = 0.998). As the line slope value (dE_p/dpH) was close to 0.029 V per pH unit, the number of exchanged protons was the half of the number of electrons according to the Nernst equation.

The stoichiometric NEV:CB(7) ratio was studied by CV in 0.04 M BR buffer pH 2, at a constant CB(7) concentration and changing NEV concentration, and vice versa. Current and potential values were plotted and the slope change depending on the NEV:CB(7) ratio was indicative of successive formation of NEV-CB(7) complexes of 1:2 stoichiometry.

As it is shown in Figure 4, the effect of the concentration (0.3-1.6 μg mL^{-1}) on the signal was studied at a 1:2 NEV:CB(7) ratio in 0.04 M BR pH 2 by DPV. Therefore, what it was done was to augment the analyte concentration but also the macrocycle one in the measured solutions. The peak current and the concentration were directly proportional up to 1 μg mL^{-1}, data points that were fitted to I_p / μA = -0.038 + 4.300c / μg mL^{-1}, R^2=0.999. For higher concentrations the analytical signal was relatively constant probably due to surface saturation.

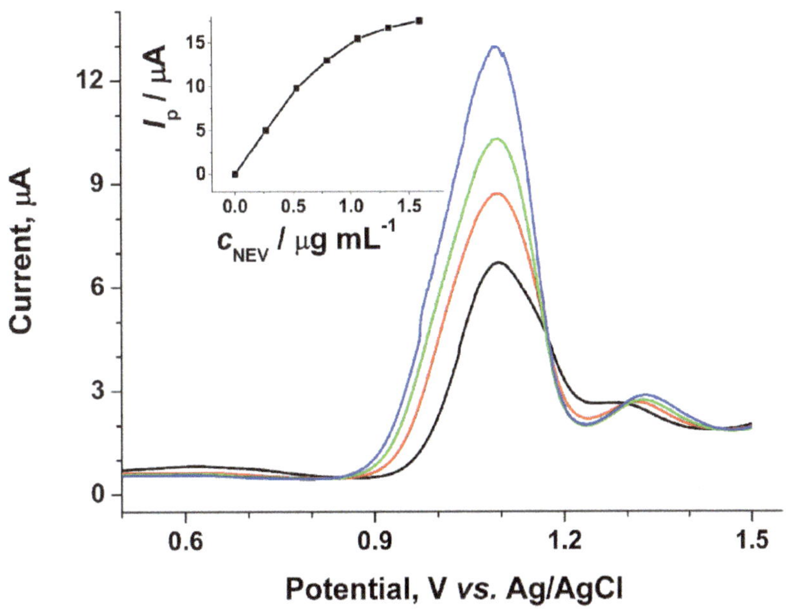

Figure 4. *Effect of the concentration of NEV on the DPV measurements, at a constant NEV:CB(7) ratio of 1:2 in 0.04 M BR pH 2 (see text). The voltammograms of the NEV concentrations 0.267, 0.534, 0.801, 1.07 μg mL^{-1} are shown. The inset graph gives the peak current* vs. *NEV concentration.*

CV measurements of approximately 200 μg mL^{-1} stavudine and lamivudine solutions in 0.04 M BR were obtained at different pH. DPV measurements of these two interferences at pH 6 are shown in Figure 5 and it can be seen that the lamivudine reduction signal was at -0.2 V (red line) and the stavudine one was at -0.1 V (black line). They both were well defined when CB(6) or CB(7) were present in the solution. The signal could be a product of the possible formation of inclusion complexes. These compounds did not show oxidation signals so did not interfere in NEV determination by DPV when they were in the analysed sample.

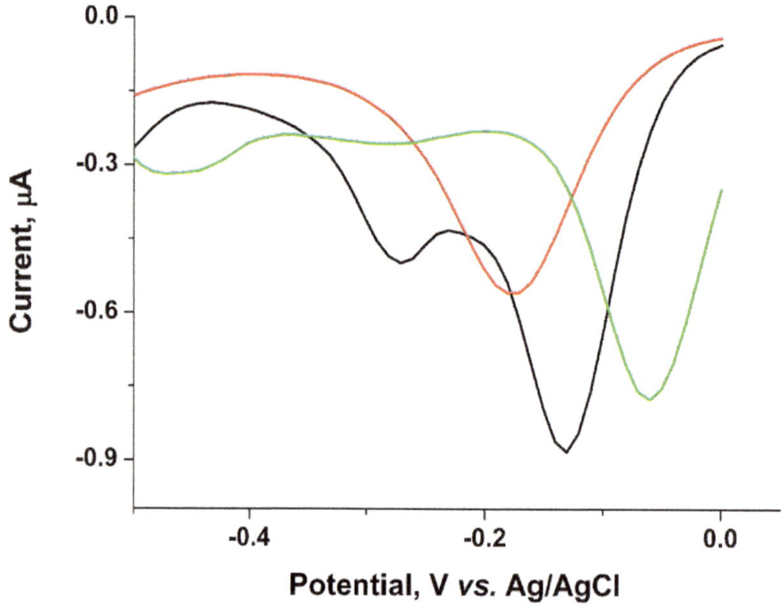

Figure 5. *DPV measurements of 4 μM stavudine and lamivudine and with CB(7) (1:2) in pH 6 BR buffer as electrolyte. Black line, stavudine; red line, lamivudine; green line, electrolyte.*

Determination in tablet

A crushed tablet was left in contact with methanol for 24 hours, the suspension was filtered, the resulting solution was transferred to a 100 mL volumetric flask and the volume was completed with methanol. It contained 2.5 mg mL^{-1} of NEV and the working solutions were prepared from this one. Voltammograms of sample solutions were recorded and, as it can be seen in Figure 6 and as previously shown, the gold oxidation and NEV waves were overlapped if no CB(7) had been added but they were separated if the macrocycle was present (1:2 NEV:CB(7) ratio). The results obtained by means of the standard addition method showed that the content of NEV/tablet was 80.4% of the nominal one (n = 4) in the presence of stavudine and lamivudine.

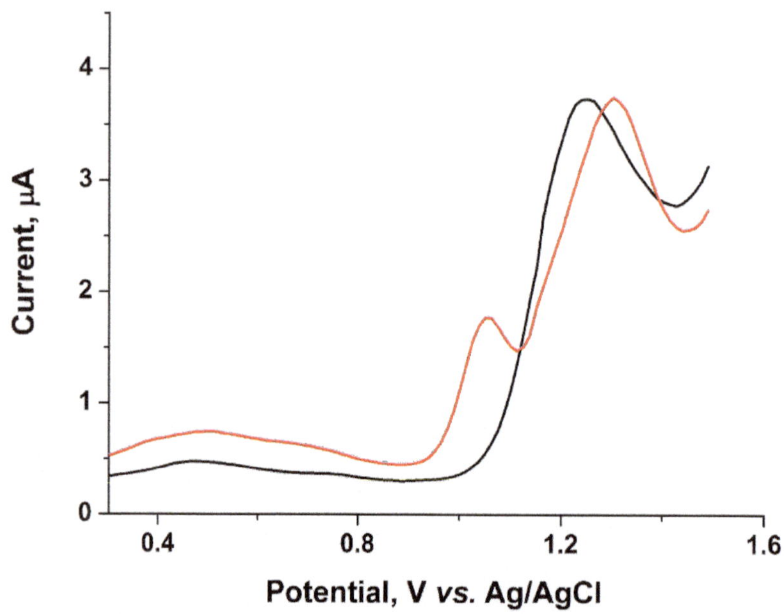

Figure 6. *DPV measurements of solutions of the pharmaceutical sample in 0.04 M BR buffer pH 2. Black line, no CB(7) in solution; red line, with CB(7) (ration 1:2 NEV:CB(7)).*

Determination in urine

A previous treatment of liquid-liquid extraction was necessary given the complexity and characteristics of the urine sample. 2 mL of urine of a healthy individuals were spiked with NEV to reach a final concentration of 26 µg mL^{-1} and were subjected to liquid-liquid extraction with 10 mL of diethyl ether. After shaking, the liquid was left for 3 min, the aqueous phase was discarded and the organic one was evaporated. 10 mL of 0.04 M BR buffer pH 2 were used to dissolve the residue, the solution was introduced in the electrochemical cell and DPV measurements of increasing concentrations of NEV were performed to analyse the sample, voltammograms that are shown in Figure 7. A recovery of 98.4 % (n = 3) was obtained for the spiked urine with 26 µg mL^{-1} NEV in presence of CB(7).

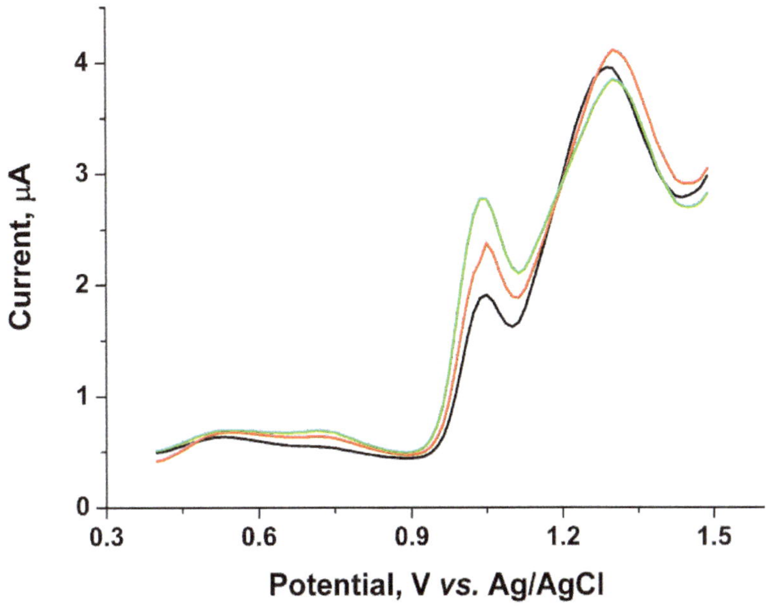

Figure 7. *DPV measurements of NEV doped urine with successive additions of the drug in presence of CB(7), maintaining the 1:2 ratio. Black line, 0 µg mL^{-1} NEV added; red line, 5.32 µg mL^{-1}; green line, 10.6 µg mL^{-1}.*

Conclusions

An electroanalytical method was developed for the analysis of NEV in pharmaceutical formulations in the presence of stavudine and lamivudine and in urine by means of a gold electrode and DPV. NEV was electroactive between pH 2 and 6 and CV measurements showed that NEV oxidation wave was very close to the gold oxides formation one but if CB(7) was added to a NEV solution, both waves were separated and an increase in the analyte peak current was observed. Measurements in 0.04 M BR buffer pH 2 were performed to get the stoichiometry of the NEV-CB(7) complex behind this electrochemical behaviour and it was found that one NEV molecule interacted with two CB(7) molecules, ratio which was kept constant in every calibration or analyzed sample. The calibration of the response was performed and found the equation I_p / µA = -0.038 + 4.300c / µg mL^{-1}, R^2=0.999. The analysis of tablets gave an 80.4 % recovery (n = 4) just dissolving the sample in methanol. 2 mL of urine were doped with NEV at a concentration of 26 µg mL^{-1} and were subjected to liquid-liquid extraction due to the complex matrix and a 98.4 % recovery (n = 3) was found.

Acknowledgements*: Authors thank to Spanish Agency for International Development Cooperation (AECID, A/030784/10) and Comunidad de Madrid (S2009/PPQ-1642, AVANSENS).*

References

[1] D. Burch, *Martindale - The complete drug reference*, Pharmaceutical Press , London, United Kingdom, 2006

[2] V. Kabra, V. Agrahari, C. Karthikeyan, P. Trivedi, *Tropical Journal of Pharmaceutical Research* **8** (2009) 79-86

[3] S. Notari, C. Mancone, T. Alonzi, M. Tripodi, P. Narciso, P. Ascenzi, *Journal of Chromatography B-Analytical Technologies In the Biomedical and Life Sciences* **863** (2008) 249-257

[4] R. Sekar, S. Azhaguvel, *Chromatographia* **67** (2008) 389-398

[5] B. Dogan-Topal, S. A. Sibel, B. Uslu, *The Open Chemical and Biomedical Methods Journal* **3** (2010) 56-73

[6] A. A. Castro, R. Q. Aucelio, N. A. Rey, E. M. Migueland, P. A. M. Farias, *Combinatorial Chemistry & High Throughput Screening* **14** (2011) 22-27

[7] N. L. Teradal, S. N. Prashanth, J. Seetharamappa, *J. Electrochem. Sci. Eng.* **2** (2012) 67-75

[8] F. F. Zhang, L. Li, L. Q. Luo, Y. P. Ding, X. Liu, *J. Appl. Electrochem.* **43** (2013) 263-269

[9] J. Lagona, P. Mukhopadhyay, S. Chakrabarti, L. Isaacs, *Angew. Chem. Int. Ed.* **44** (2005) 4844-4870

[10] E. Masson, X. X. Ling, R. Joseph, L. Kyeremeh-Mensah, X.Y. Lu, *RSC Advances* **2** (2012) 1213-1247

19

Anodic formation of nanoporous crystalline niobium oxide

LEONID SKATKOV✉, LARISA LYASHOK*, VALERIY GOMOZOV*, IRINA TOKAREVA*, BORIS BAYRACHNIY*

Technical Division, PCB "Argo" Ltd., 4/23 Shaul ha-Melekh str., 84797 Beer Sheva, Israel
**Electrochemistry Department, National Technical University "KhPI", 21 Frunze str., 61002 Kharkov, Ukraine*
✉Corresponding author

Abstract
The research results of anodic deposition of crystalline niobium oxide are presented in this work. The factors that have an impact on crystalline phase nucleation and its primary growth are revealed. Dependence of morphology and properties of nanoporous niobium oxide on modes of its formation is shown.

Keywords
Oxidation, anodic film, niobium, HF, crystallisation

1. Introduction

Increasing attention is being paid nowadays to the creation and studying of material properties that have nanometer structures. Anodic oxidation of valve metals (Al, Ti, Nb, Ta) is widely used for the formation of nanostructured oxide films. Self-organisation of nanoporous structures during electrochemical processing is completely revealed during the formation of porous anodic oxides of aluminium and titanium. These are distinguished by a high degree of sequence in the arrangement of pores and the possibility to operate in a variation of surface morphologies and thickness of the oxide film [1 2].

Anodic niobium oxide deposition in the fluoride-containing electrolytes differs significantly from the porous oxide in aluminium and titanium dioxide nanotubes. Under certain conditions of anodisation, anodic oxide films (AOF) Nb_2O_5 are formed alongside the crystalline structure in the form of microcones [3,4]. This leads to the common use of similar layers in various devices and designs, for example, in gas sensors [5], catalysts [6], electric capacitors and electrochromic devices [7], as well as in thin-film lithium ion batteries [8], etc.

The researchers' views on the mechanism of nucleation and the formation of crystalline structure of porous AOF niobium are inconsistent. Therefore, it is noted [9] that formation of the crystalline phase takes place under the influence of the internal tension that arises with a growth in oxide thickness. It is proposed in the work of Oikawa et al. [10] that formation of niobium oxide microcones has to be connected with non-uniform chemical dissolution of the anodic film during anodisation.

From our point of view, it is necessary to consider the chemical nature of niobium, which belongs to d-type elements, for understanding the nucleation and growth of the crystalline phase mechanism. When one electron has been released, it turns into an ion with empty external d-levels. As a result, the formed ion tries to achieve a stable electronic configuration, especially with oxygen and other non-metals. It is the existence of incomplete configurations of d-electrons that causes niobium to display a wide range of the valence states. The possibility of their existence is proved by thermodynamic calculations [11–13]; it is also shown that low valence oxides have to be on a metal/high valence oxide boundary. According to the authors [11–13], the transition zone cannot be considered as a certain plane-parallel layer between a metal and the high valence oxide. Local inhomogeneity on the metal surface, the border of grains, dislocation, admixture atoms and other structural and chemical defects are considered to be the centres of increased surface energy. At these centres, oxygen diffusion into metal is simplified and the primary formation of the lowest valence oxides is possible exactly here. As was shown in [7], films generated in potassium nitrate melt instead of forming an entire niobium pentoxide layer (as had first been suspected), creating a «sandwich» of Nb_2O_5, NbO_2 and NbO phases (in the direction from oxide surface to niobium).

During discussion of the research results in this article, it was taken into account [13] that, in contrast to thermal crystallisation, no transformation of an amorphous into crystalline film occurred at niobium anodising.

The present work is aimed to identifying the factors defining primary formation of nanoporous crystalline in niobium oxide solutions containing the activator.

2. Experiment

2.1. Electrochemical investigations

AOF was formed in solutions of 1 M H_2SO_4 with the addition of various HF concentrations (0.5 – 2 M). Anodisation was carried out at room temperature in a volt-static mode and varying voltage in the range of 60 to 80 V. Platinum was used as a cathode electrode. Polarisation was carried out on potentiostat PI 50-1.1 with a scan rate of 10^{-2} V/s in potentiodynamic mode [14]. The copper-decoration method was used for the research of synthesised AOF conductivity. For this purpose, the cathode samples were polarised in a solution of 220 g/L $CuSO_4$ + 60 g/L H_2SO_4. The copper plate worked as an anode, while niobium with the deposited oxide film functioned as a cathode.

2.2. SEM and X-ray investigation

Composition of the synthesised coatings was defined using a scanning electronic microscope JSM-7001F with the x-ray energy-dispersive micro-analyser Oxford INCA PentaFET-x3. Morphology of the received coatings was studied by applying the scanning electronic microscopy (SEM) method alongside the use of microscopes JSM-7001F, JSM-6390LV. The X-ray structure analysis of films was carried out using diffractometer DRON-2 with CuK_α radiation.

3. Results and Discussion

Initially, niobium surface structure was considered as defective and containing natural oxide film that inherited defects after preparatory operations. In this regard, it was assumed that along the niobium surface at potential imposing, distribution of the electric field is non-uniform and velocity of the process' electrochemical growth and oxide dissolution on various sites of the surface are not identical, and depend on the degree of its deficiency and the composition of liquid facing a electric double layer. In defective sites, the dissolution process has to proceed at a high-speed rate followed with formation of pore seeds when field density is small, and crystalline phase seeds when a strong field is available.

For identification of the factors that have an impact on the generation of the crystalline phase at Nb_2O_5 formation, we considered the received anodic polarisation curves (Fig. 1). The size of the stationary potential of the niobium electrode in electrolyte with fluoride ions had a value that was more negative at fluoride concentration growth. Therefore, activation of the surface takes place without imposing the electric current when fluoride is present.

One maximum of current was observed on curves; this confirmed irreversibility of niobium oxide formation. Current peak value increased with the growth of fluoride ion concentration in electrolyte. A sharp rise of anode current and system transition into a passive condition was connected to the formation of the oxide monolayer of the higher oxidation state at the interface with electrolyte.

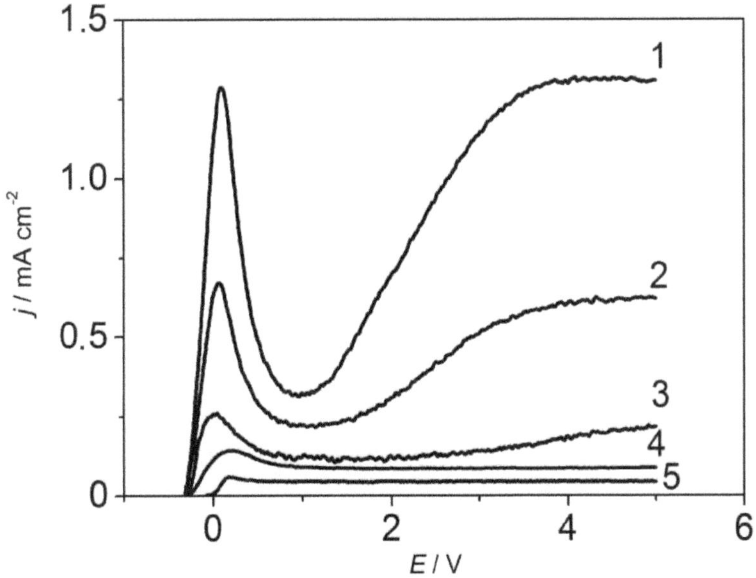

Figure 1. *Polarisation curves of Nb in solutions:* ***1*** *- 1 M H_2SO_4 + 1M HF;* ***2*** *- 1 M H_2SO_4 + 0,5 M HF;* ***3*** *- 1 M H_2SO_4 + 0,25 M HF;* ***4*** *- 1 M H_2SO_4 +0,1M HF;* ***5*** *- 1 M H_2SO_4.*

The higher niobium oxides were insulators with a wide band gap and therefore had bad electronic conductivity; the low oxides and hydroxides functioned like semiconductors. Only stoichiometrical oxides acted as a barrier, with higher oxides being formed by means of full oxidation of the lowest oxides [12].

The results showed that processing of polarisation curves on a site of current rise (Fig. 1) in semi-logarithmical coordinates transformed into a straight line in log $(1 + j_a/j_{lim})$ coordinates *vs.* ΔE; therefore, the process rate of monolayer formation of the higher oxide was determined by diffusion kinetics.

The multilayer niobium oxide formed when potentials were more positive than 1 V. If there was no activator in the solution (Fig. 1, curve 5), current practically did not depend on potential; thus, the AOF of the barrier type that had an amorphous structure was formed on niobium. The current grew in the system when concentration of the activator (fluoride ion) increased; thus, velocities of oxide formation and dissolution processes in active centres defined the geometry of porous amorphous anodic niobium oxide.

Being based on the analysis of theoretical regularities and experimental data on AOF formation on niobium, the argument can be made that its ionisation proceeds on the solid-phase multi surface mechanism [12] by formation of niobium oxides of the lowest valence. Outer layers, which were enriched with more oxygen, were the oxides of highest valence. Rate of niobium ionisation was controlled by the rate of mass transfer of oxygen-containing particles. The process of nanoporous oxide formation was defined by technological parameters of anodisation.

Results of the X-ray energy dispersive microanalyser (Fig. 2) confirmed that the synthesised films corresponded to the stoichiometric composition of Nb_2O_5 oxide, as concentration of elements (at. %) satisfied the ratio 2:5 in the received samples (Table 1).

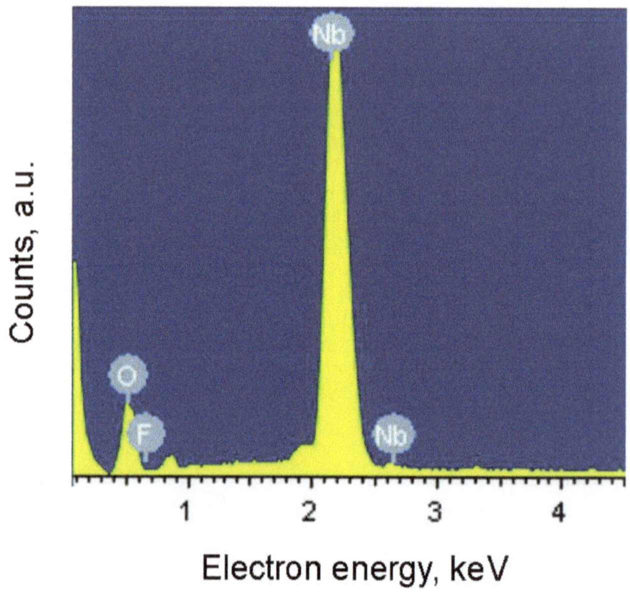

Figure 2. *EDS analysis of the niobium oxide film's surface.*

Table 1. *Analysis of the chemical composition of the niobium oxide film.*

Element	Weight content, %	Atomic content, %
Oxygen	29.39	70.02
Fluoride	0.65	1.31
Niobium	69.96	28.67
Totals	100.00	100.00

Formation of crystalline oxide required the availability of the low oxides in a transitional layer on a metal/oxide boundary and a strong electric activity field. It is known that, in contrast to high oxides, in low oxides the link metal-oxide is mainly ionic, and it exists in a transitional layer in the form of the crystalline phase. It can be assumed that these oxides appear the elements that will form the future crystal grain. In paper [12], it is shown that crystallisation velocity grows with an increase in low valent cation concentrations of oxidised metal.

As mentioned above, AOF crystalline formation on niobium in electrolyte at room temperatures required an active and strong electric field, which was necessary for carrying out the process in a volt-static mode. Crystals were formed from grains under an amorphous film in an upper metal surface layer. Kinetics of crystalline phase formation in a volt-static mode was defined by the composition of electrolyte and temperature in the reaction zone.

While the crystal grew under the film, without destroying it, gradual current decreases were applied at the volt-static mode of AOF formation on niobium (Figs. 3, 4). After separate crystals broke the amorphous oxide, the current increased as long as the area occupied with a crystal phase grew while crystallisation developed. Crystals connected amongst themselves while they grew and then the current gradually began dropping.

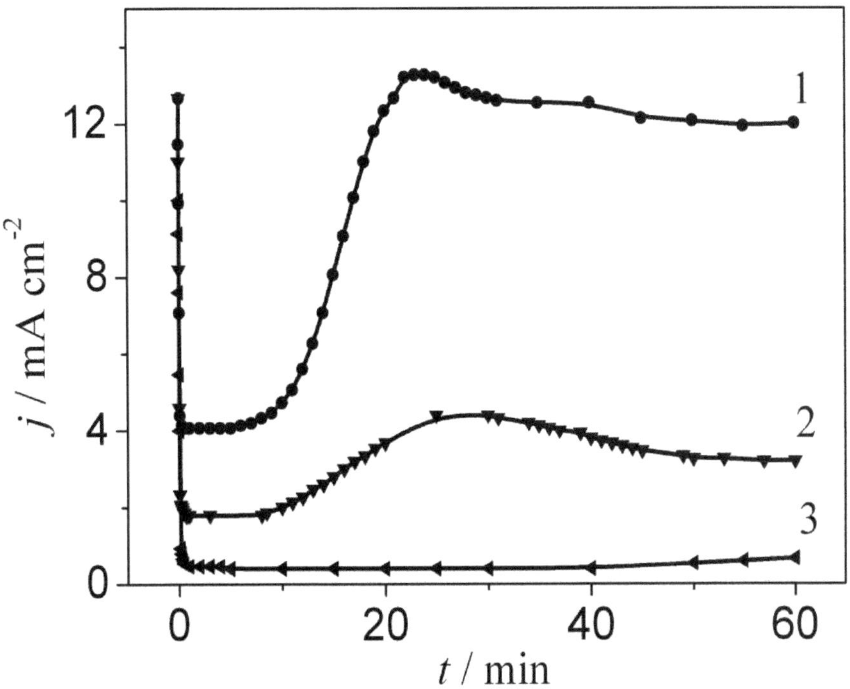

Figure 3.*Current vs. time during anodisation of Nb at 60 V in solution 1 M H_2SO_4 with addition of HF:* ***1*** *- 2 M;* ***2*** *- 1 M; 3 - 0.5 M*

Velocity of the crystalline phase formation and growth processes increased with the activator concentration jump (Fig. 3). The authors of [12] offered to call the time from the start of anodisation to the current growth using the Nb-Nb_2O_5-electrolyte (MOE) system, as the incubatory period was connected with nucleation and the growth of crystals under the amorphous film of Nb_2O_5. After completion of the incubatory period in a chain of the MOE system, rapid increase of the current was observed (Figs. 3, 4), which can be explained due to contact of electrolyte with the crystal surface. While the crystallisation process was developing, the area occupied with the crystal phase increased and current growth in the MOE system chain slowed down.

Growing crystals had the wrong polyhedron form (Figs. 5, 6). The authors of [4] called similar formations microcones of niobium oxide. Such crystalline formations consist of needle crystals that radiate from the crystallisation centre. Wrongly formed crystals are connected with sectors that grow at various speeds. All microcones (Figs. 5, 6) consisted of strongly furcated nanofibres of niobium oxide and created a well-developed oxide film surface.

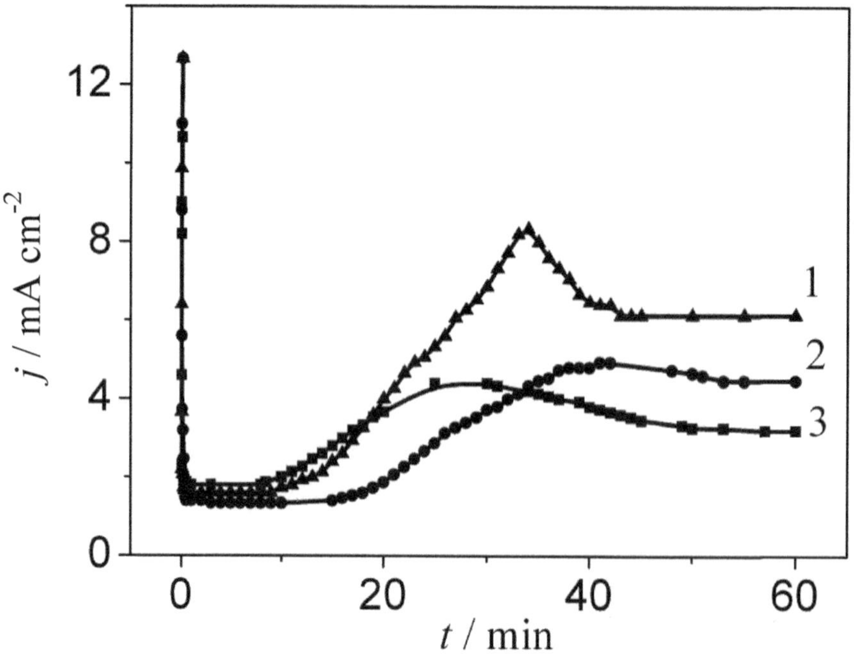

__Figure 4.__ Current vs. time during anodisation of Nb in solution 1 M H_2SO_4 + 1 M HFat:
__1__ - 80 V; __2__ - 70 V; __3__ - 60 V.

__Figure 5.__ SEM images of the surface (A) and cross-section (B) of anodic niobium oxide synthesised during 5 h at 60 V in solution1 MH_2SO_4 + 0.5 MHF.

It can be seen that crystalline oxide increased electric conductivity (Fig.6, curve 3) in comparison with the amorphous porous structure of niobium oxide (Fig. 7) and the solid barrier oxide using the decoration method. As stated above, anodic niobium oxides of the crystalline structure had to possess electronic conductivity, in contrast to amorphous films. As can be seen in Fig. 6, allocation of copper on films of the amorphous structure practically did not occur in the range of 0.2 – 0.4 V. When cathode polarisation was 0.4 V and above, copper allocation on a sample with crystalline AOF (Fig. 6, curve 3) had a larger velocity. Therefore, resistance on the interfaces of solution $CuSO_4$ – AOF – niobium varied significantly for the received coatings.

Figure 6. *SEM images of the surface of anodic niobium oxide synthesised in solution 1 M H_2SO_4+ 1 M HFat 60 V:* ***A*** *- 3 h;* ***B*** *- 5 h.*

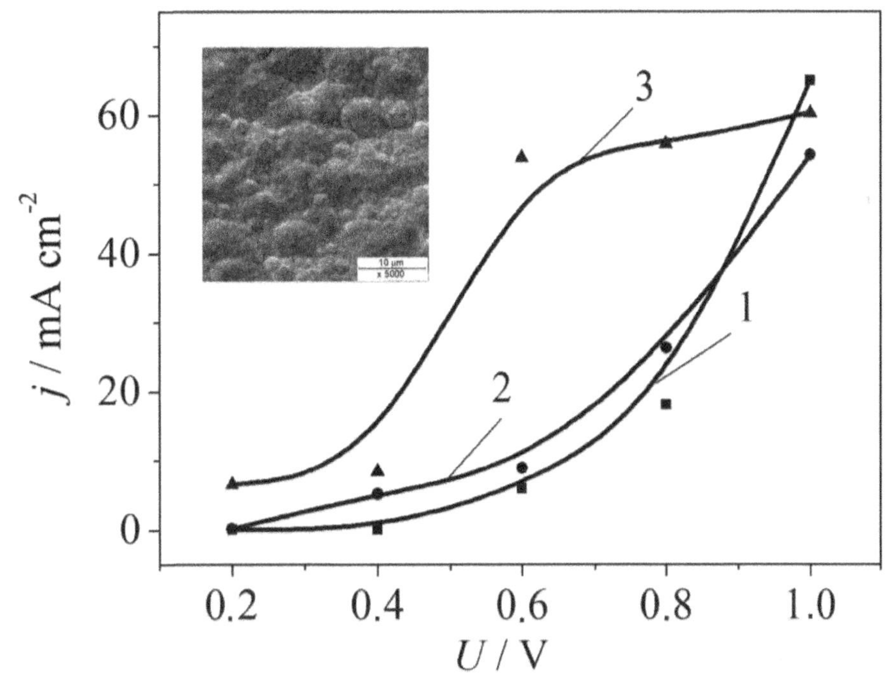

Figure 7. *Current vs. potential during deposition of copper on anodic niobium oxide synthesised in:* ***1*** *- 1 M H_2SO_4, 60 V, 1 h;* ***2*** *- 1 M H_2SO_4 + 0.25 M HF, 20 V, 1 h;* ***3*** *- 1 M H_2SO_4 + 0.5 M HF, 60 V, 5 h.*

By means of the X-ray diffraction analysis method it was confirmed that niobium AOF, having been synthesised during 1 h was X-ray amorphous (see Fig. 8, curve 2); in the same electrolyte, the crystalline oxide was formed on long anodisation. On the received roentgenogram there were peaks that corresponded to crystalline Nb_2O_5 (Fig. 8, curve 1).

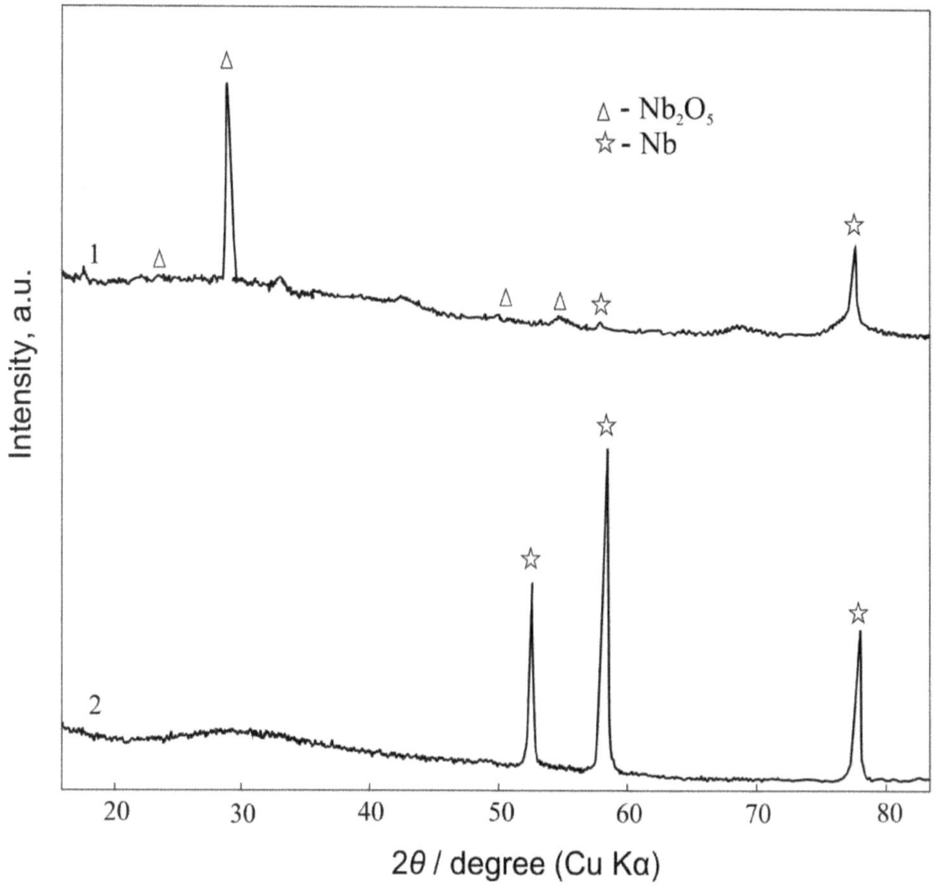

Figure 8. *XRD patterns from niobium AOF synthesised in solution 1 M H_2SO_4 + 0.5 MHF at 60 V during:* ***1*** *– 5 h;* ***2*** *- 1 h.*

4. Conclusions

Based on the research conducted it can therefore be assumed that AOF formation on niobium proceeds on the solid-phase poly-surface mechanism by formation of low valence oxides, which are grains of the crystalline phase at formation of oxide monolayer. Growth of poly-layer crystalline niobium oxide was observed to take place under the amorphous film of Nb_2O_5. After completion of the incubatory period, growing crystals broke the barrier film and the crystallisation process developed on the entire surface of the sample. Predominant formation of nanoporous crystalline oxide of niobium took place in a volt-static mode and at strong field activity in electrolyte containing the activator. The superficial crystalline structure had a nanoporous structure. Its morphology depended on the fluoride ion concentration in the solution, voltage and anodisation time.

References

[1] S. Minagar, C. C. Berndt, J. Wanga, E. Ivanova, C. Wen, *Acta Biomater.* **8** (2012) 2875–2888.

[2] P. Schmuki, *Nanostructure Sci Tech.* **3** (2009) 435–466.

[3] R. L. Karlinsey, *Electrochem. Com.* **7** (2005) 1190–1194.

[4] S. Yanga, H. Habazaki,T. Fujii, Y. Aoki, P. Skeldonb, G. E. Thompson, *Electrochim. Acta* **56** (2011) 7446–7453.

[5] R. A. Rani, A. S. Zoolfakara, J. Z. Oua, M. R. Fieldb, M. Austina, K. Kalantar-Zadeh, *Sens. Actuators B* **176** (2013) 149–156.

[6] K. Tanabe, *Catal. Today* **78** (2003) 65–77.

[7] L. Skatkov, V. Gomozov, *Niobium: Chemical Properties, Applications and Environmental Effects,* Nova Science Publ., New York ,USA, 2013, 123–136.

[8] J. E. Yoo, J. Park, G. Cha, J. Choi, *Thin Sol. Films* **531** (2013)583–587.

[9] J. Zhao, X. Wang, R. Xu, Y. Mi, Y. Li, Electrochem. *Sol. State Lett.* **10** (2007) C31–C33.

[10] Y. Oikawa, T. Minami, H. Mayama, K. Tsujii, K. Fushimi, Y. Aoki, P. Skeldon, G. E. Thompson, H. Habazaki, *Acta Mat.* **57** (2009) 3941–3946.

[11] R. Collongues, *La Non-Stoechiometrie,* Masson et Cie, Paris, France, 1971, p. 285.

[12] L. Odynetz, Anodic Oxide Films, *Nauka*, Leningrad, USSR, 1990, p. 200.

[13] Ya. Kolotyrkin, *Vestnik AN USSR* **7** (1977) 73–80.

[14] V. Gomozov V., L. Skatkov, L. Liashok, B. Bayrachny, *Sov. J. Appl. Chem.* **62** (1989)1284–1287.

20

Synthesis of $CuInSe_2$ thin films from electrodeposited $Cu_{11}In_9$ precursors by two-step annealing

TSUNG-WEI CHANG, SHAO-YU HU, WEN-HSI LEE✉

Department of Electrical Engineering, National Cheng Kung University, Tainan 701, Taiwan, R.O.C

✉Corresponding Authors

Abstract

In this study, copper indium selenide (CIS) films were synthesized from electrodeposited Cu-In-Se precursors by two-step annealing. The agglomeration phenomenon of the electrodeposited In layer usually occurred on the Cu surface. A thermal process was adopted to turn Cu-In precursors into uniform $Cu_{11}In_9$ binary compounds. After deposition of the Se layer, annealing was employed to form chalcopyrite CIS. However, synthesis of CIS from $Cu_{11}In_9$ requires sufficient thermal energy. Annealing temperature and time were investigated to grow high quality CIS film. Various electrodeposition conditions were investigated to achieve the proper atomic ratio of CIS. The properties of the CIS films were characterized by scanning electron microscopy (SEM), X-ray Diffraction (XRD), and Raman spectra.

Keywords

$CuInSe_2$, CIS, Annealing, Electrodeposition

Introduction

The solar cell has emerged as a very important non-conventional energy source. Copper indium selenide ($CuInSe_2$) is a I–III–VI group semiconductor compound offering good possibilities for thin-film photovoltaic (PV) applications because it has a energy gap of 1.02 eV [1–5]. Electrochemical deposition is a low cost method of producing thin CIS films because it has several advantages for large-area non-vacuum thin film production and little material waste. However, the crystallinity of CIS film grown by single-step electrodeposition is inferior because its growing temperature is much lower than that of the physical vapour deposition (PVD) method. The grains were small and loose. The CIS film can also be synthesized from co-sputtered $Cu_{11}In_9$-Se precursors by the thermal annealing process [6,7]. Large CIS grains can be grown due to the gas–liquid reaction during the annealing process [8]. The Cu-In alloys are usually co-sputtered by PVD. The PVD technology is excellent for good quality film growth but difficult to scale up because of the high manufacturing costs.

In this study, Cu-In precursors were prepared by multi-step electrodeposition. However, the agglomeration phenomenon of the electrodeposited In layer usually occurred on the Cu surface. The surface was non-uniform and discontinuous. An annealing process was adopted to transform Cu-In to $Cu_{11}In_9$ compound and create a uniform surface structure. After deposition of the Se layer on the annealed $Cu_{11}In_9$, another annealing process is required to synthesize the CIS structure. Various annealing temperatures and times were adopted to investigate the proper annealing conditions and the mechanism of CIS synthesis. The deposition conditions were adjusted to achieve a better atomic ratio. The properties of the CIS films were characterized by scanning electron microscopy (SEM), X-ray Diffraction (XRD), and Raman spectra.

Experiment

The aqueous solution for the Cu deposition contained 0.75 M $CuSO_4$, 4 mM H_2SO_4, and 0.5 mM HCl. The aqueous solution for the Se deposition contained 17 mM H_2SeO_3 and 0.5 mM HCl. The aqueous solution for the In deposition contained 50 mM $InCl_3$ and 30 mM HCl. The electrodepositions were carried out with AUTOLAB PGSTAT302, a conventional three-electrode potentiostat, and the deposition conditions listed in Table 1. A thin slice of 99.99% pure Pt electrode measuring 1 × 4 cm was employed as the counter electrode, and an Ag/AgCl electrode served as the reference electrode. Glass substrates with sputtered Mo film were used as the working electrodes. The electrodeposition area was a square measuring 1 x 1 cm. The substrates were cleaned by ultrasonication in acetone, 99.5 % pure ethanol, and water before sputtering and electrodeposition. A magnetic stirrer was used for the stirring procedure. The rotation speed of the magnetic stirrer was set at 50 rpm.

Table 1. *Electrodeposition and annealing condition of samples a - f.*

	Cu	In	Se		$T_{annealing}$ / °C
	i / mA : τ / s	τ / s (i = 5 mA)	i / mA	τ / s	(τ = 5 min)
Sample a	60 : 20 + 20 : 10	350	4	1450	550
Sample b	60 : 20 + 20 : 10	350	4	1450	600
Sample c	60 : 20 + 20 : 10	350	4	1450	650
Sample d	70 : 20 + 20 : 10	350	4	1450	650
Sample e	60 : 10 + 30 : 10	375	4	1450	650
Sample f	50 : 10 + 20 : 10	350	4	1450	650

The In deposition was carried out after the two-step growth of Cu. The Cu-In layers were treated by rapid thermal annealing (RTA) at 500 °C for 5 min. The Se layer was deposited on the Cu-In layer after annealing. The CIS film was synthesized by annealing the Cu-Se and In-Se precursors. The deposition conditions and RTA listed in Table 1 were found to achieve better proportions and structures.

The surface morphology and chemical composition of the films were characterized by SEM (Philips XL-40FEG) and EDS, respectively. The Raman spectra were produced with a backscattering configuration at room temperature with unpolarized light using a DILOR XY 800 spectrometer and an Ar laser with a 514.5 nm wavelength as the light source. The phase composition and the crystallographic structure were analysed by XRD using a Bruker D8 SSS multipurpose thin-film x-ray diffractometer.

Results and discussion

In this study, $CuInSe_2$ thin films were synthesized from electrodeposited Cu, In, and Se thin-film precursors. However, the electrodeposition of the In layer on the copper surface with a current of 60 mA in 20 s + 20 mA in 10 s induced serious agglomeration phenomenon. Figure 1 shows the SEM image of the deposited In on the Cu layer. It was deposited with a current density of 3-6 mA cm^{-2}. The agglomeration phenomenon made it difficult for the In layer to cover the whole surface. The grains of In were separated. Deposition with a high current density can lead to a better distribution of In. However, it also leads to large over potential, which would cause bubbles and a rough structure. A suitable value of 5 mA was employed for the deposition of In. After the deposition of Se, the Cu-In-Se precursor was annealed by RTA to form the CIS structure. The XRD results in Figure 2 show that the CIS structure can be synthesized at 450-550 °C. However, the SEM images in Figure 3 show that the film has a rough surface and a non-uniform grain size. The CIS structure came from the non-uniform In layer and the miscellaneous precursor type. The precursors not only contain Cu, In and Se, but also some binary compounds, including $Cu_{11}In_9$, Cu_xSe and In_2Se_3, which were produced in the electrodeposition and annealing process. Different precursors have different reactions and temperature requirements to form CIS, and other reactions lead to a uniform structure [9].

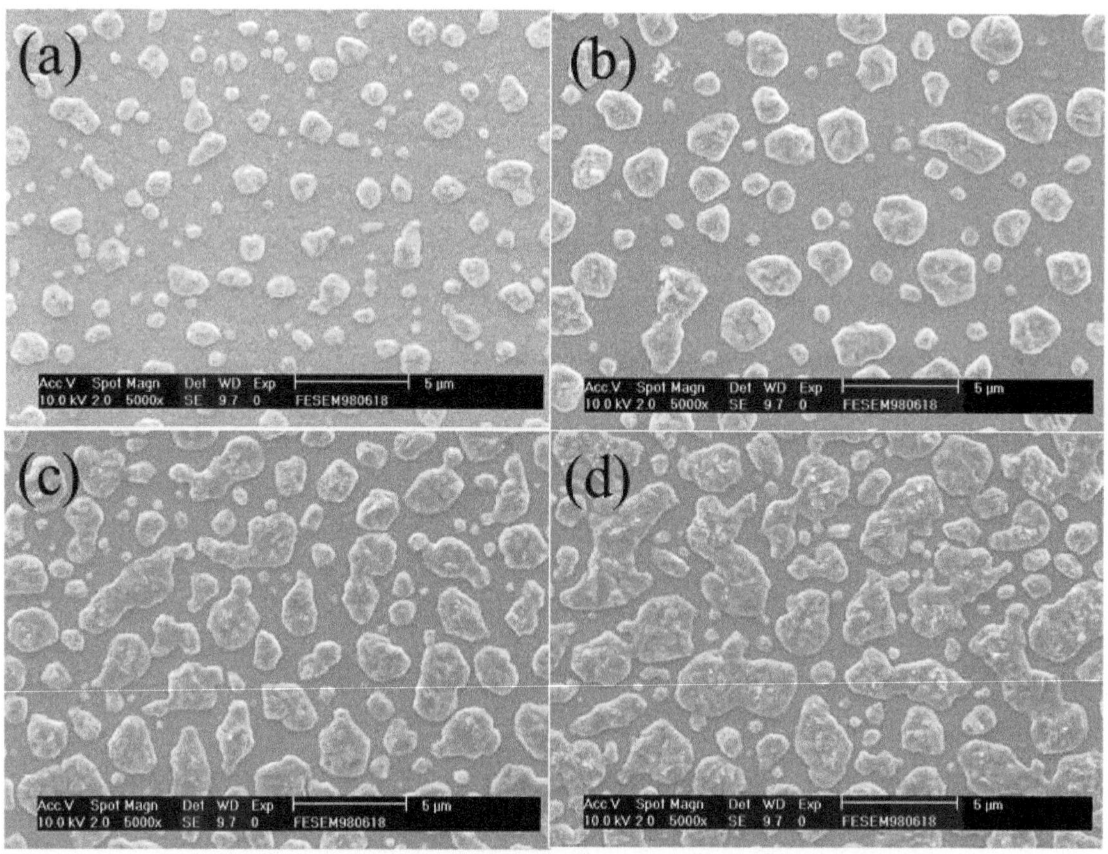

Fig. 1. SEM images of deposited In on Cu surface with current densities (a) 3 mA cm^{-2} (b) 4 mA cm^{-2} (c) 5 mA cm^{-2} (d) 6 mA cm^{-2} for 200s.

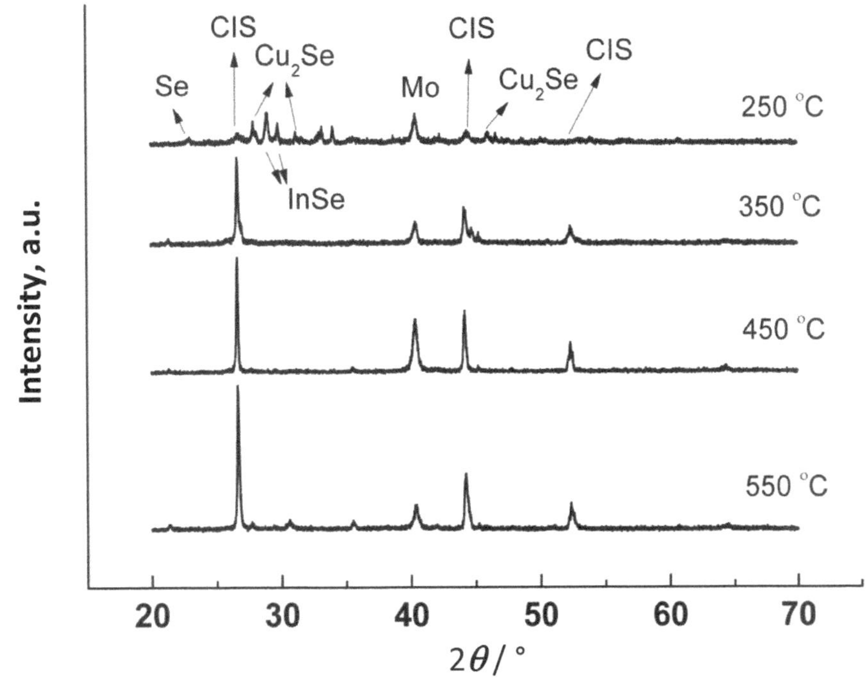

Fig. 2. XRD of annealed CIS film synthesized from Cu-In-Se precursor with temperature.

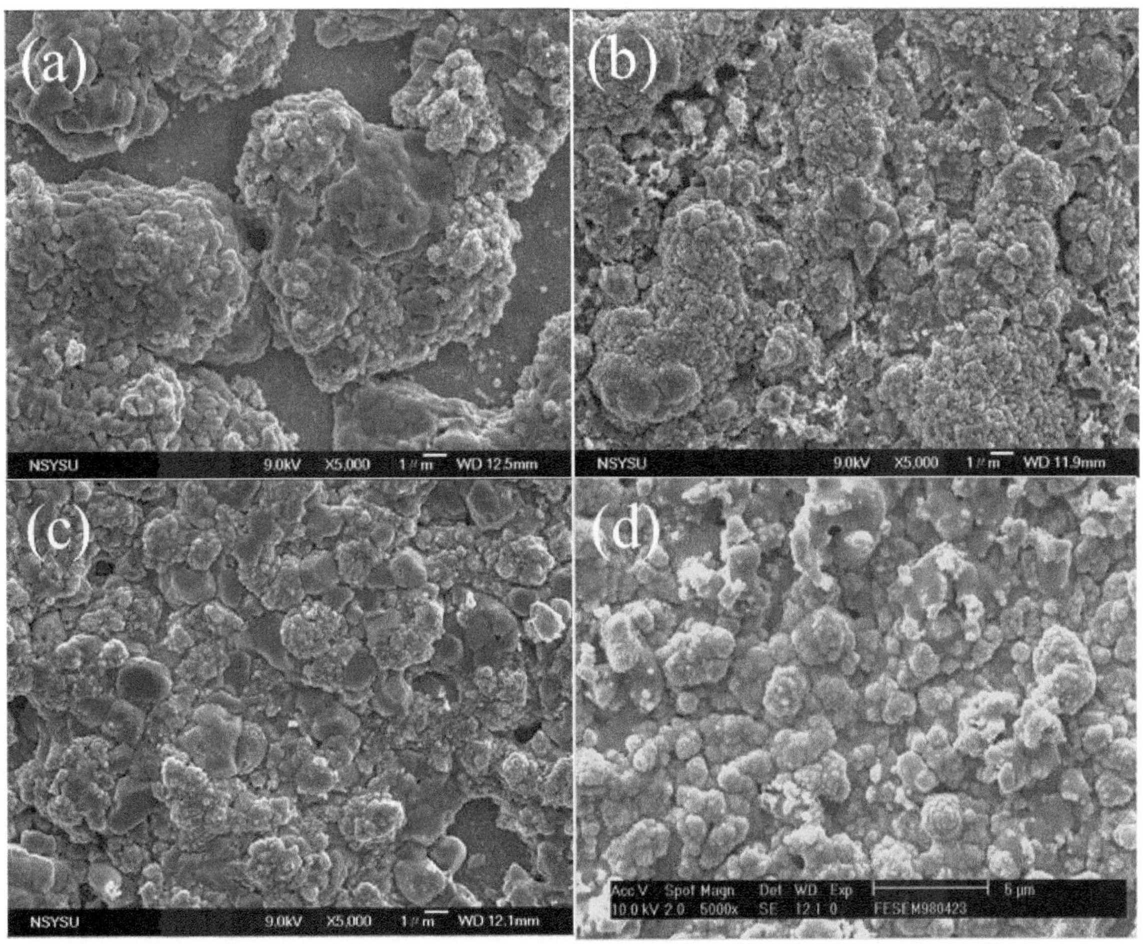

Fig. 3. SEM images of annealed CIS film synthesized from bilayer electrodeposition of Cu-In-Se precursors with temperature (a) 250 °C, (b) 350 °C, (c) 450 °C, (d) 550 °C.

In order to create a smooth In layer distribution and uniform CIS synthesis, a thermal pre-treatment at 500 °C for 5 min was employed to improve the topography of the Cu-In layer. Figure 4 shows the SEM image of the Cu-In thin film after annealing at 500 °C for 5 min. The film re-grows continuously and covers the whole Cu surface. Figure 5 shows the XRD pattern of the annealed Cu-In precursor. Most of the precursors were turned to $Cu_{11}In_9$ and small amounts to $CuInSe_2$.

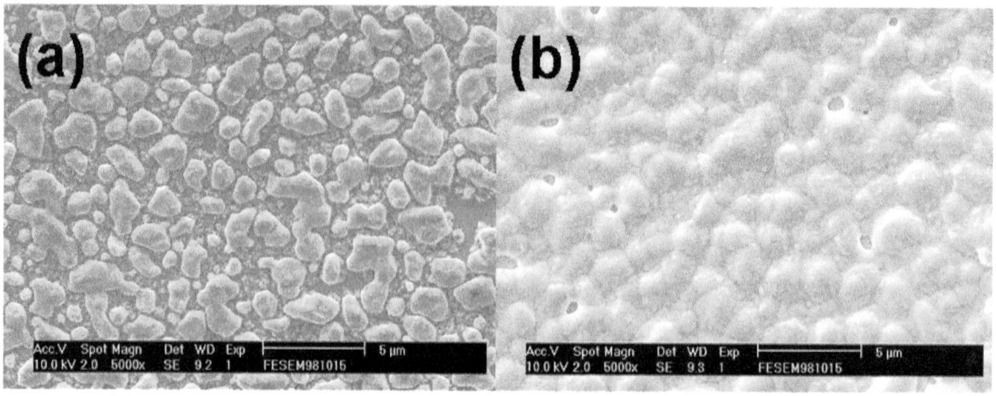

Fig. 4. *SEM images of electrodeposited In layer on Cu surface (a) as-deposited film (b) annealed film.*

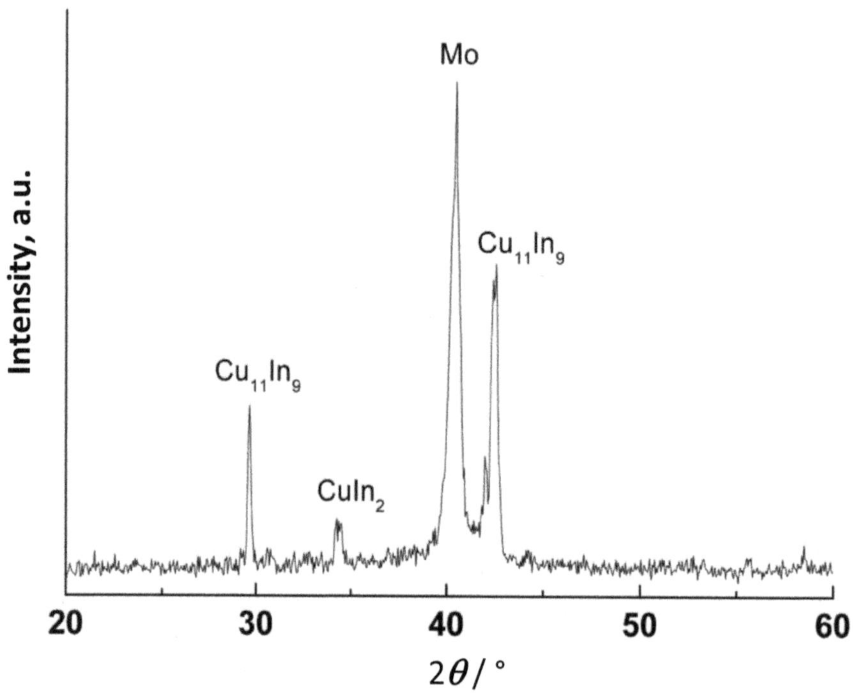

Fig.5. *XRD pattern of annealed Cu/In precursors.*

The Se layer was deposited on the annealed Cu-In surface at 4 mA cm^{-2} current density for 1200 s. All of the precursors were recrystallized by RTA at 600 °C for 3 min and observed by SEM and EDS. Figure 6(a) shows the microstructure of the annealed CIS film at 600 °C for 3 min. The grain size of the annealed CIS film synthesized from $Cu_{11}In_9$-Se precursors is much larger than that of CIS film annealed by co-electrodeposition. During the thermal annealing, the Cu-In became liquid phase and reacted with the gas phase Se. CIS grain growing and diffusion were easier in the gas-liquid reaction. However, many voids were observed in the SEM image. This was because CIS synthesis from the reaction of $Cu_{11}In_9$-Se precursors requires a higher annealing temperature.

Figure 6(b) shows the microstructure of the CIS film with an annealing temperature of 630 °C for 3 min. A higher synthesis temperature certainly reduces the voids in the film.

However, the EDS results shown in Table 2 indicate that the composition of the film is not optimum for CIS. Se gas would easily dissipate during annealing. In Samples a, b, and c listed in Table 1, the deposition times of Se was increased to 1450 s and the annealing temperature was adjusted to find the proper value. The EDS results shown in Table 2 indicate that the atomic ratio of Se increases to an appropriate value of 50 %.

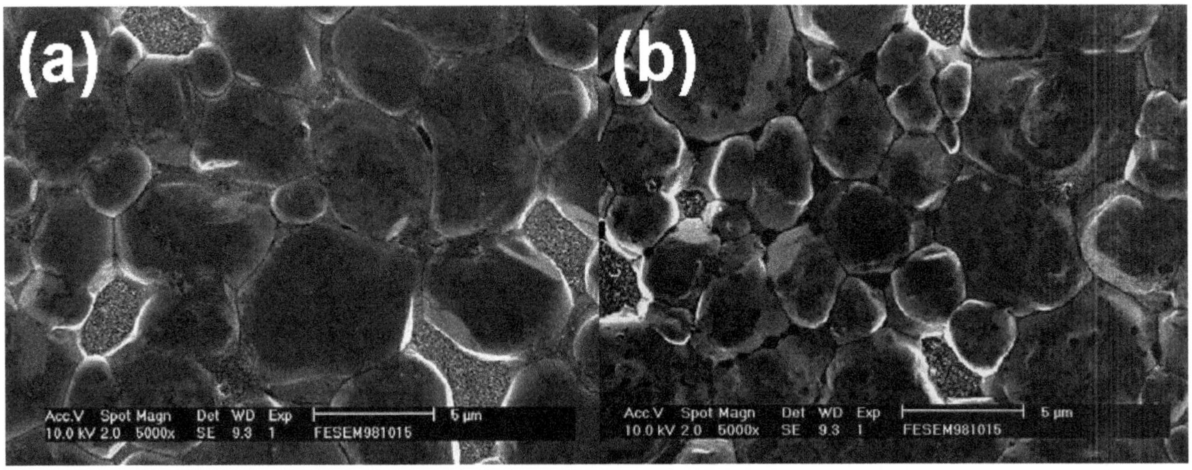

Fig. 6. *SEM images of CIS film after annealing at (a) 600 °C, (b) 630°C for 3min.*

Table 2. *The EDS analysis of atomic percent of CIS film after increasing the deposition time for 1200 s and 1450 s of Se*

at. % of Cu	at. % of In	at. % of Se	Cu/In ratio	Se deposition time, s
29.1	24.7	46.2	1.17	1200
26.6	23.2	50.2	1.14	1450

Figure 7 shows the SEM images of Samples a, b, and c. It is observed that increasing the annealing temperature can increase the grain size of the CIS film. The voids on the surface were also reduced with the higher temperature. A higher annealing temperature could provide sufficient energy for the CIS film to diffuse and react more completely. The deposition condition was adjusted in Samples d, e, and f to achieve a proper Cu-In ratio. The EDS results are shown in Table 3. The Cu-In ratio of Sample f achieved nearly 1:1.

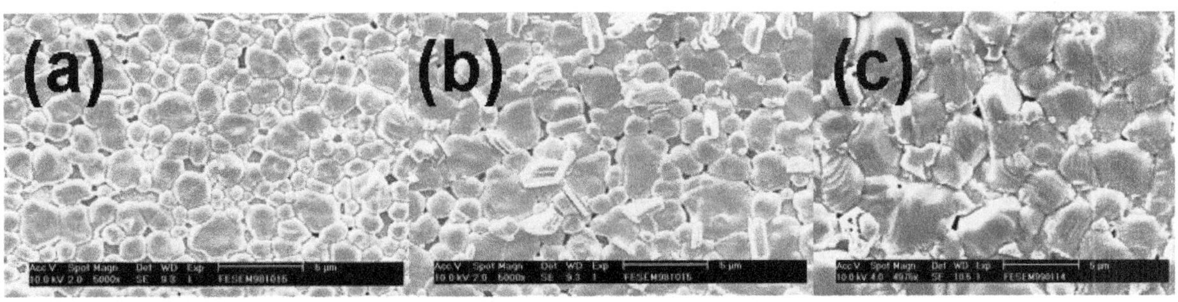

Fig. 7. *SEM images of CIS film with the Se layer deposited on the annealed Cu/In surface at 4 mA cm^{-2} current density for 1200 s after annealing at (a) 600 °C (b) 630°C for 3min.*

Table 3. *The EDS analysis of atomic percent in samples d, e, and f.*

	at. % of Cu	at. % of In	at. % of Se	Cu/In ratio
Sample d	25.6	24.3	50.1	1.05
Sample e	25.4	22.8	51.8	1.11
Sample f	24.9	24.5	50.6	1.01

Figure 8 shows the SEM images of Sample f after annealing at 630 °C for 5 min and 10 min. The grain of CIS film became large and dense, and the voids almost disappear in Sample f. The cross-section images of Sample f are shown in Figure 9. Large dense grains could be clearly observed and the thickness of the CIS approached 2 μm.

Fig. 8. *SEM images of sample f, annealed at 630 ℃ for (a) 5 min (b) 10 min.*

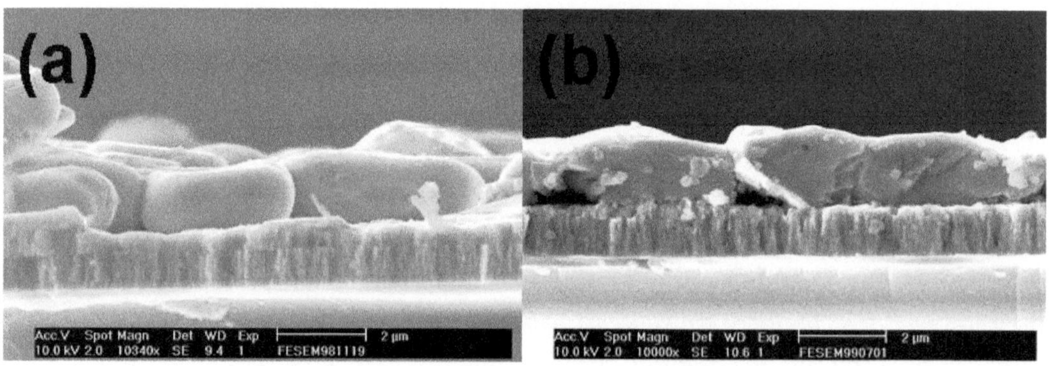

Fig. 9. *SEM images of cross-section of sample f, annealed at 630 °C for (a) 5 min, (b) 10 min.*

Figure 10 shows the XRD patterns of the CIS films with various annealing conditions. The main (112) peak confirmed the existence of chalcopyrite CIS. The (112) main peaks of the CIS films annealed at 550 °C for 5 min and at 600 °C for 5 min contained some small peaks of impure phase. However, the peaks were too close to be differentiated clearly by XRD, but the Raman analysis found $Cu_{1-x}Se_x$ or In_xSe.

A Raman spectrum was employed to analyse the film composition. Figure 11 shows the Raman spectra of the CIS films with various annealing conditions. Cu_2Se was found in the samples with the lower annealing temperature or shorter annealing time. This is because Cu_2Se was formed before the synthesis of CIS. If the annealing does not provide sufficient energy or reaction time, the precursors cannot completely transform to CIS. Figure 12 shows the XRD patterns of the CIS film with the lower annealing temperature. Cu-Se and In-Se were found at 300 °C and CIS (112) was found at 350 °C. This indicates that the precursors would turn into Cu-Se and In-Se binary compounds before the synthesis of the chalcopyrite CIS [10].

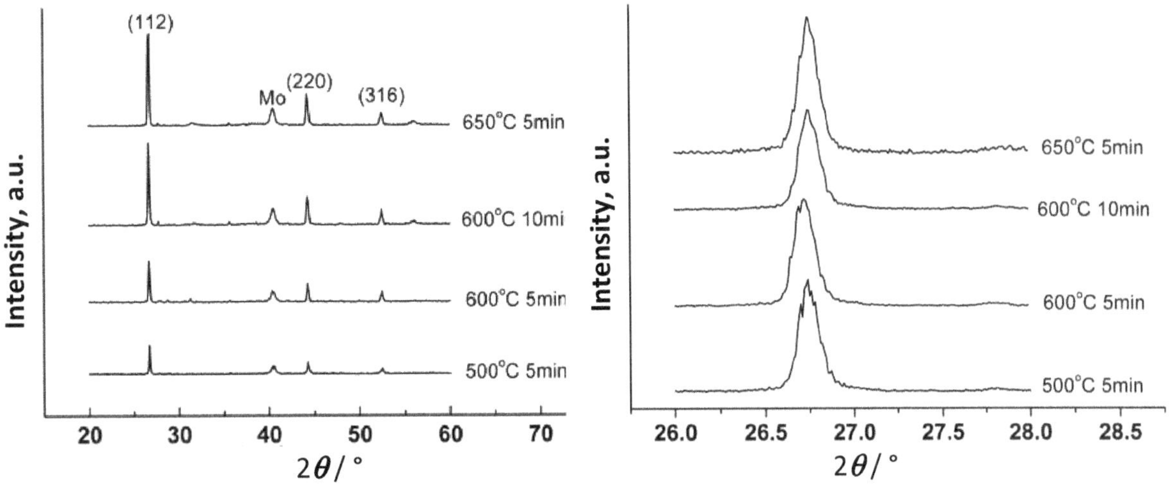

Fig.10. *XRD patterns of CIS films with conditions.*

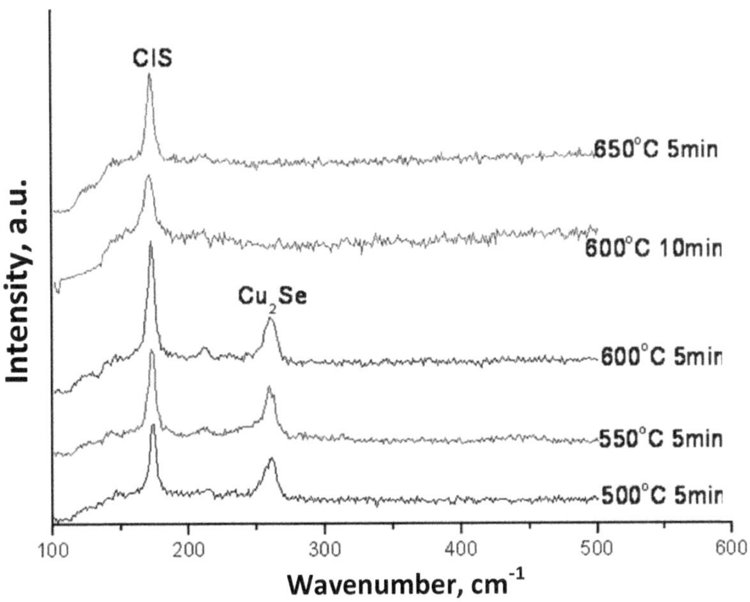

Fig.11. *Raman spectra of CIS films obtained with conditions.*

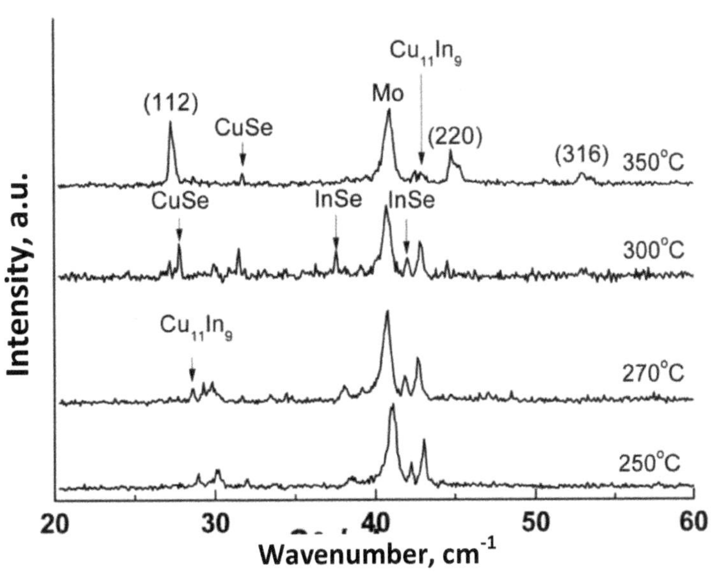

Fig.12. *XRD patterns of CIS films after annealing at low temperature 250 ℃~350 ℃.*

However, synthesized CIS from co-sputtered $Cu_{11}In_9$ can produce high quality film with large grains. In this study, synthesis of the CIS from the electrodeposited Cu-In precursors was investigated. A thermal process was adopted to eliminate the agglomeration phenomenon of electrodeposited In and to form $Cu_{11}In_9$ compound. The electrodeposition conditions of Cu, In and Se, were adjusted to achieve the preferred atomic proportion. However, the annealing temperature of the synthesized CIS from $Cu_{11}In_9$ is critical. The XRD patterns and Raman spectra show that the residue of the Cu_2Se compound is due to an incomplete reaction at lower annealing temperatures. Large dense grains could be grown at 650 °C for 5 min. Finally, we produced a high quality CIS film with large grains from a cheap method of electrodeposition of Cu-In precursors.

Conclusions

Electrodeposition is a cheap and efficient method of producing CIS film. The crystallinity of the co-electrodeposited film is inferior because of the low growing temperature. Synthesizing CIS from co-sputtered $Cu_{11}In_9$ can produce high quality film with large grains. In this study, synthesizing CIS from electrodeposited Cu-In precursors was investigated. A thermal process was adopted to eliminate the agglomeration phenomenon of electrodeposited In to form $Cu_{11}In_9$ compound. The electrodeposition conditions of Cu, In and Se, were adjusted to achieve the preferred atomic proportion. However, the annealing temperature of synthesized CIS from $Cu_{11}In_9$ is critical. The XRD patterns and Raman spectra show that the Cu_2Se compound residue is due to an incomplete reaction at lower annealing temperatures. Large dense grains could be grown at 650 °C for 5 min.

References

[1] K. Siemer, J. Klaer, I. Luck, J. Bruns, R. Klenk, D. Bräunig, *Sol. Energy Mater. Sol. Cells* **67** (2001) 159.

[2] J. Klaer, I. Luck, A. Boden, R. Klenk, I. Gavilanes Perez, R. Scheer, *Thin Solid Films* **432** (2003) 534.

[3] D. Lincot, J. F. Guillemoles, S. Taunier, D. Guimard, J. Sicx-Kurdi, A. Chaumont, O. Roussel, O. Ramdani, C. Hubert, J. P. Fauvarque, *Sol. Energy* **77** (2004) 725.

[4] R. N. Bhattacharya, J. F. Hiltner, W. Batchelor, M. A. Contreras, R. N. Noufi, J. R. Sites, *Thin Solid Films* **361** (2000) 396.

[5] K. Singh, R. Tanveer, *Sol. Energy Mater. Sol. Cells* **36** (1995) 409.

[6] F.O. Adurodija, J. Song, S.D. Kim, S.H. Kwon, S.K. Kim, S.H. Yoon and B.T. Ahn, *Thin Solid Films* **338** (1999) 13.

[7] F. Adurodija, J. Song, S. K. Kim, K. H. Kang and K. H. Yoon, *Journal of the Korean Physical Society* **32** (1998) 87.

[8] T. L. Chu, Shirley S. Chu, J. Yue, *Solid-State Science and Technology* **131** (1984) 2182-2185.

[9] O. Volobujeva, M. Altosaar, J. Raudojaa, E .Mellikov, M. Grossberg, L. Kaupmees, P. Barvinschi, *Solar Energy Materials & Solar Cells* **93** (2009) 11.

[10] S. D. Kim and H. J. Kim, *Journal of the Korean Physical Society* **35** (1999) 403.

21

The effect of cetyltrimethylammonium bromide on size and morphology of ZnO and CuO

Anantha N. Subba Rao, Venkatesha T. Venkatarangaiah✉

Department of P.G. studies and Research in Chemistry, School of Chemical Sciences, Kuvempu University, Shankaraghatta-577451, Karnataka, India

✉*Corresponding author*

Abstract

The nanoparticles (NP) ZnO and CuO were synthesized by electrochemical-thermal method. The influence of cetyltrimethylammonium bromide (CTAB) on size and morphology of NP was evaluated. They were characterized by powder X-ray diffraction spectroscopy (XRD), scanning electron microscopy (SEM), UV-Visible absorption spectroscopy. The average crystallite size and the average grain size of NP decreased with CTAB concentration. The CTAB significantly affected the morphology of CuO and ZnO NP. The regular spindle shape of CuO transformed into irregular spherical shape and the homogeneity in the morphology of spherical ZnO NP was lost with increase in CTAB concentration. The effect of morphology and size of ZnO on its photocatalytic activity was evaluated by subjecting methylene blue (MB) dye to photocatalytic degradation under the irradiation of UV light. The color removal of MB dye during electrolysis was monitored by UV-Visible spectroscopy. The highest photocatalytic activity was noticed for ZnO 10 mM CTAB.

Keywords

Electrolysis; Photocatalyst; Morphology; Methylene blue; Degradation; Surfactant; Nanoparticles; Color removal; Absorbance; Band gap

Introduction

The TiO_2, SnO_2, Fe_2O_3, PdO, WO_3, Ga_2O_3, In_2O_3, ZrO_2, CuO and ZnO are amongst the extensively studied transition metal-oxide semiconducting nanomaterials and have potential applications in optics, optoelectronics, microelectronics, biosensors, gas sensors, chemical sensors, switching, magnetic, luminescent, electrical and acoustic devices, solar cells, catalysis and powder metallurgy [1-4,5]. The CuO and ZnO nanomaterials are particularly of special interest as they are cheap, easy to prepare by simple, low cost methods at normal temperatures [6]. They exhibit diverse nanostructure configurations and show superior performance in many applications.

CuO is a p-type semiconductor with band gap of 1.8 to 2.5 eV [2,7]. It is an excellent solar absorber with low emittance, which makes it a suitable candidate for solar cells as thermal collector [8]. The suspension of CuO NPs in the heating or cooling fluids substantially increases the thermal conductivity [9]. CuO is highly responsive in non-enzymatic glucose sensing and H_2S gas sensing [1,10-12]. CuO on ceria shows high catalytic activity for the oxidation of NO, CO and dehydrogenolysis of CH_4 [13].

ZnO on the other hand, is extensively studied and widely recognized for its photocatalytic activity. ZnO, with a high surface reactivity owing to large number of active sites, has emerged to be an efficient photocatalyst [14,6]. The nanocrystalline ZnO is non-toxic and has the ability to kill or restrain bacteria [4]. It has a wide band gap of 3.2 eV and large exciton binding energy of 60 meV, which makes it a preferable material in low-voltage and short wavelength optoelectronic devices such as light emitting diodes and diode lasers [5]. The applicability of ZnO is widening with the extending knowledge of its physico-chemical properties. ZnO nanomaterials are also used in solar cells, luminescent, electrical and acoustic devices, chemical sensors, catalysis, electronics, gas sensor devices, optoelectronics, transducers [3,4].

The synthesis of ZnO and CuO thus acquires special interest. The common chemical synthetic methods adopted in the preparation of these nanomaterials are precipitation, sol-gel, hydrothermal, solvothermal, solution combustion, thermal decomposition, simple hydrolysis, wet-chemical, microemulsion, non-microemulsion, microwave assisted solvothermal, ultrasonic radiation precipitation, mechanical milling, electrochemical, chemical vapor deposition (CVD), laser vaporization, exfoliation, laser ablation methods and so on [2]. However, these methods involve organic solvents, rigorous reaction conditions, demand sophisticated instruments and few of these techniques require longer time duration affording less yield than expected [4].

The organic solvents can be replaced completely with aqueous medium using electrochemical method. Electrochemical method is fast, cheap, and easy to control and provides appreciable yields even at normal temperature and mild conditions. It was Reetz and co-workers who proposed the sacrificial anode electrochemical route for the synthesis of nano metal clusters and colloids and showed that it is possible to control the particle size by adjusting the applied current density [15-17]. Gao-Qing *et al.* demonstrated that the size and shape of the synthesized CuO nanocrystals could be controlled by changing the solvent system [12]. Mahamuni and co-workers prepared ZnO quantum dots using the sacrificial anode electrochemical method [18]. The ZnO [19], Cu_2O [20], CuO [21], Au/AgI [22] have been successfully synthesized by electrochemical method. We have recently synthesized hexagonal shaped CuO NP in bulk [2], cubic shaped Zn_2SO_4 [23] and large size ZnO NP [3,4] by hybrid sacrificial anode electrochemical-thermal route.

There are many factors to be monitored in the electrochemical synthesis to control the size and shape of NPs. The current density, distance between the electrodes, solvent system, temperature, supporting electrolyte and stabilizing agent are the most critical parameters which dictate the size and morphology of resulting NPs. The role of stabilizing agent is to bind to the surface of NP, decrease the surface energy, control growth and shape evolution and prevent the agglomeration of NPs [24]. Cetyl trimethylammonium bromide (CTAB) a cationic surfactant is the most commonly used stabilizing agent in controlling the shape and size of NPs [24-29]. Oscar *et al.* [24] reported that in the synthesis of gold nanorods, CTAB dictates the crystal growth along the faces by binding to the surfaces along the sides of nanorod. Reetz and co-workers have demonstrated the role of surfactant (CTAB) in the formation of nanoclusters and nanocolloids [15-17]. In this paper, we have systematically studied the effect of CTAB concentration in the electrolyte bath, on the morphology and size of ZnO and CuO metal oxide NPs prepared by hybrid sacrificial anode electrochemical-thermal route. Furthermore, the photocatalytic degradation of methylene blue dye on ZnO NP was studied in order to evaluate the effect of size and shape of the NPs on its photocatalytic activity.

Experimental procedure

Materials and synthesis of ZnO and CuO NP

Zinc and copper metal plates (99.99 % pure) were purchased from Sisco research laboratories, Mumbai. AR grade $NaHCO_3$, $NaNO_3$, CTAB and methylene blue (MB) dye purchased from S. D Fine Chemicals Ltd. India, were used without further purification. The electrolyte solution was prepared in Millipore water (specific resistance > 18 MΩ at 25 °C, Millipore Elix 3 water purification system, France). The electrolysis was carried out under constant current supply from potentiostat / galvanostat source (Model PS-618 chemilink systems, Navi Mumbai, India).

The experimental procedure for the electrochemical-thermal synthesis of ZnO NP was followed as mentioned in our previous work [4]. The two Zn metal plates with dimension 4 x 4 x 0.1 cm^3, were dipped in 1 M HCl for 30 seconds and washed thoroughly with Millipore water. These plates were then immersed in a rectangular electrolytic glass cell containing 250 cm^3 of 60 mM $NaHCO_3$ with or without CTAB. Three different concentrations of CTAB (5 mM, 10 mM and 15 mM) were used. The NP prepared in presence of 'x' mM concentration of CTAB is hereafter represented as MO/x, where M is Zn or Cu. The distance between the two electrodes was maintained to be 5 cm. The electrolysis was carried out for 90 minutes under constant current density of 4.9 mA cm^{-2} with constant stirring at the rate of 500 rpm. The electrolytic bath temperature was maintained around 26 ± 1 °C. The pH of the electrolytic bath was noted before and after electrolysis. The resulting colloidal solution was filtered and the solid residue thus obtained was calcined at 300 °C for 1 hour. The same procedure was followed in the preparation of CuO NP, except that the electrodes were pretreated Cu plates and the electrolyte was 60 mM $NaNO_3$. As a pretreatment, the Cu metal plates were dipped in 5 % HNO_3 for 1 minute.

Characterization of NP

The crystallite size of the NPs was evaluated by powder X-ray diffraction (XRD) pattern using X'pert Pro diffractometer (Phillips, Cu-K_α radiation, λ_{Cu}= 1.5148 Å) working at 30 mA and 40 kV recorded in the 2θ range between 10° and 90° (scan rate 1° min^{-1}). The Scherrer equation was made use in estimating the average crystallite size. The scanning electron microscope (Philips XL 30) fitted with an energy dispersive X-ray (EDAX) analyzer recorded the morphology and composition of the synthesized NP. The band gap of the synthesized NPs was evaluated by the UV-Visible absorption spectrum of solid ZnO and CuO NPs using UV-Visible spectrophotometer (HR 4000 UV-Vis spectrophotometer, UV-Vis-NIR light source, DT-MINI-2-GS, Jaz detector).

Photocatalytic degradation of MB dye

To evaluate the photocatalytic activity of ZnO NP, 50 mg L^{-1} MB dye solution was prepared using Millipore water. 30 mg ZnO nanoparticle was added to 30 mL MB solution and stirred for 2 hours in dark to establish equilibrium between adsorption/desorption of MB on NPs. Then, this solution was exposed to continuous flash of UV light for duration of 10 minutes. A high pressure mercury vapor lamp (HPML) of 125 W (Phillips, India) was jacketed in a quartz tube provided with inlet and outlet for water circulation to avoid rise in temperature of the solution. The UV-lamp radiated predominantly at 365 nm (3.4 eV) and photon flux of 5.8×10^{-6} mol of photons sec^{-1}. 0.5 mL of the irradiated sample was removed, diluted with 1 mL Millipore water and subjected to centrifugation for 5 minutes. The supernatant liquid was taken for UV-Visible spectroscopic analysis. After 2 minutes, the UV light irradiation was again continued for 10 minutes and then sample was removed. The total time of UV light irradiation on the sample was 60 minutes.

Results

XRD analysis

The XRD patterns obtained for ZnO NPs prepared in presence of different concentrations of CTAB are given in Fig. 1.

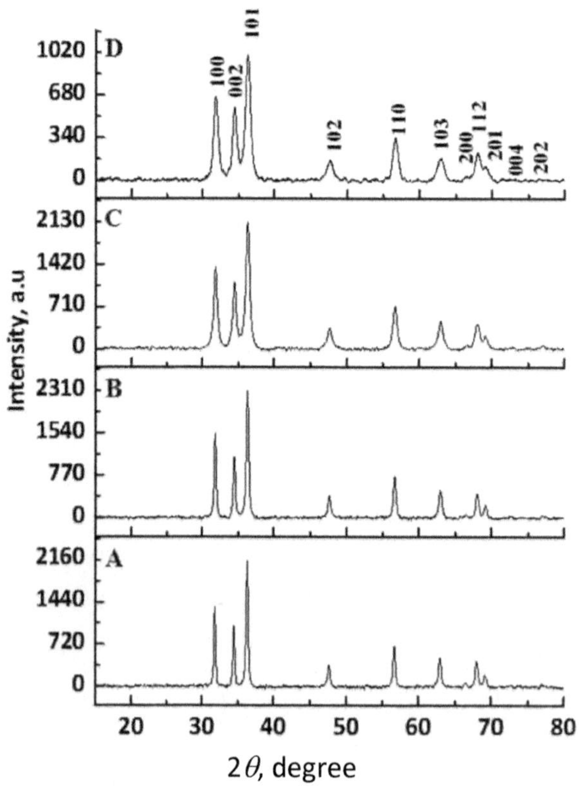

Figure 1. *The XRD pattern of A) ZnO/0; B) ZnO/5; C) ZnO/10; D) ZnO/15*

The positions and intensities of the diffraction peaks of ZnO NPs are in good agreement with the literature values of hexagonal phase ZnO (JCPDS card No. 36-1451) with primitive lattice structure. The crystallite size of the NPs was calculated from full wave half maximum intensity (FWHM) of each peak using Scherrer's formula and the average crystallite size was determined.

$$D = \frac{K\,\lambda}{\beta\,\cos\theta}$$

In the above formula, *D* is the diameter of the crystallite size, *K* - the shape factor (the typical value is 0.9), λ - the wavelength of the incident beam, β - the broadening of the diffraction line measured in radians at FWHM and θ is the Bragg's angle. The crystallite size of ZnO NPs calculated using the XRD data is given in Table 1.

Table 1. *The average crystallite size of ZnO NPs prepared in presence of CTAB*

Zinc oxide NP	Crystallite size, nm									
	Family of crystallographic planes {hkl} and corresponding 2θ values									Average
	{100}	{002}	{101}	{102}	{110}	{103}	{200}	{112}	{201}	
	31.8°	34.5°	36.3°	47.6°	56.7°	62.9°	66.4°	68.1°	69.2°	
ZnO/0	44	98	45	28	106	55	23	40	47	**48**
ZnO/5	44	44	41	23	66	27	35	31	24	**34**
ZnO/10	25	31	23	16	39	18	24	21	24	**23**
ZnO/15	16	20	27	13	11	20	23	20	24	**18**

The crystallite size of NP markedly decreased with increase in concentration of CTAB in the electrolyte bath. The width of the peaks appreciably increased suggesting the reduction in the crystallinity of ZnO with increase in CTAB concentration.

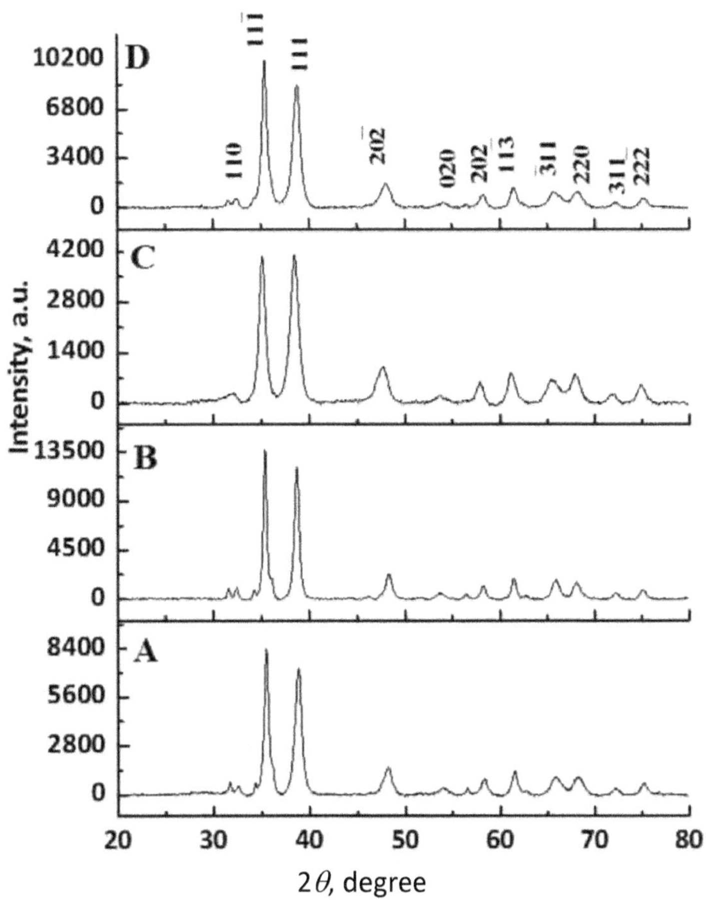

Figure 2. *The XRD pattern of A) CuO/0; B) CuO/5; C) CuO/10; D) CuO/15*

The XRD patterns of synthesized CuO (Fig. 2) are in close match with the XRD pattern of JCPDS card no. 48-1548 with monoclinic (tenorite) phase crystallite structure. The crystallite size of CuO NPs calculated using the XRD data is given in Table 2. The crystallite size of CuO NPs also decreased with increase in CTAB concentration.

Table 2. *The average crystallite size of CuO NP prepared in presence of CTAB.*

Copper oxide NP	Crystallite size, nm											
	Family of crystallographic planes {hkl} and corresponding 2θ values											Average
	{110} 32.1°	{111} 35.1°	{111} 38.5°	{202} 47.8°	{020} 53.7°	{202} 57.9°	{113} 61.1°	{311} 65.3°	{220} 68.1°	{311} 71.9°	{222} 74.9°	
CuO/0	22	49	44	14	15	24	24	25	11	22	18	**31**
CuO/5	36	26	36	16	14	20	20	20	30	26	15	**26**
CuO/10	09	22	22	13	10	24	15	18	18	13	13	**16**
CuO/15	18	33	11	17	16	14	33	17	09	12	22	**19**

SEM/EDAX analysis

The changes in the morphology of ZnO (Fig. 3) and CuO (Fig. 4) NPs with CTAB concentration are evident from the SEM images. From Fig. 3 and Fig. 4, it can be noticed that the change in the morphology of both CuO and ZnO crystals is not obvious within 5 mM CTAB, but the tendency from regular shapes to irregular ones increases with the increase of CTAB concentration.

The spherical shape of ZnO crystals was though retained, the irregularity in the shape increased with increase in CTAB concentration. The shape of CuO crystals changed from nanospindles (Fig. 4A and Fig. 4B) to irregular nanospheres (Fig. 4C and Fig. 4D). Similar changes in the morphology of CuO crystals were observed by Gao-Qing Yuan *et al.* [12] on varying the solvent system. The average grain size of ZnO and CuO NPs as depicted from SEM images is given in Table 3.

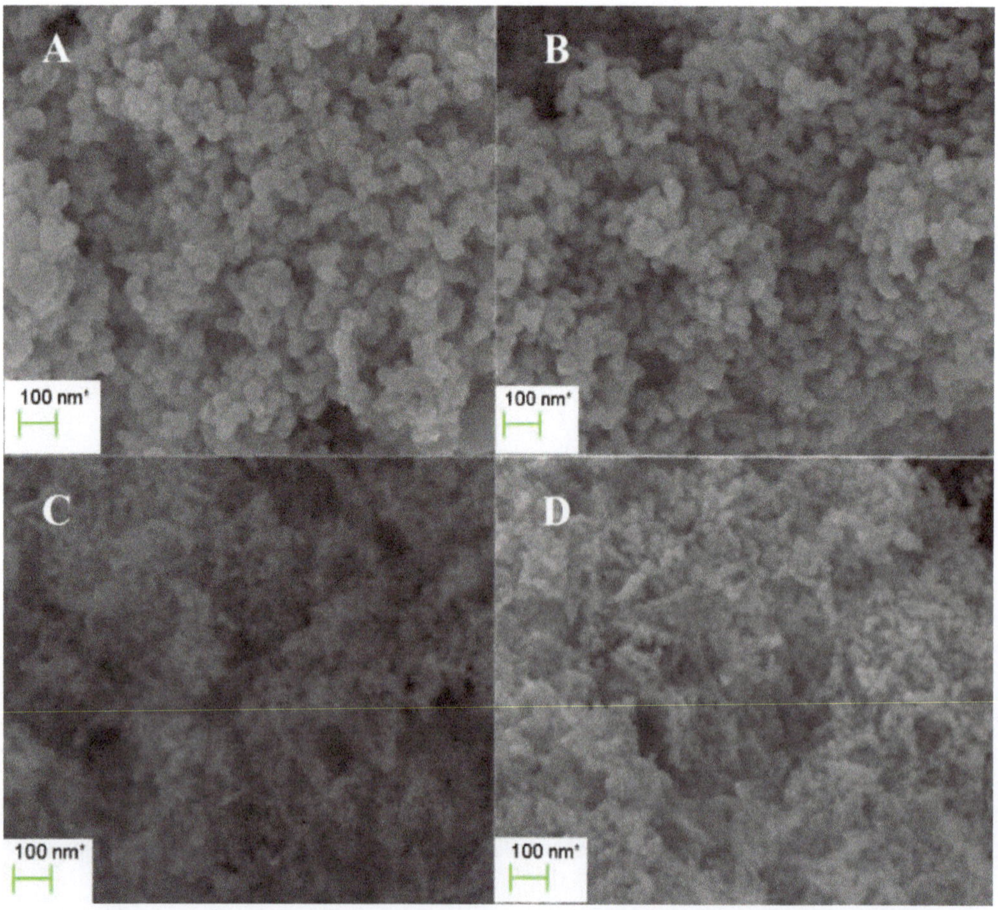

Figure 3. *SEM images of ZnO NP; A) ZnO/0; B) ZnO/5; C) ZnO/10; D) ZnO/15*

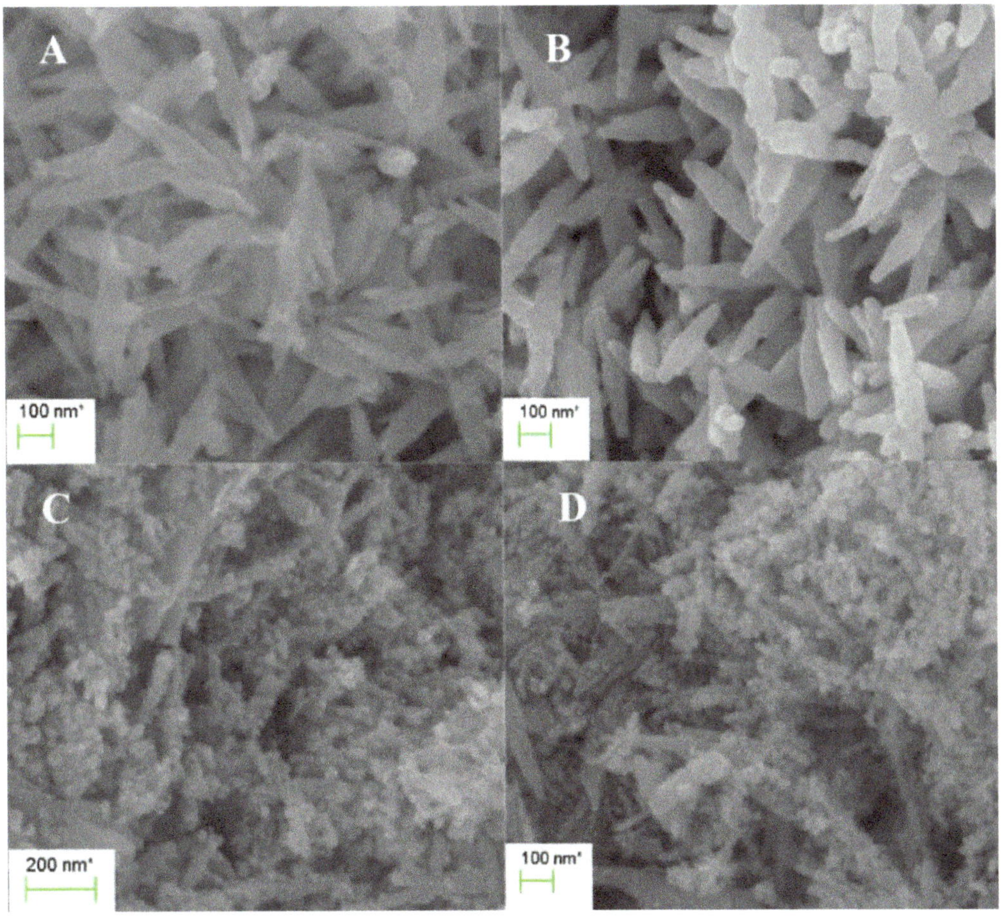

Figure 4. *SEM images of CuO NP; A) CuO/0; B) CuO/5; C) CuO/10; D) CuO/15*

Table 3. *The average grain size of ZnO and CuO NPs prepared in presence of CTAB*

Metal oxide NP	Average grain size, nm			
	0 mM CTAB[a]	5 mM CTAB	10 mM CTAB	15 mM CTAB
ZnO	32-42	31-40	30-69	25-29
CuO	60-65	85-95	18-40	27-64

[a] The CTAB concentration in electrolyte bath

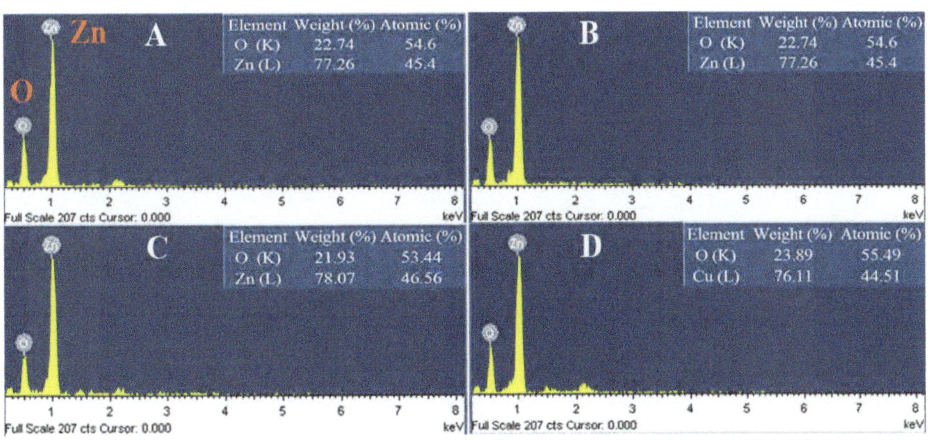

Figure 5. *EDAX analysis of ZnO NP; A) ZnO/0; B) ZnO/5; C) ZnO/10; D) ZnO/15*

The shape and size of CuO/0 and CuO/5 nanospindles was homogeneous. The length of the nanospindles ranged between 430 nm to 580 nm and the rigidity enhanced from CuO/0 to CuO/5.

The CuO/10 and CuO/15 NPs showed irregular morphologies (nanospindle, nanosphere, nanoplate). In case of ZnO NPs, the homogeneity in the spherical shape was lost as the CTAB concentration was increased in the electrolyte bath. There was only slight reduction in the grain size from ZnO/0 to ZnO/15.

The stoichiometry of metal ion and oxygen in the synthesized ZnO and CuO nanomaterials was determined by EDAX analysis (Fig. 5 and Fig. 6). No peaks corresponding to impurities were found suggesting that the synthesized ZnO and CuO nanomaterials are pure.

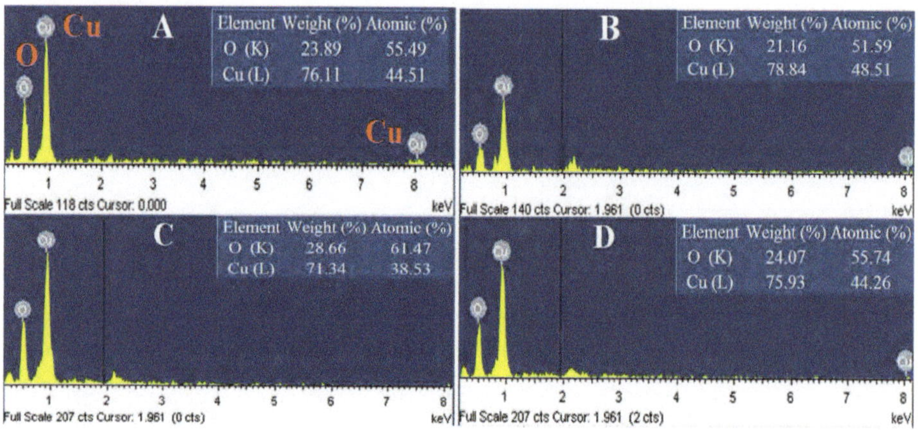

Figure 6. *EDAX analysis of CuO NP; A) CuO/0; B) CuO/5; C) CuO/10; D) CuO/15*

UV-Visible spectroscopy

To better understand the optoelectronic properties of NP, the band gap energy was calculated from their absorption curves using Tauc plots. The UV-Visible absorption spectra of ZnO and CuO NPs are given in Fig. 7A and Fig. 7B respectively. Firstly, the absorption coefficient (α) was calculated using Eq. (1):

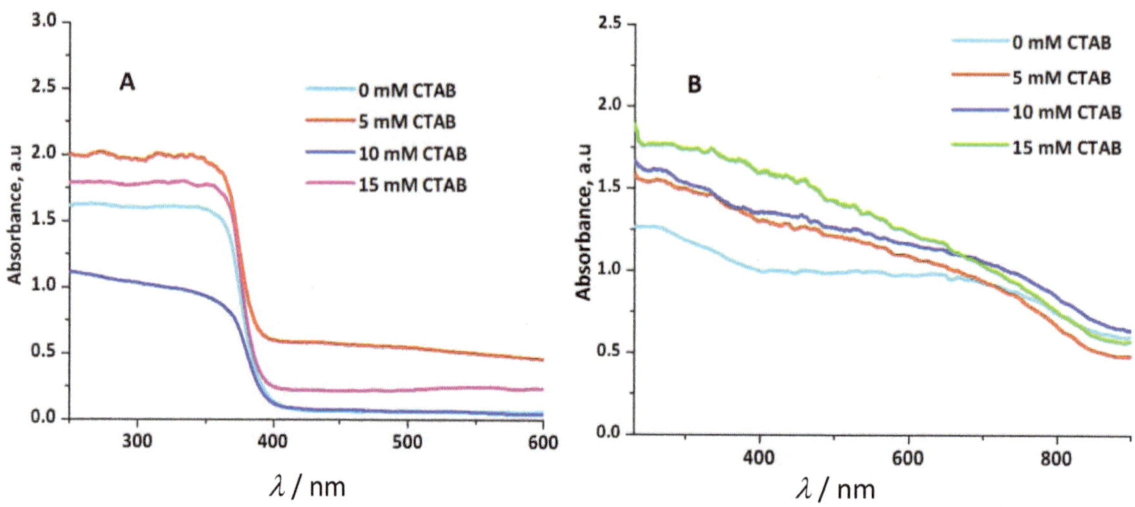

Figure 7. *UV-Visible absorption spectra of A) ZnO NPs and B) CuO NPs synthesized in presence of different concentrations of CTAB*

$$\alpha = \frac{2.303\,A}{d} \qquad (1)$$

where *A* is absorbance, *d* is the thickness of sample. To determine the band gap of NP, a plot of $(\alpha h\nu)^n$ *vs.* *Eg* was made (Tauc plot). The value of 'n' is dependent on the mode of electronic transition; n = 2 for direct band transitions and n = 1/2 for indirect transitions. For both ZnO and CuO, it is known that the electronic transition is direct. Therefore, Eq. (2) is valid for both ZnO and CuO NP:

$$(\alpha hv)^2 = B\,(hv - Eg) \tag{2}$$

where, *B* is a constant, *hv* is the photonic energy. The band gap (*Eg*) was obtained by extrapolating the linear part of the graphics to the axis of the abscissa [30].

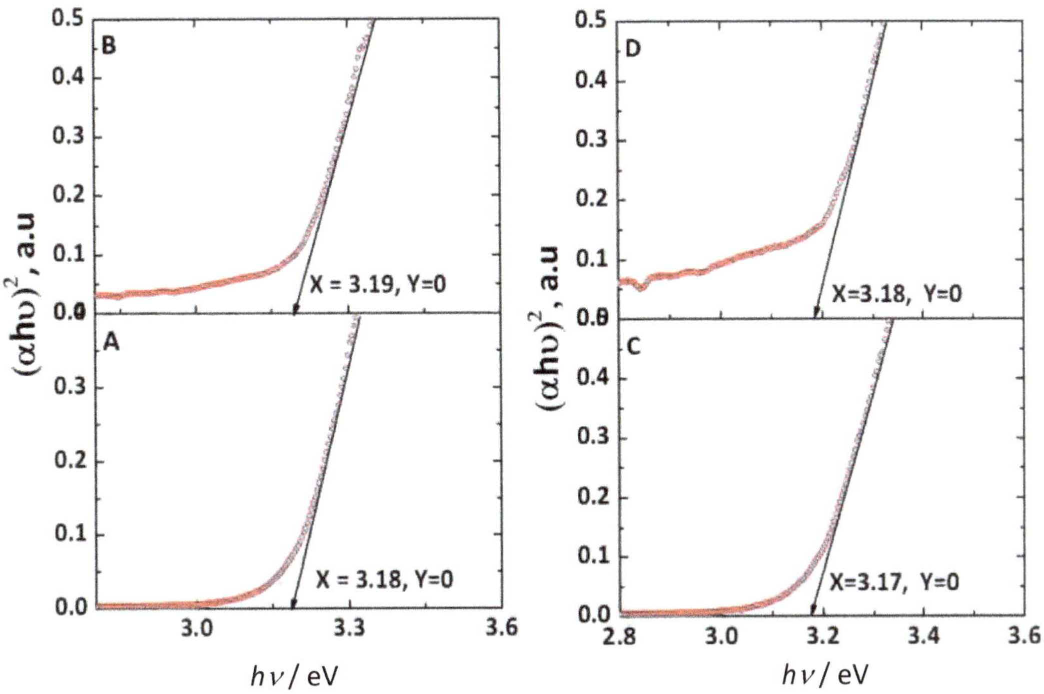

Figure 8. *Tauc plots for the evaluation of band gap of samples A) ZnO/0; B) ZnO/5; C) ZnO/10; D) ZnO/15*

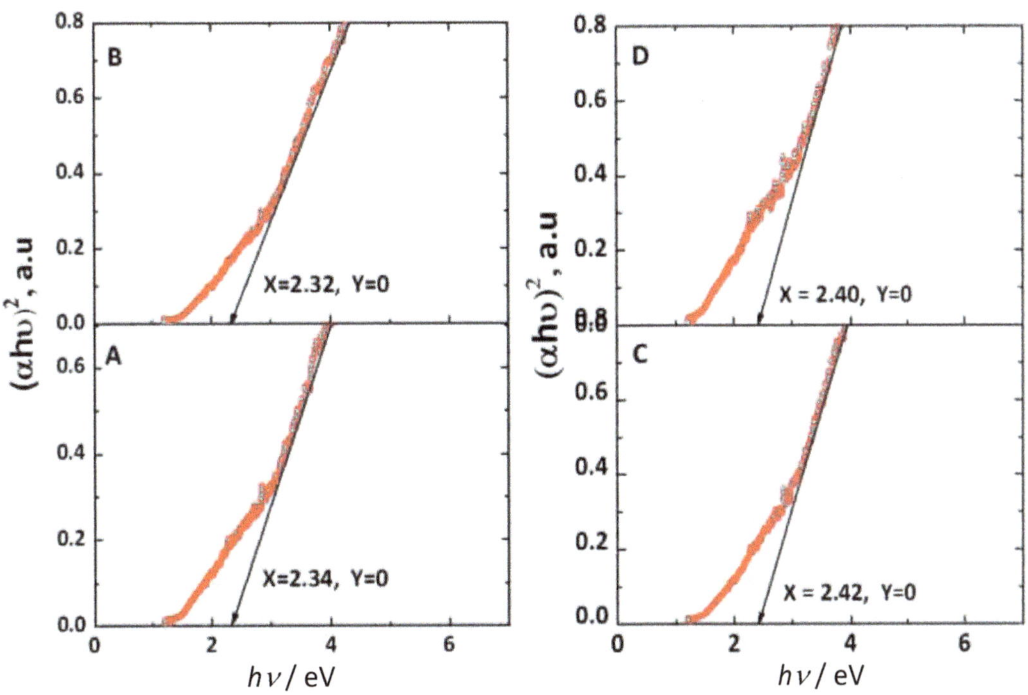

Figure 9. *Tauc plots for the evaluation of band gap of samples A) CuO/0; B) CuO/5; C) CuO/10; D) CuO/15*

Table 4. *Eg values for different ZnO and CuO NPs obtained from Tauc plot*

c (CTAB) / mM	*Eg* / eV	
	ZnO	CuO
0	3.18	2.34
5	3.19	2.32
10	3.17	2.42
15	3.18	2.40

The Tauc plots for different ZnO and CuO NPs has been presented in Fig. 8 and Fig. 9 respectively and corresponding band gap values have been summarized in Table 4. The band gap of CuO NPs increased with increase in the concentration of CTAB in the electrolytic bath during synthesis and it was found to be in between 2.32 – 2.42 eV, which is higher than that reported by Wang *et al.* [31]. However, such notable variation in the band gap of ZnO NPs was not observed. The average band gap of ZnO NP was 3.18 eV; where the literature reports a 3.2 eV band gap [3].

Photocatalytic degradation of MB dye

The photocatalytic degradation of methylene blue dye was monitored by UV-Visible spectroscopy. The variation in the absorption intensity of the methylene blue dye on UV irradiation in presence of 30 mg of ZnO NPs is presented in Fig 10.

The degradation of methylene blue was expressed in terms of % color removal by recording the changes in absorption intensity at the characteristic peak position (661 nm) in the visible region.

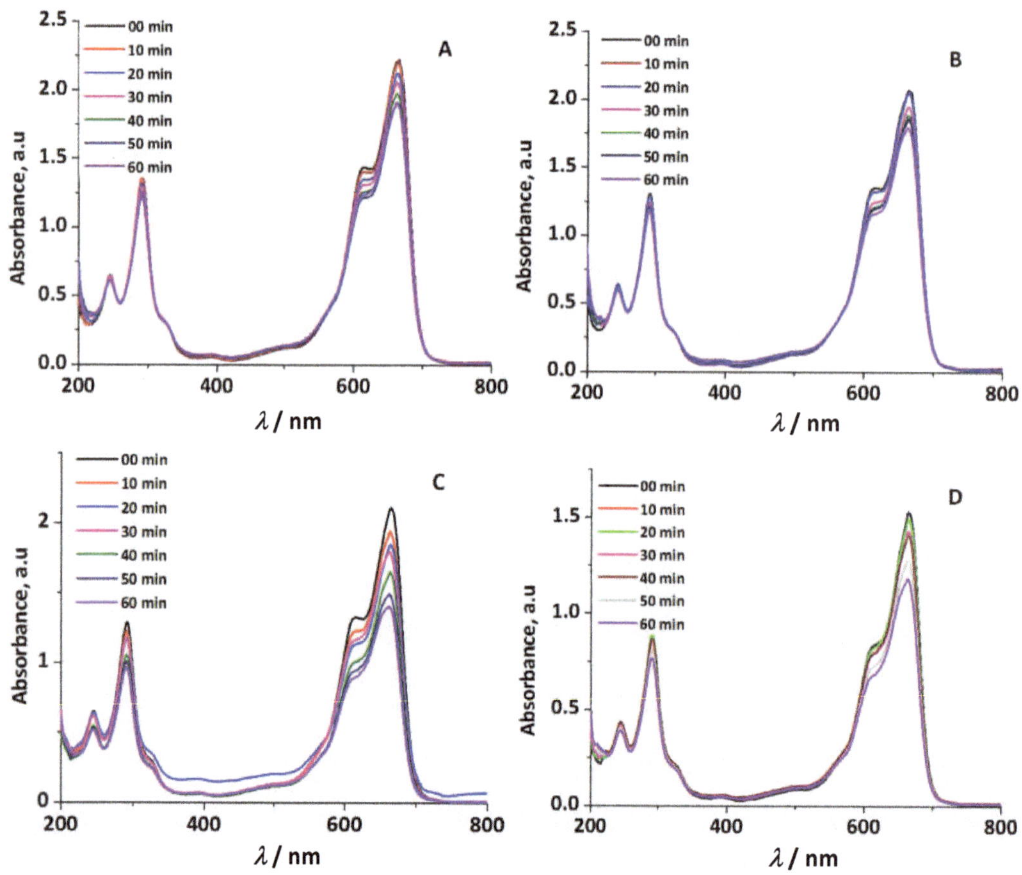

Figure 10. *The variation in the UV-Visible absorption intensity of methylene blue dye with time on electrolysis. UV irradiation in presence of 30 mg ZnO NPs; A) ZnO/0; B) ZnO/5; C) ZnO/10; D) ZnO/15*

The Eq. (3) was used to calculate the color removal:

$$\text{color removal, \%} = \frac{[A_o - A_t]}{A_o} \times 100 \qquad (3)$$

where, A_o and A_t are the absorbance at 661 nm at 0 minutes and ***t*** minutes respectively. The color removal of methylene blue achieved in presence of ZnO photocatalyst was in the following order; ZnO/10 > ZnO/15 > ZnO/0 = ZnO/5. The color removal achieved after 60 minutes of UV irradiation was evaluated and given in Table 5.

Table 5: *Photocatalytic degradation of methylene blue dye*

ZnO NP	color removal, %[a]
ZnO/0	13
ZnO/5	13
ZnO/10	33
ZnO/15	22

[a] color removal achieved after 60 minutes irradiation

The CuO was also tested for its photocatalytic activity. The CuO NP under the irradiation of UV light (predominantly at 365 nm) was ineffective as a photocatalyst to degrade the MB dye. The color removal of MB achieved in presence of CuO NP and UV irradiation was insignificant.

Discussion

It is crucial to understand the mechanism of formation of ZnO and CuO in order to control their size and shape. The electrolytes used in the electrolyte bath should enhance the conductivity and assist the formation of metal hydroxides. The $NaHCO_3$ and $NaNO_3$ have been used in the synthesis of ZnO and CuO NP respectively [2-4,12]. These electrolytes facilitate the formation of metal hydroxides along with enhancing the conductivity of the electrolyte bath. The following reactions take place during the electrolysis which leads to the formation of ZnO NP in the electrolyte bath containing $NaHCO_3$ electrolyte [Eqs. (4) to (7)];

$$Zn \rightarrow Zn^{2+} + 2e^- \text{ (on anode), } E° = 0.76 \text{ V} \qquad (4)$$

$$NaHCO_3 + H_2O \rightarrow NaOH + CO_2 \text{ (in electrolyte solution)} \qquad (5)$$

$$Zn^{2+} + 2^-OH \rightarrow Zn(OH)_2 \text{ (in electrolyte solution)} \qquad (6)$$

$$Zn(OH)_2 \rightarrow ZnO + H_2O \qquad (7)$$

In presence of $NaNO_3$, the CuO is generated according to the following reactions [12] [Eqs. (8) to (11)];

$$Cu \rightarrow Cu^{2+} + 2e^- \text{ (on anode), } E° = 0.34 \text{ V} \qquad (8)$$

$$NO_3^- + H_2O + 2e^- \rightarrow NO_2^- + 2^-OH \text{ (on cathode), } E° = 0.1 \text{ V} \qquad (9)$$

$$Cu^{2+} + 2^-OH \rightarrow Cu(OH)_2 \text{ (in electrolyte solution)} \qquad (10)$$

$$Cu(OH)_2 \rightarrow CuO + H_2O \qquad (11)$$

The mechanism of NP crystal growth is usually explained as oriented aggregation in the electrolyte solution. The metal ions generated at the anode react with the ^-OH ions in the electrolyte bath to form corresponding nano-sized amorphous metal hydroxide precipitate

($Cu(OH)_2$ or $Zn(OH)_2$). A part of the precipitate undergoes dehydration to give metal oxide in the electrolyte bath (CuO or ZnO) [2-4]. However, complete conversion of all metal hydroxide into metal oxide is ensured by calcinating the sample at elevated temperatures (in this case, 300 °C). The metal oxide thus generated acts as seed for further directional aggregation of metal oxides to form precisely oriented nanostructures [12].

The advantage in electrochemical synthesis is that the precursor metal ion is slowly released from the anode metal. The current density determines the number of ions released per unit time and their availability in the electrolyte bath for further reaction. However, in chemical methods (like solvothermal and precipitation) the entire precursor metal ion is made available for reaction. The nucleation and crystal growth are the two processes which decide the particle size. If the nucleation is fast, the crystal growth is hindered and the particle size eventually remains small. These two processes occur at controlled rate in electrochemical method. The crystal growth and shape evolution is controlled by the stabilizing agent when it is present in the electrolyte bath. In this case, CTAB binds to the surface of the nuclei and prevents agglomeration or dictates the orientation of incoming species [12]. In the present study, the concentration of CTAB was more than 6 times its critical micelle concentration (CMC) *i.e.*, 0.8 mM [24]. The CTAB micelles existing at this concentration of CTAB tend to act as template for the shape evolution of NPs. The effect of CTAB concentration on the morphology and size of both ZnO and CuO NPs is clearly evident from the SEM images and the crystallite size calculated using XRD data. As mentioned earlier, from the SEM images (Fig. 3 and Fig. 4), the change in morphology of both CuO and ZnO crystals is not obvious within 5 mM CTAB. But, the tendency from regular shapes to irregular ones increased beyond the CTAB concentration ≥ 10 mM. The morphology of CuO NPs was much affected by the CTAB concentration as compared to ZnO NPs. The spindle shape of CuO NPs increased in length and gained rigidity in presence of 5 mM CTAB as noticed in the SEM images. However, when the CTAB concentration was raised to 10 mM and 15 mM, the spindle shape of CuO NPs transformed into irregular spherical shape with reduced grain size. On the other hand, the ZnO NPs retained their spherical shape irrespective of the CTAB concentration. However, the homogeneity in the shape of ZnO NPs was disrupted. The grain size of both ZnO and CuO NPs decreased with increase in CTAB concentration. From these observations, it can be deduced that the minimum concentration of CTAB required to influence the shape and size of ZnO and CuO NP crystal is ≥ 10 mM *i.e.*, more than 10 times its CMC value.

The effect of CTAB concentration on the band gap energy and absorption edge in the UV region for ZnO NPs was insignificant. However, there was a hike of 0.1 eV in the band gap of CuO NPs on increasing the concentration of CTAB from 0 mM to 15 mM. The decrease in size of CuO NPs led to the increment in band gap. The absorption edge in the UV region for CuO NPs altered with the grain size which is intern dependent on the CTAB concentration.

The influence of changes in the morphology and size of ZnO NPs reflected on its photocatalytic activity. The electrons from the valence band (VB) of ZnO NPs leap to the conduction band (CB) on UV light irradiation. Vacancies created in the valence band (holes) are represented as h_{VB}^+ and the excited electrons are given the symbol e_{CB}^-. The photogenerated positive vacancies in the VB have high oxidizing potential. These holes can react with the adsorbed water to produce hydroxyl radical and the photogenerated electrons lead to the formation of other oxidizing agents such as superoxide anions ($\cdot O_2^-$), hydrogenperoxide (H_2O_2), hydroxyl radicals ($\cdot OH$), hydrogen dioxide anion (HO_2^-) and hydroperoxy radicals ($\cdot HO_2$) as given in Eqs. (12) to (18) [7];

$$MO + hv \rightarrow MO\ (h_{VB}^+ + e_{CB}^-) \quad (12)$$

$$MO\ (h_{VB}^{+}) + H_2O \rightarrow MO + \cdot OH + H^{+} \quad (13)$$

$$MO\ (e_{CB}^{-}) + O_2 \rightarrow MO + \cdot O_2^{-} \quad (14)$$

$$MO\ (e_{CB}^{-}) + \cdot O_2^{-} + 2H^{+} \rightarrow MO + H_2O_2 \quad (15)$$

$$MO\ (e_{CB}^{-}) + H_2O_2 \rightarrow MO + \cdot OH + {}^{-}OH \quad (16)$$

$$O_2^{-} + H_2O_2 \rightarrow \cdot OH + {}^{-}OH + O_2 \quad (17)$$

$$O_2^{-} + H^{+} \rightarrow \cdot OH_2 \quad (18)$$

These oxidants are powerful enough to completely incinerate the organics into CO_2 and H_2O. The photocatalytic reaction efficiency increases with increase in surface area available for the adsorption of target molecule. As the size of ZnO NPs decreased with increase in CTAB concentration, the surface area available for the adsorption of dye molecules also increased. The number of MB molecules adsorbed on the ZnO/15 and ZnO/10 NPs was higher than that on ZnO/0 and ZnO/5 NPs. The % color removal of MB dye achieved after 60 minutes UV light irradiation was thus highest with ZnO/10 and ZnO/15 NPs.

Conclusion

The ZnO and CuO NPs of varying size and morphology were successfully synthesized by sacrificial anode electrochemical-thermal method. The crystallite size of both ZnO and CuO reduced with increase in concentration of CTAB in the electrolyte bath during electrolysis. The crystallite size of ZnO NPs was between 48-18 nm and for CuO it varied between 31-19 nm. The ZnO/0 and ZnO/5 NPs exhibited homogeneous size and morphology. The homogeneity of ZnO NPs was lost with increase in CTAB concentration in the electrolyte bath during electrolysis and led to the formation of irregular sized nanostructures of reduced size in presence of 10 mM and 15 mM CTAB. Similar effects were observed for the CuO NPs. The CuO NPs of uniform size and morphology were obtained in the absence of CTAB and in presence of 5 mM CTAB. The rigidity of CuO NPs enhanced in presence of 5 mM CTAB. However, on increasing the CTAB concentration to 10 mM and 15 mM the spindle shape of CuO NPs disrupted to form irregular nanostructures. The CTAB adsorbs on the surface of the metal oxide nuclei and restrains the growth of NPs by reducing the surface energy and preventing further agglomeration. CTAB tends to gain spherical structure in aqueous solution at these concentrations and act like template for the evolution of spherical NPs of irregular morphology. It can be deduced from the results that the minimum concentration of CTAB influencing the evolution of shape and size of the NP is ≥ 10 mM (greater than 10 times the CMC value of CTAB).

The effect of size and shape of ZnO NPs on the absorption edge and band gap was insignificant. The band gap of ZnO NPs lied between 3.17 – 3.19 eV. The band gap of CuO NPs showed considerable changes with size. A hike of 0.1 eV was observed due to the reduction in size of CuO NPs with increase in CTAB concentration. The photocatalytic activity of ZnO NPs was tested for the degradation of MB dye. The ZnO/10 NPs showed comparatively higher photocatalytic activity and a color removal of 33 % was achieved using this NPs with the UV light irradiation for 60 minutes.

Acknowledgement: *This work was supported by University Grants Commission, New Delhi, Govt. of India under the grant Ref: F. No. 41-231/2012 (SR) Dated 16/07/2012]; and Department of Chemistry, Kuvempu University. Authors thank Mr. Manu Kumar K N, Mr. Kiran Kumar K S and Mr. Nagaraja H N for their valuable assistance.*

References

[1] J. Song, L. Xu, C. Zhou, R. Xing, Q. Dai, D. Liu, H. Song, *Appl. Mater. Interfaces.* **5** (2013) 12928–12934.

[2] K.G. Chandrappa, T.V. Venkatesha, *J. Exper. Nanosci.* **8** (4) (2013) 516–532.

[3] K. G. Chandrappa, T. V. Venkatesha, *Nano-Micro. Lett.* **4** (1) (2012)14-24.

[4] K. G. Chandrappa, T. V. Venkatesha, K. Vathsala, C. Shivakumara, *J. Nanopart. Res.* **12** (2010) 2667–2678.

[5] H. Zhang, D. Yang, Y. Ji, X. Ma, J. Xu, D. Que, *J. Phys. Chem. B.* **108** (13) (2004) 3955-3958.

[6] F. Liu, Y. H. Leung, A. B. Djurišić, A. M. Ching, W. K. Chan, *J. Phys. Chem. C.* **117** (2013) 12218–12228.

[7] K. Kannaki, P.S. Ramesh, D. Geetha, *Int. J. Sci. Eng. Res.* **3** (9) (2012) 1-4.

[8] T. Maruyama, *Sol. Energy Mater. Sol. Cells.* **56** (1998) 85-92.

[9] Y .J. Hwang, Y. C. Ahn, H. S. Shin, C. G. Lee, G. T. Kim, H. S. Park, J. K. Lee, *Curr. Appl. Phys.* **6** (2006) 1068–1071.

[10] J. Tamaki, K. Shimanoe, Y. Yamada, Y. Yamamoto, N. Miura, N. Yamazoe, *Sens. Actuators B.* **49** (1998) 121–125.

[11] G. L. Luque, M. C. Rodrìguez, G. A. Rivas, *Talanta.* **66** (2005) 467–471.

[12] G.Q.Yuan, H.F. Jiang, C. Lin, S.J. Liao, *J. Cryst, Growth.* **303** (2007) 400–406.

[13] B. Skårman, T. Nakayama, D. Grandjean, R. E. Benfield, E. Olsson, K. Niihara, L. R. Wallenberg, *Chem. Mater.* **14** (2002) 3686-3699.

[14] S. Baruah, S. K. Pal, J. Dutta, *Nanosci. Nanotechnol-Asia.* **2** (2) (2012) 90-102.

[15] M. T. Reetz, W. Helbig, S. A. Quaiser, *Chem. Mater.* **7** (1995) 2227-2228.

[16] M. T. Reetz, W. Helbig, *J. Am. Chem. Soc.* **116** (1994) 1401-1402.

[17] M. T. Reetz, M. Winter, R. Breinbauer, T. Thurn-Albrecht, W. Vogel, *Chem. Eur. J.* **7**(5) (2001) 1084 – 1094.

[18] S. Mahamuni, K. Borgohain, B. S. Bendre, V. J. Leppert, S. H. Risbud, *J. Appl. Phys.* **85**(5) (1999) 2861-2865.

[19] M. Starowicz, B. Stypuła, *Eur. J. Inorg. Chem.* (2008) 869–872.

[20] A. B. Isaev, N. A. Zakargaeva, Z. M. Aliev, *Nanotechnol. Russia.* **6** (7–8) (2001) 463–467.

[21] P. Pandey, S. Merwyn, G. S. Agarwal, B. K. Tripathi, S. C. Pant, *J. Nanopart. Res.* **14** (2012) 709-722.

[22] M. El-Kouedi, M. L. Sandrock, C. J. Seugling, C. A. Foss, Jr, *Chem. Mater.* **10** (1998) 3287-3289.

[23] C. K. Govindappa, V. T. Venkatarangaiah, S. B. Abd Hamid, *Nano-Micro Lett.* **5**(2) (2013) 101-110.

[24] O. R. Miranda, N. R. Dollahon, T. S. Ahmadi, *Cryst. Growth Des.* **6** (12) (2006) 2747-2753.

[25] C. Bullen, P. Zijlstra, E. Bakker, M. Gu, C. Raston, *Cryst. Growth Des.* **11** (2011) 3375–3380.

[26] M. Grzelczak, J. Perez-Juste, B. Rodrıguez-Gonzalez, M. Spasova, I. Barsukov, M. Farle, L. M. Liz-Marzan, *Chem. Mater.* **20** (2008) 5399–5405.

[27] M. M. Husein, E. Rodil, J. H. Vera, *Langmuir*, **22** (2006) 2264-2272.

[28] K. S. Choi , E. M. P. Steinmiller, *Electrochim. Acta.* **53** (2008) 6953–6960.

[29] L. Ding, R. Zhang, L. Fan, *Nanoscales Res. Lett.* **8** (2013) 78.

[30] S. Valencia, J. M. Marín, G. Restrepo, *Open Mater. Sci. J.* **4** (2010) 9-14.

[31] H. Wang, J. Z. Xu, J. J. Zhu, H. Y. Che, *J. Cryst. Growth.* **244** (2002) 88–94.

22

Novel electroanalysis of hydroxyurea at glassy carbon and gold electrode surfaces

Keerti M. Naik and Sharanappa T. Nandibewoor✉

P. G. Department of Studies in Chemistry, Karnatak University, Dharwad 580 003, India

✉*Corresponding author*

Abstract

A simple and a novel electroanalysis of hydroxyurea (HU) drug at glassy carbon and gold electrode was investigated for the first time using cyclic, linear sweep and differential pulse voltammetric techniques. The oxidation of HU was irreversible and exhibited a diffusion controlled process on both electrodes. The oxidation mechanism was proposed. The dependence of the current on pH, the concentration, nature of buffer, and scan rate was investigated to optimize the experimental conditions for the determination of HU. It was found that the optimum buffer pH was 7.0, a physiological pH. In the range of 0.01 to 1.0 mM, the current measured by differential pulse voltammetry showed a linear relationship with HU concentration with limit of detection of 0.46 μM for glassy carbon electrode and 0.92 μM for gold electrode. In addition, reproducibility, precision and accuracy of the method were checked as well. The developed method was successfully applied to HU determination in pharmaceutical formulation and human biological fluids. The method finds its applications in quality control laboratories and pharmacokinetics.

Keywords

Hydroxyurea; voltammetry; glassy carbon electrode; gold electrode; electroanalysis; oxidation

Introduction

Synthesis, chemical analysis and testing of the drugs are important events in pharmaceutical laboratories. The many of the essential drugs has been reported by previous workers, which requires the development and validation of analytical methods for their analysis. The requirements of the quality of the drugs have increased tremendously due to the regulations led by regulatory departments like FDA and ICH. Conventional methods have some shortcomings like time consumption and the use of costly and hazardous chemicals. This in turn motivated the analysts to develop and establish newer and faster methods of analysis of the quality of the drug substance and drug products. Hydroxyurea (HU), the simplest, 1-carbon organic antitumor agent, is a member of the substituted urea group and is chemically known as hydroxycarbamide [1]. In 1981 it was reported to have antineoplastic activity against sarcoma [2]. Presently, the primary role of hydroxyurea (Scheme 1) in chemotherapy is the management of granulocytic leukemia and thrombocytosis. It has been used in combination with radiotherapy for carcinomas of the head and neck [3]. HU is used in the treatment of cancer [4], sickle cell anemia [5] and infection with the human immunodeficiency virus (HIV) [6]. HU is a potent, nonalkylating myelosuppressive agent that inhibits DNA synthesis [7].

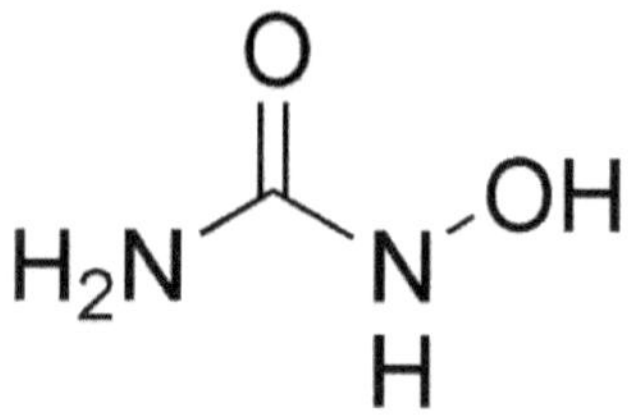

Scheme 1. *Structural formula of hydroxyurea.*

Only a few analytical procedures have been reported for the determination of HU. Nuclear magnetic resonance spectroscopy [8], liquid chromatographic (LC) procedures have been recommended by the U.S. Pharmacopeia [9] and others [10-12] for determination of hydroxyurea in pharmaceutical formulations and biological fluids. Capillary gas chromatography (GC) with thermionic (N-P) specific detection has also been reported [13]. Main problems encountered in using such methods are either the need for derivatization or the need for time-consuming extraction procedures.

Electrochemical methods may offer certain advantages, such as easier sample preparation, being less time-consuming and offering detection limits and dynamic range comparable to other analytical methods [14,15]. These methods have proven useful for the development of very sensitive and selective methods for the determination of organic molecules including drugs. Redox properties of drugs can give insights into their metabolic fate or their in vivo redox processes or pharmaceutical activity [16].

Electrochemical methods, especially differential pulse voltammetry (DPV) make it possible to decrease the analysis time as compared to the time exhaustive chromatographic methods [17]. The advantages of DPV over other electroanalytical techniques are greater speed of analysis, lower consumption of electroactive species in relation to the other electroanalytical techniques, and fewer problems with blocking of the electrode surface.

To the best of our knowledge, there is no report on the electroanalytical method for the determination of HU using glassy carbon (GCE) and gold (GE) electrodes until now. The aim of this study is to establish the suitable experimental conditions, to investigate the voltammetric behavior and oxidation mechanism of HU at GCE and GE by cyclic, linear sweep and differential pulse voltammetric methods for the direct determination of HU in real samples like pharmaceuticals and human biological fluids.

Experimental

Reagents and chemicals

Hydroxyurea (HU) was obtained from Sigma Aldrich and used without further purification. A stock solution of HU (1.0 mM) was prepared in water and stored in a refrigerator at 4 $^{\circ}$C. Standard working solutions were prepared by diluting the stock solution with the selected supporting electrolyte. The phosphate buffers from pH 3.0 – 10.4 were prepared according to the method of Christian and Purdy [18]. The HU containing pharmaceutical product, HYDROX-L, was purchased from a local pharmacy. Other reagents used were of analytical grade. All solutions were prepared with millipore water.

Instrumentation

Electrochemical measurements were carried on a CHI 630D electrochemical analyzer (CH Instruments Inc., USA). The voltammetric measurements were obtained in a 10 ml single compartment three-electrode glass cell with Ag/AgCl as a reference electrode, a platinum wire as counter electrode and a 2-mm diameter glassy carbon electrode and gold electrode as working electrodes. All the potentials are given against the Ag/AgCl (3 M KCl). pH measurements were performed with Elico LI120 pH meter (Elico Ltd., India). All experiments were carried at an ambient temperature of 25 ± 0.1 °C.

Analytical procedure

Polishing of the glassy carbon electrode (GCE) and gold electrode (GE) was done on microcloths (Buehler) glued to flat mirrors. Al_2O_3 (0.3 μm) was used for polishing before each experiment. Before transferring the electrode to the solution, it was rinsed thoroughly with doubly distilled water. After this mechanical treatment, the GCE and GE were placed in 0.2 M phosphate buffer solution, and various voltammograms were recorded until a steady-state baseline voltammogram was obtained.

The parameters for differential pulse voltammetry (DPV) were initial potential: 0.0 V; final potential: 1.2; increase potential: 0.004 V; amplitude: 0.05 V; frequency: 15 Hz; quiet time: 2 s; sensitivity: 1×10^{-5} A/V.

Area of the electrodes

The area of the electrode was obtained by the cyclic voltammetry method using 1.0 mM $K_3Fe(CN)_6$ as a probe at different scan rates. For a reversible process, the following Randles-Sevcik formula can be used [19].

$$I_{pa} = 0.4463(F^3 / RT)^{1/2} n^{3/2} A D_0^{1/2} C_0\, \upsilon^{1/2} \qquad (1)$$

where I_{pa} refers to the anodic peak current, n is the number of electrons transferred, A is the surface area of the electrode, D_0 is diffusion coefficient, υ is the scan rate, and C_0 is the concentration of $K_3Fe(CN)_6$. For 1.0 mM $K_3Fe(CN)_6$ in 0.1 M KCl electrolyte, T = 298K, R = 8.314 J K^{-1} mol^{-1}, F = 96480 C mol^{-1}, n = 1, $D_0 = 7.6\times10^{-6}$ $cm^2\, s^{-1}$, then from the slope of the plot of I_{pa} *vs.* $\upsilon^{1/2}$ relation, the electroactive area was calculated. In our experiment the slope was 2.46×10^{-6} μA $(V\, s^{-1})^{-1/2}$ and 2×10^{-5} μA $(V\, s^{-1})^{-1/2}$ and the area of electrodes were calculated to be 0.033 cm^2 and 0.0269 cm^2 for GCE and GE.

Sample preparation

Two pieces of HU containing tablets were weighed and ground to a homogeneous fine powder in a mortar. A portion equivalent to a stock solution of a concentration of about 1.0 mM was accurately weighed and dissolved in water. The contents were sonicated for 20 min to affect complete dissolution. The excipient was separated by filtration and the residue was washed three times with water. The filtrate was diluted to 1.0 mM. Appropriate solutions were prepared by taking suitable aliquots from this stock solution and diluting them with the phosphate buffer solutions. Each solution was transferred to the voltammetric cell. The differential pulse voltammograms were subsequently recorded following the optimized conditions. The content of the drug in tablet was determined referring to the calibration graph or regression analysis. To study the accuracy of the proposed method and to check the interferences from excipients used in the dosage form, recovery experiments were carried out. The concentration of HU was calculated using standard addition method.

Plasma sample preparation

Human blood samples were collected in dry and evacuated tubes (which contained saline and sodium citrate solution) from a healthy volunteer. The samples were handled at room temperature and were centrifuged for 10 min at 1500 rpm for the separation of plasma within 1 h of collection. The samples were then transferred to polypropylene tubes and stored at 20 °C until analysis. The plasma samples, 0.2 mL, were deproteinized with 2 mL of methanol. After vortexing for 15 min, the mixture was then centrifuged for 15 min at 6000 rpm, and supernatants were collected. The supernatants were spiked with known amounts of HU. Appropriate volumes of this solution were added to phosphate buffer as supporting electrolyte and the voltammograms were then recorded.

Results and discussion

Cyclic voltammetric behavior of hydroxyurea

The electrochemical behavior of HU at GCE and GE were studied using cyclic voltammetry (CV) at physiological pH = 7.0. The cyclic voltammograms obtained for 1.0 mM HU solution at a scan rate of 50 mV s^{-1} exhibit well-defined irreversible anodic peaks at 0.59, 0.78 and 0.91 V at glassy carbon electrode and 0.32, 0.84 and 1.13 V at gold electrode. The cathodic peak was appeared at 0.59 V corresponding to reduction of gold oxides [20]. The results are shown in Figure 1. However, no peak was observed in the reverse scan, suggesting that the oxidation process is an irreversible one. There are two possibilities, either the charge transfer kinetics are slow at surfaces of GCE and GE or products of the electron transfer were unstable and the reaction is accompanied by the fast chemical follow-up reaction resulting in electrochemically inactive products.

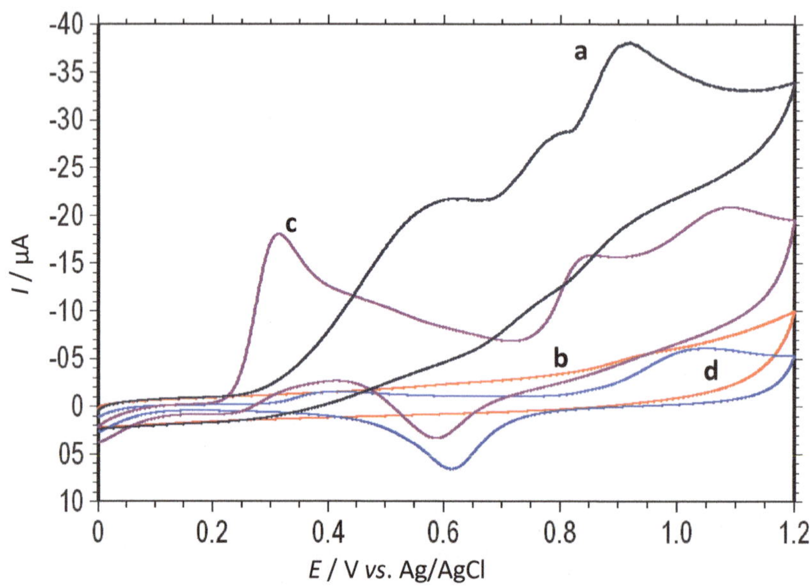

Figure 1. *Cyclic voltammograms obtained for 1.0 mM HU on glassy carbon electrode (GCE) and gold electrode (GE): (a) HU on GCE, (b) blank run of GCE, (c) HU on GE and (d) blank run of GE in pH 7.0, 0.2 M buffer at v = 50 mV s^{-1}.*

Influence of pH

The electrode reaction might be affected by pH of the medium. The electrooxidation of 1.0 mM HU was studied over the pH range of 3.0 - 10.4 in phosphate buffer solution by cyclic voltammetry. A well-defined sharp oxidation peaks were appeared only in the pH range 7.0 - 10.4 (Figures 2A and 2B). Below the pH 7.0, the oxidation peak was not observed; hence the pH study was restricted only in the range from 7.0 to 10.4. However, with the increase in the pH of the solution, the oxidation peak current decreased continuously from 7.0 - 10.4. A stability study showed that the HU was unstable in aqueous solutions even at 4 °C [13] since it degrades with the production of hydroxylamine. The peak potential (E_p) shifted towards less positive potentials with an increase in pH, suggesting the involvement of protons in the chemical process. From the plot of I_{pa} vs pH of GCE and GE it is clear that the peak height decreased with pH. Since the best sensitivity was achieved at pH 7.0, this pH was selected for further experiments.

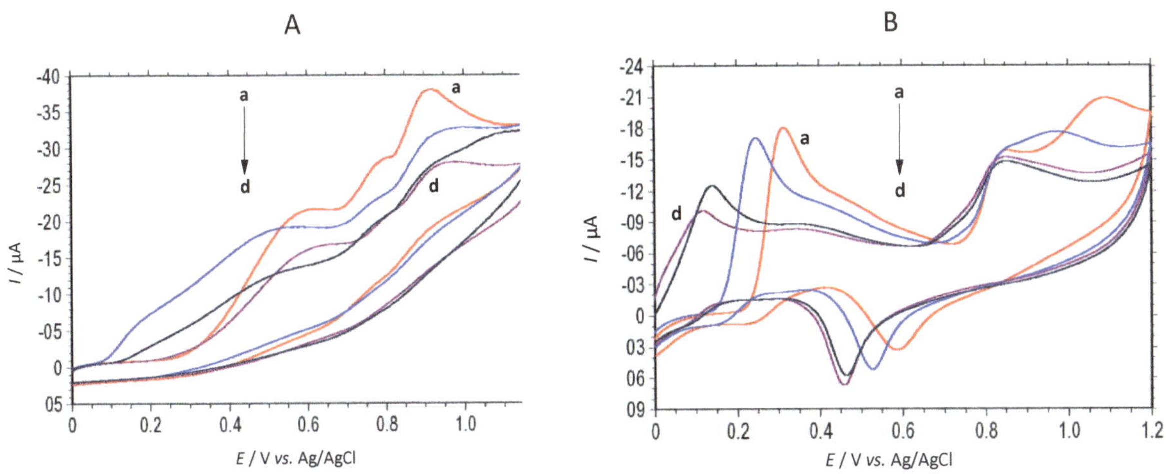

Figure 2. *Influence of pH on the shape of the peak in phosphate buffer solution at (a) 7.0, (b) 8.0, (c) 9.2 and (d) pH 10.4. For 1.0 mM HU on **A** - GCE, **B** - GE.*

Influence of scan rate

Useful information involving electrochemical mechanism can be acquired from the relationship between peak current and scan rate. Therefore, the voltammetric behavior of HU at different scan rates from 10 to 50 mV s^{-1} was also studied using linear sweep voltammetry (Figures 3A and 3B). Scan rate studies were carried out to assess whether the processes on GCE and GE were under diffusion or adsorption-controlled.

The plot of square root of scan rate with the peak current showed a linear relationship in the range of 10 to 50 mV s^{-1} which is of diffusion controlled process [21].

A plot of logarithm of anodic peak current vs. logarithm of scan rate gave a straight line with a slope of 0.424 and 0.516 (Figure 3C), which are close to the theoretical value of 0.5 for a purely diffusion-controlled process [22] which in turn confirms that the processes are diffusion controlled.

The E_p of the oxidation peak was also dependent on scan rate. The peak potential shifted to more positive values on increasing the scan rate, which confirms the irreversibility of the oxidation process, and a linear relationship between peak potential and logarithm of scan rate (Figure 3D).

For an irreversible electrode process, according to Laviron [23] E_p is defined by the following equation:

$$E_p = E^{o'} + \frac{2.303RT}{\alpha nF}\log\frac{RTk^o}{\alpha nF} + \frac{2.303RT}{\alpha nF}\log \upsilon \qquad (2)$$

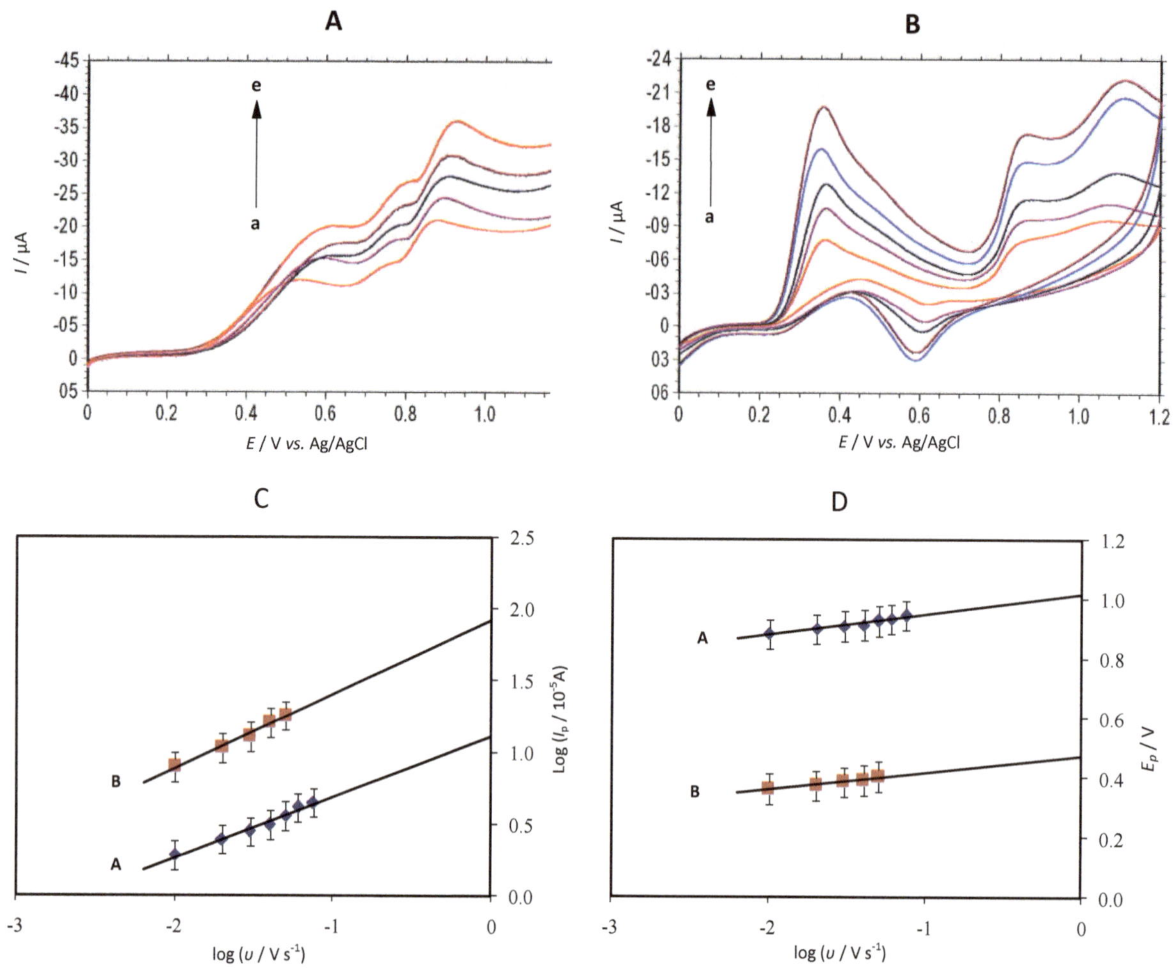

Figure 3. *Linear sweep voltammograms of 1.0 mM HU on*
A *– GCE and* ***B*** *– GE, with different scan rates,* ***a - e*** *were 10, 20, 30, 40 and 50 mV s^{-1}, respectively;*
C *- Dependence of the logarithm of peak current $I_p/10^{-5}$ A on log of scan rate (υ / V s^{-1}); 1.0 mM HU on*
(A) *GCE (log (I_p / μA) = 0.424 log (υ / V s^{-1}) + 1.110; r = 0.9810) and*
(B) *GE (log (I_p / μA) = 0.516 log (υ / V s^{-1}) + 1.914; r = 0.9890);*
D *- Relationship between peak potential E_p/ V and logarithm of scan rate log (υ / V s^{-1}); 1.0 mM HU on*
(A) *GCE (E_p / V = 0.059 log (υ / V s^{-1}) + 0.998; r = 0.9860) and*
(B) *GE (E_p / V = 0.057 log (υ / V s^{-1}) + 0.474; r = 0.9720).*

where α is the transfer coefficient, k^{o} is the standard heterogeneous rate constant of the reaction, n is the number of electrons transferred, υ is the scan rate, and E^{o} is the formal redox potential. Other symbols have their usual meanings. Thus, the value of αn can be easily calculated from the slope of E_p *vs.* log υ. In this system, the slope is 0.059 and 0.057 for GCE and GE, taking T = 298 K and substituting the values of R and F, αn was calculated. According to Bard and Faulkner [24] α can be given as

$$\alpha / \mathrm{mV} = \frac{47.7}{E_p - E_{p/2}} \tag{3}$$

where $E_{p/2}$ is the potential where the current is at half the peak value. So, from this we obtained the value of α. Further, the number of electrons (n) transferred in the electrooxidation of HU was also calculated using linear sweep voltammetry. The value of k^{o} can be determined from the intercept of the above plot if the value of $E^{o'}$ is known. The value of $E^{o'}$ in Equation (2) can be obtained from the intercept of E_p versus υ curve by extrapolating to the vertical axis at $\upsilon = 0$ [24]. All the values of αn, α, n, $E^{o'}$ and k^o obtained from linear sweep voltammetry are tabulated in Table 1.

Table 1. *The calculated values of* αn, α, n, $E^{o'}$ *and* k^o *for the electro-oxidation of HU by linear sweep voltammetry (LSV) at GCE and GE*

Parameters	Linear sweep voltammetry	
	Glassy carbon electrode	Gold electrode
αn	1.0024	1.0375
α	0.5690	0.5860
n	1.76	1.78
$E^{o'}$ / V	0.8740	0.3520
k^o / min^{-1}	4.910×10^3	5.530×10^3

Calibration curve and detection limit

To develop a rapid and sensitive voltammetric method for the determination of HU, differential pulse voltammetric method was adopted as the peaks obtained are better defined at lower concentration of HU than those obtained by cyclic voltammetry. According to the obtained results, it was possible to apply this technique to the quantitative determination of HU. The phosphate buffer solution of pH = 7.0 was selected as the supporting electrolyte for the quantification of HU as it gave a maximum peak current at pH = 7.0 for both GCE and GE. Differential pulse voltammograms obtained with increasing amounts of HU showed that the peak current increased with increasing concentration, as shown in Figures 4A and 4B. The concentration of HU was varied from 0.01 to 1.0 mM. Figure 4C shows that the graph of anodic peak current vs. concentration of HU shows two linear relationships in the range 0.01 to 0.08 and 0.2 to 1.0 mM.

Above 1.0 mM, deviation from linearity was obtained which might be due to the adsorption of HU or its oxidation products on the electrode surface. The decrease of sensitivity (slope) in the second linear range is likely to the kinetic limitations [25]. Related statistical data of the calibration curves were obtained from the six different determinations. The detection limits (LOD) and quantification (LOQ) in the lower range regions were given in Table 2.

Table 2. *The Values of LOD and LOQ for HU at GCE and GE by using differential pulse voltammetric method*

	Glassy carbon electrode (GCE)	**Gold electrode (GE)**
Linearity range, mM	0.01 to 1.0	0.01 to 1.0
Number of data points	06	06
Limit of detection (LOD), μM	0.46	0.92
Limit of quantification (LOQ), μM	1.54	3.08
Repeatability – RSD, %	1.14	2.02
Reproducibility – RSD, %	1.46	2.48

During the actual analysis, the analytical response was checked through the peak potential and its height. No change in peak potential was observed within an hour, while its height changed about ±1 % for five different quantitative determinations. This proposed method was better as compared with other reported methods [11,26].

To ascertain the repeatability of the analysis, six measurements of 1.0 mM HU solution were carried out using GCE and GE at intervals of 30 min. The RSD value of peak current was found to be 1.14 % and 2.02 % respectively, which indicated that the methods had good repeatability. As to the reproducibility between days, it was similar to that of within day repeatability if the temperature was kept almost unchanged.

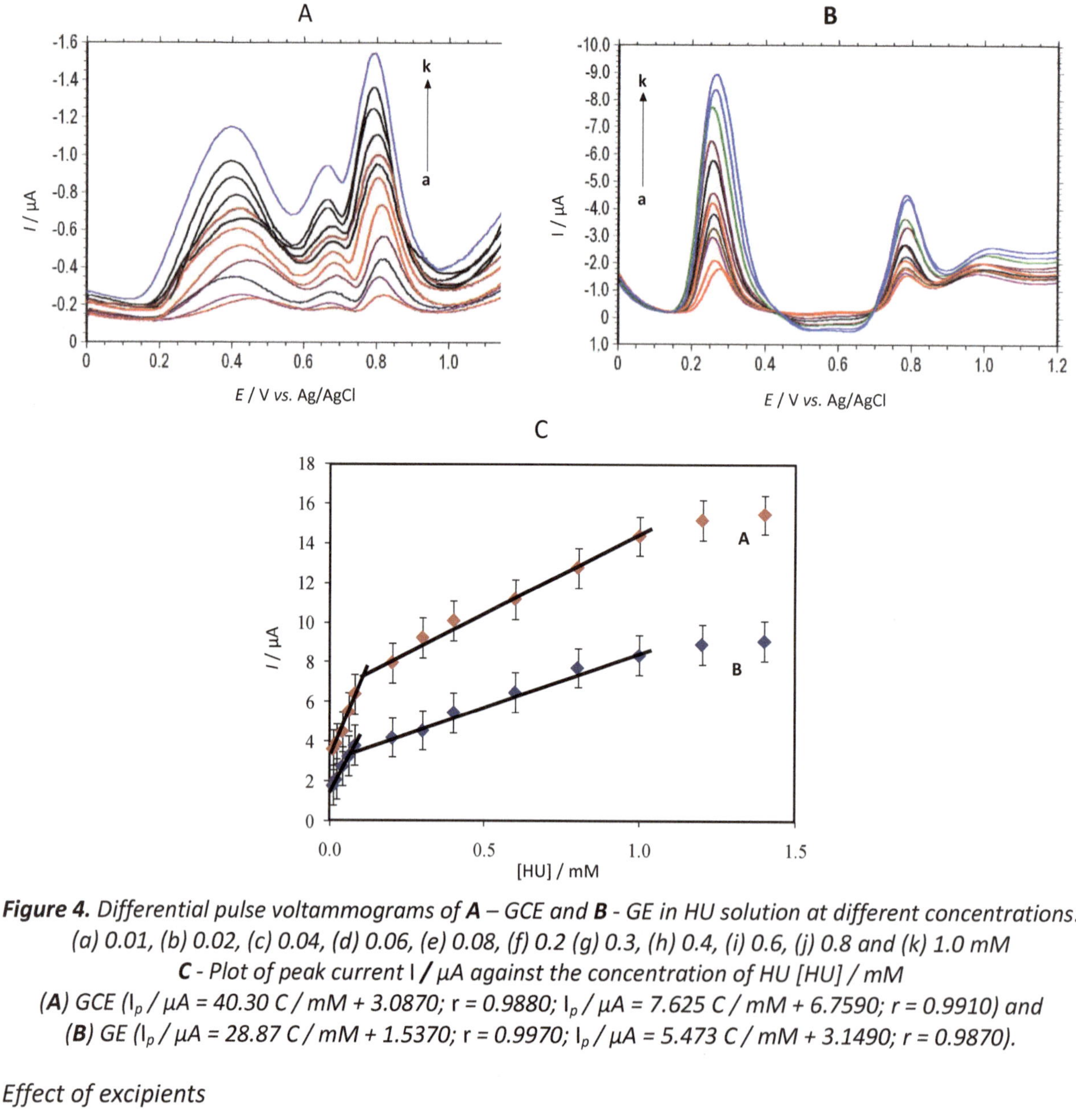

Figure 4. *Differential pulse voltammograms of **A** – GCE and **B** - GE in HU solution at different concentrations: (a) 0.01, (b) 0.02, (c) 0.04, (d) 0.06, (e) 0.08, (f) 0.2 (g) 0.3, (h) 0.4, (i) 0.6, (j) 0.8 and (k) 1.0 mM*
C *- Plot of peak current I / µA against the concentration of HU [HU] / mM*
*(**A**) GCE (I_p / µA = 40.30 C / mM + 3.0870; r = 0.9880; I_p / µA = 7.625 C / mM + 6.7590; r = 0.9910) and (**B**) GE (I_p / µA = 28.87 C / mM + 1.5370; r = 0.9970; I_p / µA = 5.473 C / mM + 3.1490; r = 0.9870).*

Effect of excipients

For the possible analytical application of the proposed method, the effect of some common excipients used in pharmaceutical preparations was examined. The tolerance limit was defined as the maximum concentration of the interfering substance that caused an error less than ±5 % for determination of HU. Under the optimum experimental conditions, the effects of potential excipients on the voltammetric response of 1.0 mM HU as a standard were evaluated. The experimental results showed that hundred-fold excess of citric acid, dextrose, glucose, gum acacia, lactose, starch, and sucrose did not interfere with the voltammetric signal of HU. Thus, the procedures were able to assay HU in the presence of excipients, and hence it can be considered specific.

Determination of HU in pharmaceutical preparations and recovery test

The proposed method was validated for the determination of HU in pharmaceutical preparations in tablets as a real sample by applying DPV using the standard addition method. The procedure for the tablet analysis was followed as described in sample preparation section. The results are in good agreement with the content marked in the label (Table 3). Recovery studies were carried out after the addition of known amounts of the drug to various pre-analyzed formulations of HU. The recovery in the sample was found to be 98.83 % with RSD of 1.37 %.

Table 3. *Determination of HU in pharmaceutical formulation samples using differential pulse voltammetric method*

	Labeled claim 500.0, mg		Added 20.0 mg	
	GCE	**GE**	**GCE**	**GE**
Amount found, mg[a]	499.2	492.5	19.5	19.2
Recovery, %	98.83	98.50	97.78	96.11
RSD, %	1.37	1.87	3.51	3.54
Bias, %	-1.16	-1.50	-2.22	-3.88

[a] average of six determinations.

Detection of HU in spiked human plasma samples

The applicability of the DPV to the determination of HU in spiked human plasma sample was investigated. The recoveries from human plasma were measured by spiking drug free plasma with known amounts of HU. The plasma samples were prepared as described in plasma sample preparation section. A quantitative analysis can be carried out by adding the standard solution of HU into the detect system of plasma sample. The calibration graph was used for the determination of spiked HU in plasma samples. The detection results obtained for four plasma samples are listed in Table 4.

Table 4. *Determination of HU in spiked human plasma samples using differential pulse voltammetric method*

Human plasma sample	**Amount of spiked HU, 10-4 M**	**Amount of detected HU*, 10-4 M**		**Recovery, %**		**RSD, %**		**Bias, %**	
		GCE	**GE**	**GCE**	**GE**	**GCE**	**GE**	**GCE**	**GE**
1	0.3	0.3002	0.2998	100.1	99.95	2.92	3.12	0.067	-0.067
2	0.7	0.6863	0.6797	98.05	97.10	2.67	2.25	-1.96	-2.90
3	3.0	3.0407	3.0074	101.4	100.2	2.37	0.48	1.36	0.25
4	7.0	6.9620	6.9820	99.45	99.74	1.79	0.96	-0.54	-0.26

* average of six determinations

Detection of HU in urine samples

The developed differential pulse voltammetric method was also applied for the determination of HU in spiked urine samples. The recoveries from urine were measured by spiking drug-free urine with known amounts of HU. The urine samples were diluted 100 times with the phosphate buffer solution before analysis without further pretreatments. A quantitative determination can be carried out by adding the standard solution of HU into the detect system of urine sample. The calibration graph was used for the determination of spiked HU in urine samples. The detection results of four urine samples obtained are listed in Table 5. Thus, satisfactory recoveries of the analyte from the real samples and a good agreement between the concentration ranges studied and the real ranges encountered in the urine samples when treated with the drug make the developed method applicable in clinical analysis.

Table 5. *Determination of HU in urine samples using differential pulse voltammetric method*

Urine sample	Amount of spiked HU, 10^{-4} M	Amount of detected HU*, 10^{-4} M		Recovery, %		RSD, %		Bias, %	
		GCE	GE	GCE	GE	GCE	GE	GCE	GE
1	0.3	0.3025	0.3092	100.8	103.1	2.32	4.12	0.83	3.07
2	0.7	0.6995	0.7101	99.94	101.4	0.34	1.41	-0.07	1.44
3	3.0	2.9424	2.9424	98.08	98.08	1.73	1.73	-1.92	-1.92
4	7.0	7.0799	6.9966	101.1	99.95	1.35	0.82	1.14	-0.05

* average of six determinations.

Conclusions

The voltammetric oxidation of HU at glassy carbon and gold electrodes under physiological condition, *i.e.*, pH 7.0 in phosphate buffer solutions has been investigated. HU undergoes two electron-two proton change and follows the diffusion-controlled process at both the electrodes. A suitable oxidation mechanism was proposed. The proposed method can be used successfully to assay the drug in pharmaceutical dosage form as well as in spiked real samples. High percentage recovery and study of excipients showed that the method is free from the interferences of the commonly used excipients in the formulations of drugs. The proposed method is suitable for quality control laboratories as well as pharmacokinetic studies where economy and time are essential.

Acknowledgment: *Keerti M. Naik thanks UGC, New Delhi for the award of Research Fellowship in Science for Meritorious Students (RFSMS).*

Nomenclature

HU = Hydroxyurea
GCE = glassy carbon electrode
GE = gold electrode
DPV = differential pulse voltammetry
LSV = linear sweep voltammetry
CV = cyclic voltammetry
I_{pa} = anodic peak current μA
n = number of electrons transferred
A = surface area of the electrode cm^2
D_0 = diffusion coefficient, $cm^2\ s^{-1}$
υ = scan rate, mV s^{-1}
C_0 = concentration, M
E_p = peak potential, V
I_p = peak current, μA
α = transfer coefficient
k^o = standard rate constant of the reaction, cm s^{-1}
F = Faraday constant, C mol^{-1}
E^o = formal redox potential, V
R = gas constant, J K^{-1} mol^{-1}
T = temperature, K
LOD = limit of detection, M
LOQ = limit of quantification, M
s = standard deviation of the peak currents
m = slope of the calibration curve
RSD = relative standard deviation

References

[1] R. Donehower, *In Cancer Chemotherapy: Principles and Practice*, B. A. Chabner, J. M. Collins, (Eds), J. B. Lippincott Co., Philadelphia, PA, 1990.

[2] S. J. P. Van Belle, M. M. de Planque, I. E. Smith, A. T. van Oosterom, T. J. Schoemaker, W. Deneve, J. G. McVie, *Cancer Chemother. Pharmacol.* **18** (1986) 27–32.

[3] D. Heerenberg, *Basic Principles in Therapeutics*, McGraw-Hill, New York, 1992.
[4] R. C. Donehower, *Semin. Oncol.* **19** (1992) 11-19.
[5] S. Charache, G. J. Dover, R. D. Moore, *Blood* **79** (1992) 2555-2565.
[6] W. Y. Gao, A. Cara, R. C. Gallo, F. Lori, *Proc. Natl. Acad. Sci. U.S.A.* **90** (1993), 8925-8928.
[7] I. H. Krakoff, N. C. Brown, P. Reichard, *Cancer Res.* **28** (1968) 1559–1565.
[8] K. B. Main, T. Medwick, L. C. Bailey, J. H. Shinkai, *Pharm. Res.* **4** (1987) 412–415.
[9] The U. S. Pharmacopia 23rd Ed., U. S. Pharmacopeial Convention, Rockville, MD, 1995.
[10] Q. Wang, H. J. Altermatt, H. B. Ris, B.E. Reynolds, J. C. M. Stewart, R. Bonnett, C. K. Lim, *Biomed. Chromatogr.* **7** (1993) 155–157.
[11] E. W. Iyamu, P. D. Roa, P. Koposmbut, M. D. P. Aguinaga, E.A . Turner, *J. Chromatogr. B Biomed. Appl.* **709** (1998) 119–126.
[12] K. K. Manouilov, T. R. McGruire, P. R. Gwilt, *J. Chromatogr. B Biomed. Appl.* **708** (1998) 321–324.
[13] A. El-Yazigi, S. Al-Rawithi, *Pharm. Res.* **9** (1992) 115–118.
[14] C. Radovan, D. Cinghita, F. Manea, M. Mincea, C. Cofan, V. Ostafe, *Sensors* **8** (2008) 4330-4349.
[15] C. D. Souza, O. C. Braga, I. C. Vieira, A. Spinelli, *Sens. Actuators B: Chem.* **135** (2008**)** 66-73.
[16] S. A. Ozkan, B. Uslu, H. Y. Aboul-Enein, *Crit. Rev. Anal. Chem.* **33** (2003) 155-181.
[17] N. Erk, *Anal. Bioanal. Chem.* **378** (2004) 1351-1356.
[18] G. D. Christian, W. C. Purdy, *J. Electroanal. Chem.* **3** (1962) 363-367.
[19] B. Rezaei, S. Damiri, *Sens. Actuators B* 134 (2008) 324-331.
[20] C. Barus, P. Gros, M. Comtat, S. Daunes-marion, R. Tarroux, *Electrochim. Acta* **52** (2007) 7978-7985.
[21] K. M. Naik, S. T. Nandibewor, *Sensors and Actuators A* **212** (2014) 127-132
[22] D. K. Gosser, Cyclic Voltammetry: Simulation and Analysis of Reaction Mechanisms; VCH: New York, 1993.
[23] E. Laviron, *J. Electroanal. Chem.* **101** (1979) 19-29.
[24] A. J. Bard, L. R. Faulkner, *Electrochemical methods; Fundamentals and Applications*. 2nd edn, Wiley, 2004.
[25] M. Mazloum-Ardakani, H. Rajabi, H. Beitollahi, B. B. F. Mirjalili, A. Akbari, N. Taghavinia, *Int. J. Electrochem. Sci.* **5** (2010) 147-157.
[26] J. Havard, J. Grygiel, D. Sampson, *J. Chromatogr. B.* **584** (1992) 270-274.

23

Comparative voltammetric study and determination of carbamate pesticide residues in soil at carbon nanotubes paste electrodes

THOMMANDRU RAVEENDRANATH BABU, SARVAREDDY RAJASEKHAR REDDY✉, PUCHAKAYALA SUJANA

Electroanalytical Lab., Department of Chemistry, N.B.K.R. Science and Arts College, Vidyanagar, Nellore, AP, India

✉Corresponding Authors

Abstract

In this investigation, the persistence of carbamate pesticides in soil samples was investigated. A simple and selective differential pulse adsorptive stripping voltammetry was selected for this investigation. Carbon nanotubes paste electrodes were used as working electrodes for differential pulse adsorptive stripping voltammetry and cyclic voltammetry. A symmetric study of the various operational parameters that affect the stripping response was carried out by differential pulse voltammetry. Peak currents were linear over the concentration range of 10^{-5} to 10^{-10} M with an accumulation potential of -0.6 V and a 70 s accumulation time with lower detection limits of 1.09×10^{-7} M, 1.07×10^{-7}M, 1.09×10^{-7} M for chlorphropham, thiodicarb, aldicarb. The relative standard deviation (n=10) and correlation coefficient values were 1.15 %, 0.988; 1.13 %, 0.978; and 1.14 %, 0.987, respectively. Universal buffer with pH range 2.0 - 6.0 was used as supporting electrolyte. The solutions with uniform concentration (10^{-5} M) were used in all determinations. Calculations were made by standard addition method.

Keywords

Thiodicarb; Aldicarb; Chlorpropham; Differential pulse adsorptive stripping voltammetry; Cyclic voltammetry; CNTPE; Soil samples

Introduction

Pesticides are extensively and indiscriminately used in modern agricultural practices, resulting in widespread distribution in the environment and posing serious health hazards to animals and human beings. Besides inhalation from polluted environment, animals are also exposed to pesticides through the utilisation of treated feeds and fodders. Thiodicarb (dimethyl *N, N'* –thiobis (methyl imino) carbonyloxy bisethanimido thioate) is a new carbamate compound with a broad spectrum of activity that is being extensively used for crop protection. It is a class II category compound (moderately toxic) as set forth by the United States Environmental Protection Agency (USEPA) and World Health Organization (WHO). Various carbamate compounds have been reported to cause biochemical changes in different species of animals [1-5]. Little information on the effect of thiodicarb on biochemical profiles is available in dogs and rats [6-8]. However, no detailed report is available regarding the effects of thiodicarb on various biochemical parameters and blood enzymes in animals.

Chlorpropham ($C_{10}H_{12}ClNO_2$)

Chlorpropham is a plant growth regulator used for the pre-emergence control of grass weeds in alfalfa, Lima and snap beans, blueberries, cane berries, carrots, cranberries, ladino clover, garlic, seed grass, onions, spinach, sugar beets, tomatoes, safflower, soybeans, gladioli and woody nursery stock. It is also used to inhibit potato sprouting and for sucker control in tobacco. Parilla *et al.* [9] reported SPE and HPLC/DAD methods to determine pesticide residues in water. Richard [10] employed HPLC method to determine carbamate residues using post-column hydrolysis electrochemical detection. Aulakh *et al.* [11] reported solid phase microextraction HPLC for the analysis of pesticides. Tomomi *et al.* [12] developed a new analytical method for the determination of nine pesticide residues including chlorpropham in fruits and vegetables using ESI-LC/MS/MS with direct sample injection into a short column. Oosselton and Snelling [13] reported the use of GLC, HPLC/DAD and TLC for the determination of 51 common pesticides including chlorpropham.

Thiodicarb ($C_{10}H_{18}N_4O_4S_3$)

Thiodicarb is a non-systemic carbamate insecticide whose acetyl cholinesterase activity is related to its main methomyl degradation product[14]. Xu and Li [15] determined thiodicarb by reverse-phase high performance liquid chromatography.

Aldicarb ($C_7H_{14}N_2O_2S$)

Aldicarb is a carbamate insecticide which is the active substance in the pesticide Temik. It is effective against thrips, aphids, spider mites, lygus, fleahoppers, and leafminers, but is primarily used as a nematicide. Waliszewski and Szymczyński [16] reported a Simple method for the gas-chromatographic determination of aldicarb, aldicarb sulphoxide and aldicarb sulphone in soil and sugar beets. Mora *et al.* [17] determined the presence of the nematicide aldicarb and its metabolites aldicarb sulphoxide and aldicarb sulphone in soils and potatoes by liquid chromatography with photodiode array detection. Although there are reports in the literature for several methods of determinations of pesticides, there are few focused on electrochemical methods; hence, in this investigation, electrochemical determinations [18-20] were employed.

Experimental

Apparatus and electrodes

The electrochemical measurements were carried out with Metrohm model 101 potentiostat and galvanostat. The three-electrode system consisted of carbon nanotubes paste electrode as the working electrode, Ag/AgCl reference electrode and a platinum wire auxiliary electrode. The electrodes joined the cell through holes in its Teflon cover. All of the potentials given in this work were measured with respect to this reference system. Electrochemical experiments were carried out in a voltammetric cell at room temperature. A magnetic stirrer was used during the accumulation step. The Elico Li-129 model glass calomel combined electrode was employed for measuring pH values.

Preparation of carbon nanotubes paste electrode

The CNTPE was prepared by mixing multiwall CNTs powder (diameter 20-50 nm, either 1-5 mm or 5-20 mm lengths) and Castrol oil in an agate mortar at a ratio of 50.0 % (w/w) each. A portion of the resulting paste was packed firmly into the cavity (0.8 mm diameter) of a Teflon tube. The electrical contact was established via a copper wire [21].

Reagents and solutions

All reagents used were of analytical reagent grade. Double distilled water was used throughout the analysis. In the present investigation, universal buffers in the pH range 2.0 to 6.0 were used as supporting electrolytes and were prepared using 0.2 M boric acid, 0.05 M citric acid and 0.1 M trisodium orthophosphate solutions. Samples were obtained from RANKEM India, Ltd.

Result and discussion

All of the compounds exhibit well-defined voltammetric peaks at the same experimental conditions but the reduction electrode potentials are somewhat different; this is attributed to the difference in the nature of groups present in the compounds under investigation (Scheme 1).

Although all of the compounds possess electron-donating nitrogen on one or both sides of carbonyl carbon, there are some differences in the environment of carbonyl carbon.

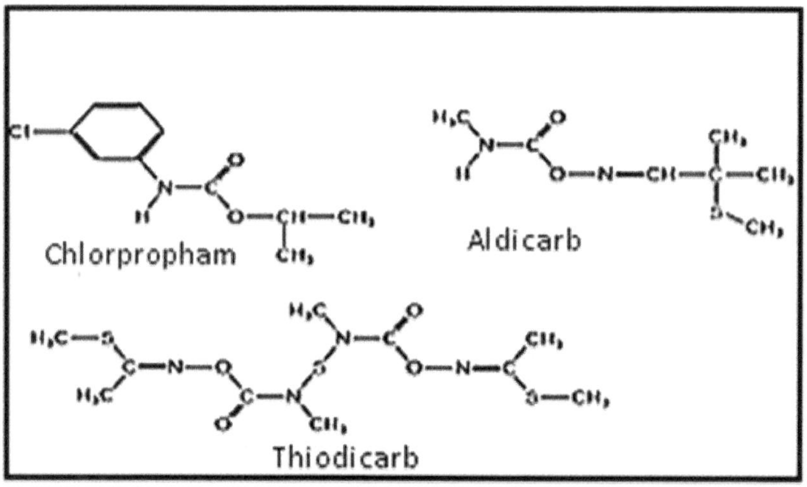

Scheme 1. Structures of the pesticides investigated in this work

In the case of chlorpropham, there is oxygen bonded with a propyl group on one side of the carbonyl carbon and on the other side nitrogen with chlorobenzene. Because the aromatic ring is closer to the electroactive group, it will experience less negative charge and undergo reduction at somewhat lower electrode potentials when compared with the other two carbonyl group-containing pesticides. Two electrons are involved in reduction of one carbonyl group into the hydroxyl group.

In the case of thiodicarb, there are two carbonyl groups with the same environments; in the case of two carbonyl groups, there is oxygen bonded with electron-donating nitrogen on one side and nitrogen bonded with electronegative sulphur and electron-donating alkyl groups on the other side along with the other carbonyl group with the same environment. In the case of thiodicarb, however, there is electron-donating nitrogen, alkyl groups with positive inductive effect; their impact on the electronic environment seems to be nil because of double bonds and electronegative groups. In the case of thiodicarb, there is a well-defined peak due to 4 electron reduction of two carbonyl groups.

In the case of aldicarb, there is only one carbonyl group on one side with nitrogen, while there is electronegative oxygen bonded with nitrogen on the other side. Because of the electro rich nitrogen being directly bonded with a carbonyl group, the environment around the electroactive species seems to be more negative and reduction will take place at greater negative potentials compared with the remaining two pesticides. Two electron reductions will take place.

Figure 1 shows DP-AdSV response for the samples (10^{-5}M) under investigation over the pH range 2.0-6.0 at CNTPE. The systematic studies of the various experimental and instrumental parameters that affect the voltammetric response were carried out in order to establish the optimum conditions. The pH of a solution is a critical factor affecting both the rate and equilibrium state of the reduction process, as well as the rate of the electrode reaction. The influence of pH on the voltammetric response was studied at CNTPE of the 10^{-5} M samples with pH between 2.0 and 6.0. The maximum peak currents were obtained with pH 4.0. Voltammograms obtained for increasing values of the scan rate showed the existence of a linear dependence of the peak current intensity on the scan rate between 10 to 60 mV s.$^{-1}$ The peak currents were directly proportional to the scan rate. The voltammetric behaviour of samples has been studied in the pH range from 2.0 to 6.0. A single well resolved peak was observed throughout the pH range and this single peak is attributed to the reduction of corresponding groups. All the compounds under investigation exhibit only one voltammetricpeak for each over the pH range 2.0 to 6.0. This wave / peak are attributed to the simultaneous reduction of carbonyl group. Typical cyclic voltammograms are shown in Fig. 2. No reduction peak is observed in basic medium ($8 \leq pH \leq 12$) for carbonyl groups due to the precipitation. The diffusion controlled nature of electrode process is evidenced from the linear plots of i_p *vs.* $V^{1/2}$ (Fig. 3).

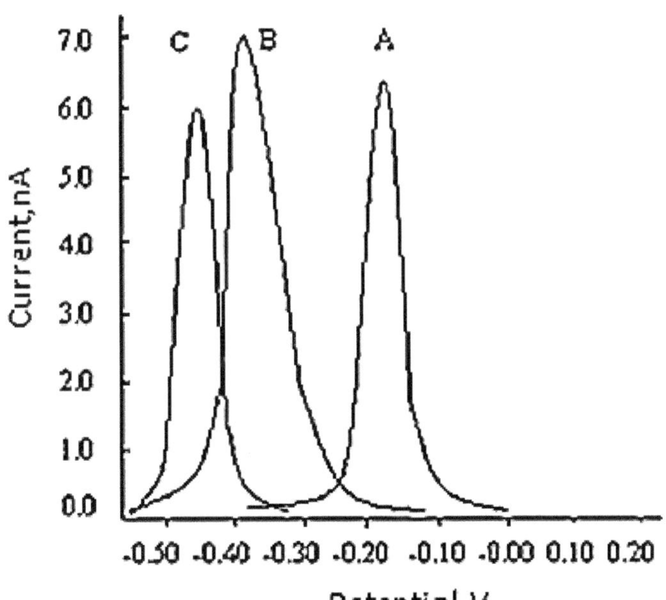

Fig. 1.*Stripping voltammograms of A - chlorpropham, B - thiodicar and C – aldicarb at CNTPE Concentration: 10^{-5} M L^{-1}, scan rate: 60 mV s^{-1}, pH 4.0*

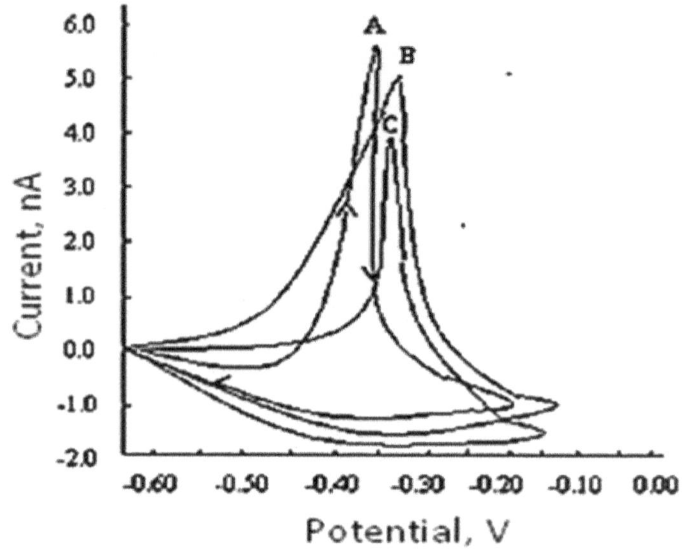

Fig. 2. *Cyclic voltammograms of A - chlorpropham, B - thiodicar and C - aldicarb at CNTPE, Concentration: 10^{-5} M L^{-1}, scan rate: 60 mV s^{-1}, pH 4.0*

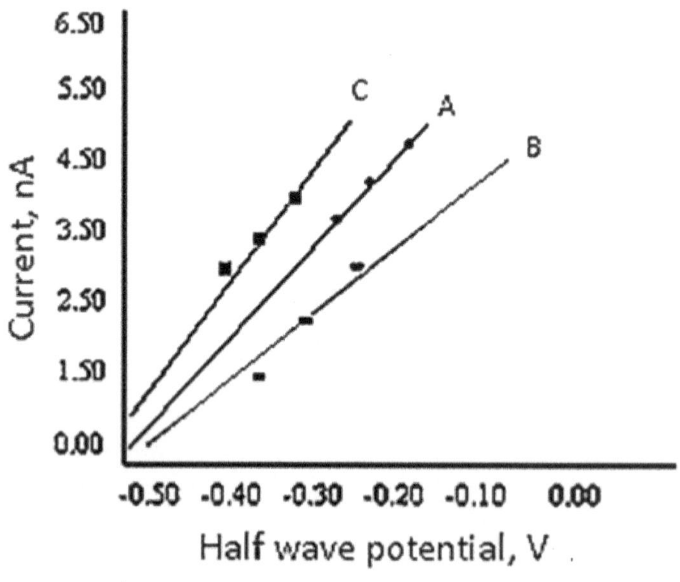

Fig. 3. I_p vs. $V^{1/2}$ *plots of A - chlorpropham, B - thiodicarb, C - aldicarb. Concentration: 10^{-5} M L^{-1}; Scan rate: 60 mV s^{-1}, pH 4.0*

Recovery experiments

Analysis

Based on the results obtained with differential pulse adsorptive stripping voltammetry and cyclic voltammetry at CNTPE, differential pulse adsorptive stripping voltammetry and cyclic voltammetry have been used for the quantitative determination of samples using both calibration and standard addition methods. The investigated compounds were found to exhibit well resolved peaks at pH 4.0, and the sharp well resolved peak was chosen for quantitative studies. Peak currents are linear over the concentration range of 10^{-5} to 10^{-10} M with lower detection limits of 1.09×10^{-7} M for chlorpropham, 1.07×10^{-7} M for thiodicarb, and 1.09×10^{-7} M for aldicarb. The relative standard deviation and correlation coefficients were found to be 1.15 %, 0.988; 1.13 %, 0.978; and 1.14 %, 0.987, respectively, for 10 replicates.

Determination of pesticide samples from their standard solutions

To check the validity of the method, a standard solution (10^{-5} M) was prepared in dimethyl formamide. 1 mL of the standard solution was transferred into a voltammetric cell and made up with 9 mL of supporting electrolyte (pH 4.0), before being deoxygenated with nitrogen gas for 10 min, and then subjected to voltammetry. After obtaining voltammograms, a small increment of the standard solution of samples was added to voltammetric cells and was deoxygenated for 10 min; voltammograms were recorded under similar conditions. In the same manner, 10 voltammograms were recorded for 10 standard additions. The optimum conditions for analytical determination were found to be at pH 4.0 and scan rate 60 mV s^{-1}. The average recovery obtained for the pesticide samples in soil samples ranged from 89.00 to 92.00 % for chlorpropham, from 97.50 to 99.33 % for thiodicarb and from 97.80 to 98.33 % for aldicarb for 10 replicates. The results are shown in Table 1.

Table 1.*Recoveries of chlorpropham, thiodicarb, aldicarb in standard solution of 1.0×10^{-5}M*

Sample	Amount added, μg mL^{-1}	Amount found, μg mL^{-1}	*Recovery, %	Standard deviation
Chlorpropham	**3.0**	**2.79**	**93.00**	**0.024**
Thiodicarb	**3.0**	**2.98**	**99.33**	**0.034**
Aldicarb	**3.0**	**2.95**	**98.33**	**0.028**

**Average of 10 replicates*

Determination of pesticide samples in spiked soil samples

The soil under investigation was spiked with known amounts of formulations and dried on filter paper at laboratory temperature. For extraction, 50 g of the dried soil was transferred into a 250 ml Erlenmeyer flask. These samples and blanks were extracted 2-5 times by acetone. The extracts were then evaporated to dryness and the resulting residues were dissolved in DMF and transferred to 50 ml voltammetric flasks. This solution was filtered through Whatman nylon membrane filter paper and voltammograms of the filtrates were recorded by following the previously mentioned procedure. The average recovery obtained for the sample in soil samples ranged from 90.00 to 93.00 % for chlorpropham (bud nip), from 93.50 to 95.66 % for thiodicarb (larvin) and from 92.70 to 95.66 % for aldicarb (aldicarb sulphone) for 10 replicates. The results are presented in Table 2.

Table 2. *Recoveries of chlorpropham, thiodicarb, aldicarb (formulations) in spiked soil samples*

Sample	Amount added, μg mL^{-1}	Amount found, μg mL^{-1}	*Recovery, %	Standard deviation
Bud Nip	**3.0**	2.76	92.00	0.015
Larvin	**3.0**	2.87	95.66	0.024
Aldicarb sulphone	**3.0**	2.88	96.00	0.018

**Average of 10 replicates*

Conclusion

In conclusion, the adopted method of differential pulse adsorptive stripping voltammetry is a less tedious and economically low consumption method; hence, this can be used satisfactorily for the determination of pesticide residues in soil. The obtained results also demonstrate the suitability of the developed DP-AdSV method for the determination of samples under investigation in soil samples. The electrochemical reduction mechanism of the carbonyl group in all three compounds was found to be irreversible. The nature of the electrode process for these compounds is found to be diffusion controlled and involves adsorption on the electrode surface without any kinetic complications. The variation of peak current with the pH of the supporting electrolyte influences the diffusion coefficient values. The slight variations in diffusion coefficient values with increasing pH may be attributed to a decrease in the availability of protons.

The heterogeneous forward rate constant values obtained for the reduction of these three pesticides are found to decrease with an increase in the pH of the solution, as expected. From the comparison of the forward rate constant values of the three compounds, it can be seen that they reduce at different electrode potentials, which is attributed to the difference in the molecular environment of the samples under investigation. Analytical procedures are described for the quantitative determination of these compounds using DP-AdSV. In the present investigation, standard addition and calibration methods were utilised for the determination of these pesticides in soil samples. From the recoveries, it has been observed that the proposed method describes the successful application of an electroanalytical technique for the analysis of these compounds. It also demonstrates that DP-AdSV at a carbon nanotubes paste electrode could conveniently be used for the quantitative determination of these pesticides in soil samples. The method shows a good reproducibility and high accuracy compared with spectrophotometric, spectrofluorimetric and chromatographic methods of analysis.

References

[1] M. Jayapragasam, I. Jasmine, V.,Thenammai, R. Kasthuri, *Madras Agric. J.* **68** (1981) 461-465

[2] R. Kiran, M. Sharma, R. C. Bansal *Pesticides* **19** (1985) 42-43.

[3] G. L. Kennedy, *J. Appl. Toxicol.* **6** (1986) 423-429.

[4] S. D. Moregaonkar, B. B .Deshpande, V. P Vadlmudi, N. M. Degloorkar, S. R. Rajurkar, *Indian. Vet. J.* **70** (1993) 945-948.

[5] Satpal, S. K. Jain, J. S. Punia, *Toxicol. Int* .**17** (2010) 30-32.

[6] *EFSA Scientific Report*, **55** (2005) 1-76

[7] H. B. Knaak, B. W .Wilson. *ACS Symposium Series*, **273** (1985) 63-79.

[8] N. N. Hamada, *Rep. No. 210-216 from Hazleton Laboratories America Inc.*, Vienna, VA to Union Carbide Agricultural Products Company Inc. Research Triangle Park, North Carolina, 1986

[9] P. Parrilla, J. L. M. Vidal, *Anal. Lett.* **30** (1997) 1719-1738.

[10] R. T. Krause, *J. Chromatogr. A* **442**(1988) 333-343.

[11] J. S. Aulakh; A. K. Malik, V. Kaur, P. Schmitt-Kopplin, *CRC Cr. Rev. Anal. Chem.* **35**_(2005) 71-85.

[12] T. Goto, Y. Ito, S. Yamada, H. Matsumoto, H. Oka and. H. Nagase, *Anal. Chim. Acta* **555** (2006) 225-232.

[13] M. D. Osselton, R. D. Snelling *J. Chromatogr. A* **368** (1986) 265-271.

[14] G. Hoizey, F. Canas, L Binet, M. L. Kaltenbach, G. Jeunehomme, M. H. Bernard, D. Lamiable, *J. Forensic Sci.* **53** (2008) 499-502.

[15] G. Xu, W. Zheng, Y. Li, S. Wang, J. Zhang, Y. Yan., *Int. Biodeter. Biodegr.* **62** (2008) 51–56.

[16] S. M. Waliszewski, G. A. Szymczyński, *Fresen. J. Anal. Chem.* **338** (1990) 75-76

[17] N. Unceta, A. Ugarte, A. Sanchez, A. Gómez-Caballero, M. A .Goicolea, R. J. Barrio, *J. Chromatogr. A* **1061** (2004) 211-216

[18] S. Rajasekharreddy, K. Chandramohan and, NY. Sreedhar, *Int. J. Sci. Eng. Res*, **2(10)** (2011) 1-4.

[19] S. Rajasekhar Reddy, T. Raveendranath Babu, B. SreenivasuluInt, *J. Res. Pharm. Life Sci.* **1** (2013) 43-47.

[20] S. Rajasekhar Reddy, T. Raveendra Nath Babu, *Int. J. Nanosci.* **12** (2013) 130058.

Permissions

All chapters in this book were first published in JESE, by International Association of Physical Chemists (IAPC); hereby published with permission under the Creative Commons Attribution License or equivalent. Every chapter published in this book has been scrutinized by our experts. Their significance has been extensively debated. The topics covered herein carry significant findings which will fuel the growth of the discipline. They may even be implemented as practical applications or may be referred to as a beginning point for another development.

The contributors of this book come from diverse backgrounds, making this book a truly international effort. This book will bring forth new frontiers with its revolutionizing research information and detailed analysis of the nascent developments around the world.

We would like to thank all the contributing authors for lending their expertise to make the book truly unique. They have played a crucial role in the development of this book. Without their invaluable contributions this book wouldn't have been possible. They have made vital efforts to compile up to date information on the varied aspects of this subject to make this book a valuable addition to the collection of many professionals and students.

This book was conceptualized with the vision of imparting up-to-date information and advanced data in this field. To ensure the same, a matchless editorial board was set up. Every individual on the board went through rigorous rounds of assessment to prove their worth. After which they invested a large part of their time researching and compiling the most relevant data for our readers.

The editorial board has been involved in producing this book since its inception. They have spent rigorous hours researching and exploring the diverse topics which have resulted in the successful publishing of this book. They have passed on their knowledge of decades through this book. To expedite this challenging task, the publisher supported the team at every step. A small team of assistant editors was also appointed to further simplify the editing procedure and attain best results for the readers.

Apart from the editorial board, the designing team has also invested a significant amount of their time in understanding the subject and creating the most relevant covers. They scrutinized every image to scout for the most suitable representation of the subject and create an appropriate cover for the book.

The publishing team has been an ardent support to the editorial, designing and production team. Their endless efforts to recruit the best for this project, has resulted in the accomplishment of this book. They are a veteran in the field of academics and their pool of knowledge is as vast as their experience in printing. Their expertise and guidance has proved useful at every step. Their uncompromising quality standards have made this book an exceptional effort. Their encouragement from time to time has been an inspiration for everyone.

The publisher and the editorial board hope that this book will prove to be a valuable piece of knowledge for researchers, students, practitioners and scholars across the globe.

List of Contributors

C. MARY ANBARASI
PG Department of Chemistry, Jayaraj Annapackiam College for Women, Periyakulam-625601, India

SUSAI RAJENDRAN
Corrosion Research Centre, PG and Research Department of Chemistry, GTN Arts College, Dindigul-624005 and Department of Chemistry, RVS School of Engineering and Technology, Dindigul-624005, India

JOHN GUSTAVSSON
Applied Electrochemistry, School of Chemical Science and Engineering, KTH Royal Institute of Technology, SE-100 44 Stockholm, Sweden

CHRISTINE HUMMELGÅRD
Department of Natural Sciences, Engineering and Mathematics, Mid Sweden University, SE-851 70 Sundsvall, Sweden

JOAKIM BÄCKSTRÖM
Department of Natural Sciences, Engineering and Mathematics, Mid Sweden University, SE-851 70 Sundsvall, Sweden

INGER ODNEVALL WALLINDER
Department of Surface and Corrosion Science, KTH Royal Institute of Technology, SE-100 44 Stockholm, Sweden

SEIKH MOHAMMAD HABIBUR RAHMAN
Department of Chemical and Biological Engineering, Chalmers University of Technology, SE-412 96 Gothenburg, Sweden

GÖRAN LINDBERGH
Applied Electrochemistry, School of Chemical Science and Engineering, KTH Royal Institute of Technology, SE-100 44 Stockholm, Sweden

STEN ERIKSSON
Department of Chemical and Biological Engineering, Chalmers University of Technology, SE-412 96 Gothenburg, Sweden

ANN CORNELL
Applied Electrochemistry, School of Chemical Science and Engineering, KTH Royal Institute of Technology, SE-100 44 Stockholm, Sweden

Mahbobeh Moazampour
Department of Chemistry, Graduate University of Advanced Technology, Kerman, Iran

Fahimeh Tahernejad-Javazmi
Department of Chemistry, Graduate University of Advanced Technology, Kerman, Iran

Maryam Salimi-Amiri
Department of Physics, Sari Branch, Islamic Azad University, Sari, Iran

Hassan Karimi-Maleh
Department of Chemistry, Graduate University of Advanced Technology, Kerman, Iran

Mehdi Hatami
Polymer Research Laboratory, University of Bonab, Bonab, Iran

NAGAPPA LAXMAN TERADAL
Department of Chemistry, Karnatak University, Dharwad-580 003, India

SHANKAR NARAYAN PRASHANTH
Department of Chemistry, Karnatak University, Dharwad-580 003, India

JALDAPPAGARI
Department of Chemistry, Karnatak University, Dharwad-580 003, India

SEETHARAMAPPA
Department of Chemistry, Karnatak University, Dharwad-580 003, India

ASHOK K. SHARMA
Department of Materials Science & Nanotechnology, D.C.R University of Science & Technology, Murthal-131 039 (Haryana) India

YASHPAL SHARMA
Department of Chemistry, G. J. University of Science & Technology, Hisar-125001, India

GVOZDEN TASIĆ
University of Belgrade, Vinča Institute, Department of Physical Chemistry, 11001 Belgrade, P.O. Box 522, Serbia

BORKA JOVIĆ
Institute for Multidisciplinary Research University of Belgrade, 11030 Belgrade, P.O. Box 33, Serbia

UROŠ LAČNJEVAC
Institute for Multidisciplinary Research University of Belgrade, 11030 Belgrade, P.O. Box 33, Serbia

NEDELJKO KRSTAJIĆ
Faculty of Technology and Metallurgy University of Belgrade, 11000 Belgrade, Karnegijeva 4, Serbia

VLADIMIR JOVIĆ
Institute for Multidisciplinary Research University of Belgrade, 11030 Belgrade, P.O. Box 33, Serbia

AFZAL SHAH
Department of Chemistry, Quaid-i-Azam University, 45320, Islamabad, Pakistan and National Center of Excellence in Analytical Chemistry, University of Sindh, Jamshoro, Pakistan

ABDUR RAUF
Department of Chemistry, Quaid-i-Azam University, 45320, Islamabad, Pakistan and National Center of Excellence in Analytical Chemistry, University of Sindh, Jamshoro, Pakistan

ASAD ULLAH
Department of Chemistry, Quaid-i-Azam University, 45320, Islamabad, Pakistan and National Center of Excellence in Analytical Chemistry, University of Sindh, Jamshoro, Pakistan

AZEEMA MUNIR
Department of Chemistry, Quaid-i-Azam University, 45320, Islamabad, Pakistan and National Center of Excellence in Analytical Chemistry, University of Sindh, Jamshoro, Pakistan

RUMANA QURESHI
Department of Chemistry, Quaid-i-Azam University, 45320, Islamabad, Pakistan and National Center of Excellence in Analytical Chemistry, University of Sindh, Jamshoro, Pakistan

IFTIKHAR AHMAD
Department of Chemistry, Quaid-i-Azam University, 45320, Islamabad, Pakistan and
National Center of Excellence in Analytical Chemistry, University of Sindh, Jamshoro, Pakistan

MUHAMMAD TAHIR SOOMRO
Department of Chemistry, Quaid-i-Azam University, 45320, Islamabad, Pakistan and National Center of Excellence in Analytical Chemistry, University of Sindh, Jamshoro, Pakistan

ZIA-UR-REHMAN
Department of Chemistry, Quaid-i-Azam University, 45320, Islamabad, Pakistan and National Center of Excellence in Analytical Chemistry, University of Sindh, Jamshoro, Pakistan

Joan Stilling
Department of Chemical Engineering, McGill University, 3610 University Street, Montreal, Quebec, Canada, H3A 0C5

Nehar Ullah
Department of Chemical Engineering, McGill University, 3610 University Street, Montreal, Quebec, Canada, H3A 0C5

Sasha Omanovic
Department of Chemical Engineering, McGill University, 3610 University Street, Montreal, Quebec, Canada, H3A 0C5

KRISHNA NAIK
Sri Krishnadevaraya University, Anantapur, A.P., India

ALURU RAGHAVENDRA GURU PRASAD
ICFAI Foundation for Higher Education, Hyderabad, A.P., India

YADATI NARASIMHA SPOORTHY
Sri Krishnadevaraya University, Anantapur, A.P., India

LAKSHMANA RAO KRISHNA RAO RAVINDRANATH
Sri Krishnadevaraya University, Anantapur, A.P., India

NIRWAN SYARIF
Department of Chemistry, Faculty of Mathematics and Natural Sciences, University of Indonesia, Indonesia
Department of Chemistry, Faculty of Mathematics and Natural Sciences, Sriwijaya University, Indonesia

IVANDINI A. TRIBIDASARI
Department of Chemistry, Faculty of Mathematics and Natural Sciences, University of Indonesia, Indonesia

WIDAYANTI WIBOWO
Department of Chemistry, Faculty of Mathematics and Natural Sciences, University of Indonesia, Indonesia

JOHN GUSTAVSSON
Applied Electrochemistry, School of Chemical Science and Engineering, KTH Royal Institute of Technology, SE 100 44 Stockholm, Sweden

GONGZHUO LI
Applied Electrochemistry, School of Chemical Science and Engineering, KTH Royal Institute of Technology, SE 100 44 Stockholm, Sweden

CHRISTINE HUMMELGÅRD
Department of Natural Sciences, Engineering and Mathematics, Mid Sweden University, SE 851 70 Sundsvall, Sweden

JOAKIM BÄCKSTRÖM
Department of Natural Sciences, Engineering and Mathematics, Mid Sweden University, SE 851 70 Sundsvall, Sweden

ANN CORNELL
Applied Electrochemistry, School of Chemical Science and Engineering, KTH Royal Institute of Technology, SE 100 44 Stockholm, Sweden

DEVESH KUMAR MAHLA
Rubber Technology Centre, Indian Institute of Technology Kharagpur, Kharagpur, India

SUBHENDU BHANDARI
Rubber Technology Centre, Indian Institute of Technology Kharagpur, Kharagpur, India

MOSTAFIZUR RAHAMAN
Presently at Chemical Engineering Department, College of Engineering Sciences, King Fahd University of Petroleum and Minerals (KFUPM), Dhahran, Saudi Arabia

DIPAK KHASTGIR
Rubber Technology Centre, Indian Institute of Technology Kharagpur, Kharagpur, India

STEFANIA GIORDANO
Dipartimento di Chimica, Università degli Studi di Milano, Via Golgi, 19, 20133, Milano, Italy

MARIANGELA LONGHI
Dipartimento di Chimica, Università degli Studi di Milano, Via Golgi, 19, 20133, Milano, Italy

LEONARDO FORMARO
Dipartimento di Chimica, Università degli Studi di Milano, Via Golgi, 19, 20133, Milano, Italy

HERMES FARINA
Dipartimento di Chimica, Università degli Studi di Milano, Via Golgi, 19, 20133, Milano, Italy

GIUSEPPE DI SILVESTRO
Dipartimento di Chimica, Università degli Studi di Milano, Via Golgi, 19, 20133, Milano, Italy

CARMEN S. HERNANDEZ DOMÍNGUEZ
Departamento de Química Analítica y Análisis Instrumental, Facultad de Ciencias, Universidad Autónoma de Madrid, Cantoblanco, 28029 Madrid, Spain

PEDRO HERNÁNDEZ
Departamento de Química Analítica y Análisis Instrumental, Facultad de Ciencias, Universidad Autónoma de Madrid, Cantoblanco, 28029 Madrid, Spain

ANNABEL FERNANDES
UMTP and Department of Chemistry, University of Beira Interior, 6201-001 Covilhã, Portugal

EDITE CATALÃO
UMTP and Department of Chemistry, University of Beira Interior, 6201-001 Covilhã, Portugal

LURDES CIRÍACO
UMTP and Department of Chemistry, University of Beira Interior, 6201-001 Covilhã, Portugal

MARIA J PACHECO
UMTP and Department of Chemistry, University of Beira Interior, 6201-001 Covilhã, Portugal

ANA LOPES
UMTP and Department of Chemistry, University of Beira Interior, 6201-001 Covilhã, Portugal

NEBOJŠA D. NIKOLIĆ
ICTM-Institute of Electrochemistry, University of Belgrade, Njegoševa 12, 11000 Belgrade, Serbia

GORAN BRANKOVIĆ
Institute for Multidisciplinary Research, University of Belgrade, Kneza Višeslava 1a, Belgrade, Serbia

MIOMIR G. PAVLOVIĆ
ICTM-Institute of Electrochemistry, University of Belgrade, Njegoševa 12, 11000 Belgrade, Serbia

ANANTHA N. SUBBA RAO
Department of P.G. Studies and Research in Chemistry, School of Chemical Sciences, Kuvempu University, Shankaraghatta-577451, Shimoga, Karnataka, India

VENKATESHA T. VENKATARANGAIAH
Department of P.G. Studies and Research in Chemistry, School of Chemical Sciences, Kuvempu University, Shankaraghatta-577451, Shimoga, Karnataka, India

ANA M. ESTEVA
Departamento de Química Analítica, Facultad de Química, Universidad de La Habana, La Habana 10400, Cuba

ELÍAS BLANCO
Departamento de Química Analítica y Análisis Instrumental, Facultad de Ciencias, Universidad Autónoma de Madrid, Cantoblanco 28049, Madrid, Spain

JUAN J. PIÑA
Departamento de Química Analítica, Facultad de Química, Universidad de La Habana, La Habana 10400, Cuba

ABEL I. BALBIN
Departamento de Química Analítica, Facultad de Química, Universidad de La Habana, La Habana 10400, Cuba

CARMEN QUINTANA
Departamento de Química Analítica y Análisis Instrumental, Facultad de Ciencias, Universidad Autónoma de Madrid, Cantoblanco 28049, Madrid, Spain

PEDRO HERNÁNDEZ
Departamento de Química Analítica y Análisis Instrumental, Facultad de Ciencias, Universidad Autónoma de Madrid, Cantoblanco 28049, Madrid, Spain

LEONID SKATKOV
Technical Division, PCB "Argo" Ltd., 4/23 Shaul ha-Melekh str., 84797 Beer Sheva, Israel

LARISA LYASHOK
Electrochemistry Department, National Technical University "KhPI", 21 Frunze str., 61002 Kharkov, Ukraine

VALERIY GOMOZOV
Electrochemistry Department, National Technical University "KhPI", 21 Frunze str., 61002 Kharkov, Ukraine

IRINA TOKAREVA
Electrochemistry Department, National Technical University "KhPI", 21 Frunze str., 61002 Kharkov, Ukraine

BORIS BAYRACHNIY
Electrochemistry Department, National Technical University "KhPI", 21 Frunze str., 61002 Kharkov, Ukraine

TSUNG-WEI CHANG
Department of Electrical Engineering, National Cheng Kung University, Tainan 701, Taiwan, R.O.C

SHAO-YU HU
Department of Electrical Engineering, National Cheng Kung University, Tainan 701, Taiwan, R.O.C

WEN-HSI LEE
Department of Electrical Engineering, National Cheng Kung University, Tainan 701, Taiwan, R.O.C

Anantha N. Subba Rao
Department of P.G. studies and Research in Chemistry, School of Chemical Sciences, Kuvempu University, Shankaraghatta-577451, Karnataka, India

Venkatesha T. Venkatarangaiah
Department of P.G. studies and Research in Chemistry, School of Chemical Sciences, Kuvempu University, Shankaraghatta-577451, Karnataka, India

Keerti M. Naik
P. G. Department of Studies in Chemistry, Karnatak University, Dharwad 580 003, India

Sharanappa T. Nandibewoor
P. G. Department of Studies in Chemistry, Karnatak University, Dharwad 580 003, India

THOMMANDRU RAVEENDRANATH BABU
Electroanalytical Lab., Department of Chemistry, N.B.K.R. Science and Arts College, Vidyanagar, Nellore, AP, India

SARVAREDDY RAJASEKHAR REDDY
Electroanalytical Lab., Department of Chemistry, N.B.K.R. Science and Arts College, Vidyanagar, Nellore, AP, India

PUCHAKAYALA SUJANA
Electroanalytical Lab., Department of Chemistry, N.B.K.R. Science and Arts College, Vidyanagar, Nellore, AP, India